Natural Computing Series

Scope

The Natural Computing book series covers theory, experiment, and implementations at the intersection of computation and natural systems. This includes:

- **Computation inspired by Nature**: Paradigms, algorithms, and theories inspired by natural phenomena. Examples include cellular automata, simulated annealing, neural computation, evolutionary computation, swarm intelligence, and membrane computing.
- **Computing using Nature-inspired novel substrates**: Examples include biomolecular (DNA) computing, quantum computing, chemical computing, synthetic biology, soft robotics, and artificial life.
- **Computational analysis of Nature**: Understanding nature through a computational lens. Examples include systems biology, computational neuroscience, quantum information processing.

Yaochu Jin

Computational Evolution of Neural and Morphological Development

Towards Evolutionary Developmental Artificial Intelligence

Springer

Yaochu Jin
Faculty of Technology
Bielefeld University
Bielefeld, Germany

ISSN 1619-7127
Natural Computing Series
ISBN 978-981-99-1853-9 ISBN 978-981-99-1854-6 (eBook)
https://doi.org/10.1007/978-981-99-1854-6

This Springer imprint is published by the registered company Springer Nature Singapore Pte Ltd.
The registered company address is: 152 Beach Road, #21-01/04 Gateway East, Singapore 189721, Singapore

Preface

When I started my research journey in artificial intelligence (AI) over thirty years ago, I did not expect that this field would experience so many ups and downs in the next years, nor did I imagine that it will create such huge impacts on almost every facet of human life today. I feel very lucky to have engaged myself in this field, working in cross-disciplinary areas of expert systems, evolutionary computation, fuzzy systems, artificial neural networks, computational neuroscience, and computational biology.

Over the past thirty years, my research focus has shifted from AI for systems and control to AI for optimization, in particular data-driven evolutionary optimization and multi-objective evolutionary optimization. In the mid-2000s during my time at Honda Research Institute Europe, I had the opportunity to make an important side trip to an exciting area, aiming to understand natural evolution by integrating computational intelligence, computational neuroscience, and systems biology. This excursion has fascinated me so much simply because it touched something deep in my heart. In my Ph.D. dissertation completed in early 1996 at Zhejiang University, I put the following two items, among others, on my research wish list, i.e., to investigate the synergies between various approaches to AI including artificial neural networks and evolutionary computation, and to examine the inherent links between biological intelligence and artificial intelligence. Looking back, I feel these ideas together exactly shape the research paradigm to be elaborated in this book, namely evolutionary developmental systems, since evolution, learning, and development are three main adaptation mechanisms in nature.

This book aims to provide a basic yet unified overview of theory and methodologies for evolutionary developmental systems for postgraduate students, researchers, and practitioners who are interested in pursuing a different approach to AI. It starts with an introduction to computational algorithms for understanding and simulating biological evolution and development, including evolutionary algorithms, gene regulatory network models, multi-cellular models for neural and morphological development, and computational models of neural plasticity. Chapter 2 discusses important properties of biological gene regulatory systems, including network motifs, network connectivity, robustness, and evolvability. To analyze robustness and evolve ability, two fundamental properties of biological gene regulatory networks, quantitative measures for robustness and evolvability, are introduced. The results reveal that the developmental plasticity is indispensable for biological systems to harness both the robustness to random genetic mutations and evolvability to achieve new functions favored in evolution. To go a step further, Chap. 3 presents methods for synthesizing regulatory motifs from scratch and creating more complex regulatory dynamics by combining basic regulatory motifs using evolutionary algorithms. In addition, a case study of reverse-engineering biological regulatory networks based on gene expression data is given, showcasing the ability of evolutionary algorithms to reconstruct large-scale biological regulatory networks. Multi-cellular growth models, which can be used for simulating either neural or morphological development, are presented in Chaps. 4 and 5. In addition, various local plasticity rules are evolved. Chapter 6 examines the synergies and coupling between neural and morphological evolution and development. This is divided into three steps, namely neural and morphological evolution, co-evolution of neural controller and morphological development, and finally co-evolution of neuro-morphological development. These investigations reveal rich findings that are important for both understanding biological intelligence and developing artificial intelligence models. Different from the previous chapters that focus on addressing analyzing and synthesizing basic properties of biological evolution and development, Chap. 7 provides preliminary yet promising examples of how evolutionary developmental systems can help in self-organized pattern generation, coined morphogenetic self-organization, highlighting the great potentials of evolutionary and developmental systems. Finally, Chap. 8 summarizes the book, stressing the importance and promises of the evolutionary developmental approach to AI.

This book is basically a collection of the research work on evolutionary developmental systems that I have authored or co-authored since the mid of 2000s. Relevant published papers on which a chapter or a section is based are explicitly indicated at the beginning of the chapter or section. Therefore, I would like to thank, first and foremost, Bernhard Sendhoff for his contributions to many ideas presented in the book. Several other colleagues and collaborators of mine, and postdocs and Ph.D. students I supervised or co-supervised, have contributed to the contents of this book, including Markus Olhofer, Xin Yao, Yan Meng, Lis Schramm, Till Steiner, Ben Jones, Jens Trommler, Rubin Gruna, Hongliang Guo, Jun Yin, Spencer Thomas, Joseph Chrol-Cannon, Ataollah Ramezan Shirazi, Hyondong Oh, Xinjie Wang, and Yan Zhou.

I would like to thank Mio Sugino and Ronan Nugent, Editors at Springer, for their great patience over the years, without which the book would not have been possible. At last, I would like to acknowledge the funding of the Alexander von Humboldt Professorship for Artificial Intelligence endowed to me by the German Federal Ministry of Education and Research.

Bielefeld, Germany Yaochu Jin
February 2023

Contents

Chapter 1
Computational Models of Evolution and Development

1.1 Evolutionary Algorithms

1.1.1 A Generic Evolutionary Algorithm

Natural evolution includes all changes in the characteristics and diversity of life that occur throughout time. Evolution may occur on both large and small scales, namely microevolution, which refers to the evolution at the population level, and macroevolution, which is a form of evolution undergone at or above the level of species. According to Charles Darwin, the inherited traits of a population of organisms change from one generation to the next subject to a natural selection that is usually known as the survival of the fittest. As a result, favorable heritable traits will become increasingly common in successive generations of a population while unfavorable heritable traits will become less common. In a natural evolution, favorable traits are mostly related to the capability of survival and reproduction. It has been found later that it is DNA, a double helix of nucleotides consisting of four nucleobases, A, C, G, and T, that makes biological traits heritable. In biology, a paired length of DNA is called a chromosome, which is composed of multiple genes, each of which is a place (locus) on a chromosome that determines a particular trait. In biological evolution, two main driving forces are natural selection that shapes the adaptation to the environment and genetic drift, which is a random loss of alleles.

Evolutionary algorithms (EAs) are a class of computational algorithms that emulate the key elements of natural evolution at a high abstraction level. EAs are able to adapt themselves to the environment by applying three main operations, namely crossover (or recombination), mutation, and selection. Through these conceptually simple operations, EAs offer very attractive properties such as adaptation, optimization, and creation. Due to these properties, evolutionary algorithms have been widely applied to solving science and technological problems, ranging from basic research on understanding biological intelligence such as artificial life to real-world applications such as optimization of deep neural networks.

Y. Jin, *Computational Evolution of Neural and Morphological Development*, Natural Computing Series, https://doi.org/10.1007/978-981-99-1854-6_1

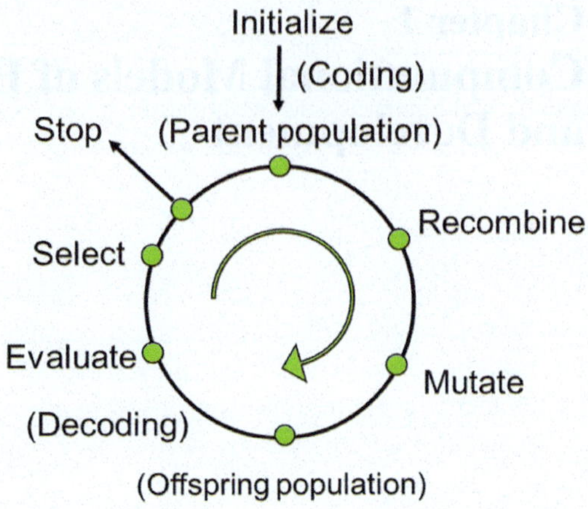

Fig. 1.1 A diagram of a generic evolutionary algorithm

The origin of evolutionary algorithms can be traced back to the 1950s [42, 118]. Three main schools of EAs, namely genetics algorithms [134, 166], evolution strategies [18, 368], and evolution programming [112, 113] were proposed independently. In the 1990s, genetic programming [230, 239], which was developed on the basis of genetic algorithms, has become another main paradigm of evolutionary computation. Despite the differences in representation, genetic operators, and selection, EAs can be described in a generic form as follows:

```
begin EA;{
t = 0;
generate initial population p(0)
evaluate p(0);
do {
        t = t + 1;
        select parents from population p(t).
        recombine p(t);
        mutate p(t);
        evaluate p(t);
        } while {stop criteria are not met};
}
```

A generic diagram of an EA is also plotted in Fig. 1.1. In artificial evolutionary systems, evolution occurs on three levels, i.e., the genotype level, phenotype level, and fitness level. In EAs, a set of chromosomes is called the genotype, simulating the DNA in biological evolution, whereas the definition of phenotype is problem-specific. For example, if an evolutionary algorithm is used to optimize the architecture of an artificial neural network, the phenotype describes all possible structures of the neural networks. The introduction of the phenotype makes it possible to perform a detailed analysis of evolutionary algorithms, since the genotype-fitness map, sometimes called the fitness landscape, is strongly problem-dependent and a quantitative analysis may be intractable. Recall that all genetic variations, such as recombination

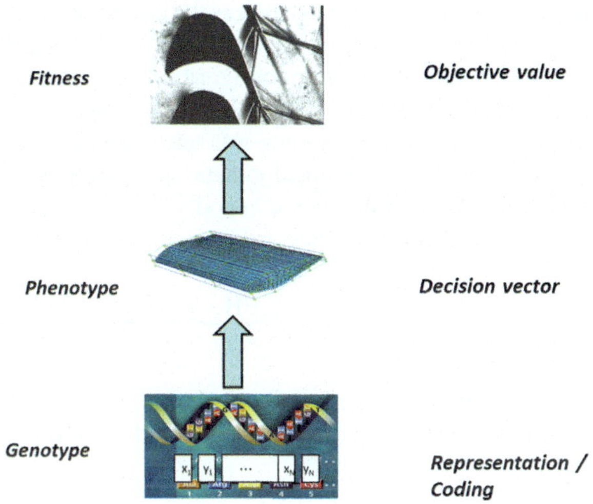

Fig. 1.2 Genotype, phenotype, and fitness of an evolutionary algorithm for optimizing the geometry of a turbine blade

and mutation, happen in the genotype space, which directly results in variations in phenotype space.

An illustrative example genotype-phenotype-fitness map in the context of an optimization problem is given in Fig. 1.2. To be specific, for geometry optimization of a turbine blade, the genotype is the representation or coding of the geometry of the blade, the phenotype is the corresponding geometry, and fitness is the aerodynamic performance of the blade, which can be evaluated using either computational fluid dynamic simulations or wind tunnel experiments.

The following main properties of a genotype-phenotype map should be considered when designing a representation of an optimization problem:

- Causality: Small changes in genotype should result in small changes in phenotype. Vice versa, large changes in genotype result in large changes in phenotype.
- Evolvability: Heritable new phenotypes can be produced to be adaptable to environmental changes.
- Robustness: Critical phenotypes are insensitive to small changes in genotype.

In the following, we introduce in greater detail the main components of EAs, including representation, recombination, mutation, and selection. Since both genetic algorithms and evolution strategies are employed in this book, the following discussions are limited to these two main algorithms.

1.1.1.1 Representation

Representation is concerned with the coding of the problem to be solved by the EA in the genotype space. In genetic algorithms, usually all decision variables are encoded, while in evolution strategies, both decision variables and some important parameters in the evolutionary algorithm, such as the population size, mutation step-sizes, mutation probability, crossover positions, and crossover probability, are encoded. In general, the performance of an evolutionary algorithm heavily depends on its parameters; therefore, it is highly desirable to evolve the parameters together with the decision variables during evolution. This is usually known as self-adaptation.

Traditionally, a canonical genetic algorithm uses binary or gray coding [52], in which decision variables are encoded into a string consisting of 0's and 1's. Suppose the decision variables are $\mathbf{x} = [x_1, x_2, ..., x_n]$, then the chromosome is in the following form:

$$\underbrace{101...1}_{x_1}\,\underbrace{110...0}_{x_2}\,...\,\underbrace{111...1}_{x_n} \tag{1.1}$$

For an l-bit binary string $s_j \in 0, 1; j = 1, 2, ..., l$ that encodes a real number x_i within the range of $[a_i, b_i]$, the decoding function is as follows:

$$x_i = a_i + (b_i - a_i)\frac{1}{2^l - 1}\left(\sum_{j=0}^{l-1} s_j 2^j\right). \tag{1.2}$$

It is noticed that the binary coding is nonlinear and suffers from the so-called Hamming cliff problem. That is, a large change in the genotype is needed for a small change in the phenotype, making it hard for exploitation. Alternatively, gray coding can also be used. The most important feature of gray coding is that the decimal value increases or decreases by 1 if only one bit is changed, which is strongly causal. In decoding, the gray-coded string can first be converted into a binary string and then Eq. (1.2) can be used.

In the above binary representation, the length of the chromosomes is fixed. It is known, however, that the length of biological genomes may change over time or over the generations, and that they may contain duplicate or competing genes [334]. To simulate this, EAs with a variable length representation have also been proposed. An early example of variable length representations is the messy GA [135], where both the value and position of each bit are encoded in the chromosome. The delta-coding was designed in [278], whose length is varied during evolution so that the genetic algorithm can strike a better balance between exploration and exploitation.

Both binary and gray representations have been shown to work well for continuous, integer, or combinatorial optimization, although they are not the most efficient for continuous optimization and inadequate in tackling more complicated tasks such as structure optimization of neural networks [371, 469, 479]. To address the above weaknesses, real-coded genetic algorithms have been proposed [160], making them much efficient for continuous optimization. Meanwhile, other more sophisticated coding schemes have also been developed. Examples include grammar coding [219], cellular coding [141], recursive coding [374], and context-dependent coding [247], among many others. These coding methods are sometimes called indirect coding or generative coding because not every property in the phenotype has an explicit representation in the genotype. For example, the context-dependent coding has two types of genes, one for instructions and the other for binary bits, where the instructions indicate which binary bits are to be activated, emulating the gene expression process in biology. Therefore, a chromosome may have some genes that are not explicitly seen in decoding. Suppose A and B are used for instruction coding, and 1 and 0 are the bits for binary coding, then a typical chromosome may look like the following:

A0BA1011AAB101B00B1B

When decoding the above chromosome, the system will first look for particular instructions, which, for instance, are AA, AB, BA, or BB. Consequently, the first two genes are ignored until the sequence BA is found. The binary genes are then checked, which is 1011 in this example. The system stops reading the binary genes once the next instruction gene is encountered. Since binary genes are expected after the instruction genes AA, B will be skipped. It should be noticed that genetic programming uses tree structures for representation, and the width and depth of the trees may also change after crossover and mutation.

By contrast, evolution strategies encode decision variables directly in the chromosome. Different from the representations of genetic algorithms, evolution strategies also encode their strategy parameters in the chromosome. Thus, the chromosome of an evolution strategy for a decision vector $\mathbf{x} = [x_1, x_2, ..., x_n]$ will be

$$x_1\, x_2 \ldots x_n\, \sigma_1\, \sigma_2 \ldots \sigma_m, \tag{1.3}$$

where $x_i, i = 1, 2, ..., n$ are the decision variables and $\sigma_j, j = 1, 2, ..., m$ are the strategy parameters representing the step-sizes. Note that the number of strategy parameters is different in various evolution strategy algorithms. Note that the real-coded representation can be changed to an integer representation if an evolution strategy is adopted for integer optimization.

From the above discussions, it is noted that genetic algorithms are generally more flexible than evolution strategies in representation, although modern evolutionary algorithms may use a hybrid representation consisting of real-valued, binary, or other coding schemes. Accordingly, different genetic operators should be applied to different representations, resulting in more complex search behaviors, mainly because different representations exhibit different causality properties between the genotype and phenotype. By causality, it is meant that the degree of variations in the genotype space should be properly reflected by the degree of variation in the phenotype space. For a strongly causal search, it is expected that the neighborhood in the genotype space should be preserved in the phenotype space [373]. Strong causality, however, may be helpful or detrimental to the search performance, depending on the search phase and the fitness landscape. Generally speaking, causality is desirable for exploitation, and required for self-adaptation in evolutionary algorithms, whereas weak causality is more helpful for exploratory search. It has been found that the representation of evolution strategies has a higher degree of causality than that of the binary coding used in genetic algorithms, and that the gray coding is more causal than the binary coding [373].

1.1.1.2 Recombination

Recombination has different implementations in different representations, providing a mechanism for exchanging genetic materials between different individuals. Recombination, better known as crossover, is the main driving force in genetic algorithms and genetic programming. By contrast, recombination is a secondary search operator in evolution strategies.

As mentioned before, recombination heavily depends on the nature of representation, i.e., genetic coding. In addition, recombination can be divided into genotypic-level and phenotypic-level schemes [397]. For example, genetic algorithms based on binary or gray coding usually apply genotypic-level recombination, and the most widely used genotypic-level recombination methods include

- *1-point recombination* [166]. Given two parents for recombination: $[AB101A001]$ and $[BA010A110]$. One position in the chromosome is randomly selected for conducting crossover, say between the third and fourth bits. Two offspring individuals are created by exchanging two pieces of the genomes cut at the selected position: $[BA001A001]$ and $[AB110A110]$.
- *n-point recombination* [80]. 1-point recombination was extended to n-point recombination. In this scheme, n 'cut-points' are introduced. For instance, assume that a two-point crossover is to be applied to the two parents that are cut (denoted by '|') at the third and sixth loci $[AB1|01A|001]$ and $[BA0|10A|110]$, which will result in the following two offspring: $[AB110A001]$ and $[BA001A110]$.
- *Adaptive Recombination*. The recombination methods described above are static. It is conceivable to adjust the position of crossover during evolution. A simple idea is to identify groups of genes that should be kept together in crossover, avoiding destroying promising building blocks according to the knowledge acquired during the optimization. For example, one can generate the following rules:
  IF the fitness of an offspring is better than that of its parents, **THEN** reward the bits coming from the parents with a probability.
  Another idea is to assign a label to a particular crossover, such as good, poor and inactive [370]. An inactive crossover means that the fitness of offspring generated using this particular crossover is the same as that of its parents.
  Self-adaptation has shown to be essential in enhancing the search performance of evolutionary algorithms, which is realized by encoding some parameters of the evolutionary algorithm in the representation so that they are also subject to variations and selection.
- *Uniform recombination*. Uniform recombination [412] does not use cut-points. As its name suggests, uniform crossover decides bit by bit whether the particular gene should be swapped between the two parents. The probability of swapping is usually set to 0.5, although other probabilities can also be used. For a swapping probability of 0.5, uniform recombination is implemented as follows. First, a binary decision string of the same length to the parents with 0 and 1 is randomly generated. If the i-th bit of the decision string is 1, then swap the i-th bit of the parent chromosomes. Given a generated decision string of [011010010], two

offspring individuals, $[AA000A011]$ and $[BB111A100]$, will be generated from the two parents $[AB101A001]$ and $[BA010A110]$. Generally speaking, uniform recombination is more disruptive than 1-point and 2-point recombinations, which can be both positive and negative to the search performance, since it can not only lead to more exploratory search but can also destroy existing good building blocks.

Note that all the above operators can be generalized to multiple parents.

Recombination in evolution strategies is usually considered to be supplementary. The widely used recombination methods are [368]

$$x_i' = \begin{cases} x_{a,i} \text{ or } x_{b,i} & \text{discrete recombination} \\ \frac{1}{2}(x_{a,i} + x_{b,i}) & \text{intermediate recombination} \\ x_{a,i} \text{ or } x_{b_j,i} & \text{global discrete recombination} \\ \frac{1}{2}(x_{a,i} + x_{b_j,i}) & \text{global intermediate recombination} \end{cases} \tag{1.4}$$

where a, b, b_j are parent individuals and i is the i-th element of the decision vector x. Notice that in global recombination methods, three individuals are involved, which can be extended to multiple individuals too, which is termed panmictic recombination [18]. In addition, the weighting for each individual can be tuned.

1.1.1.3 Mutation

Mutation plays a minor role in genetic algorithms. The main purpose of mutation is to alleviate the premature convergence problem, when all individuals in the population become very similar before the genetic algorithm converges to the global optimum. If all individuals are very similar or even the same in a population, it is hardly possible to generate new building blocks by means of recombination. Mutation inverts a bit (changes 1 to 0 or 0 to 1) if binary or gray coding representation is used.

Mutation can also be made adaptive in genetic algorithms, by, e.g., adjusting the mutation rate [111]. Similar to the recombination parameter, the mutation rate can be encoded in the chromosome and evolved. The mutation rate can also be made position-dependent [372], which offers an easy way of balancing exploitation and exploration.

Contrary to genetic algorithms, mutation is the main operator in evolution strategies. In evolution strategies, mutation is realized by adding a normally distributed random number to the decision variables and the encoded strategy parameters. It is important, however, that the variance of the normal distributions is properly tuned during the evolution.

1.1.1.4 Selection

Selection in evolutionary algorithms reflects the biological mechanism of survival of the fittest. In evolutionary algorithms, the key issue in choosing a selection method is to maintain a proper selection pressure at different search stages. Some often used selection schemes for genetic algorithms include the Roulette wheel selection [134], also known as fitness-proportionate selection, the rank-proportionate selection, and tournament selection [136]. The above selection methods are all probabilistic, meaning that the best individuals may get lost during the optimization. This problem can be fixed by introducing the elitism strategy by preserving the best individual(s). Elitism plays an important role in accelerating the convergence of GAs [343]; it may also lead to premature convergence in evolving complex systems.

There are two selection schemes in evolution strategies, namely (μ, λ)-selection and $(\mu + \lambda)$-selection, where μ, λ are the population sizes for the parents and the offspring. In (μ, λ)-selection, the best μ individuals in the offspring are chosen as the parents for the next generation, whereas in $(\mu + \lambda)$-selection, the best μ individuals from the union of parents and offspring are chosen as the parent individual for the next generation. Clearly, $(\mu + \lambda)$ is an elitism, since the best μ will always be preserved. Notice that both selection methods in evolution strategies are deterministic, and typically μ is much smaller than λ.

1.1.2 Self-Adaptation and Evolution Strategies

Self-adaptation is one most important feature in evolution strategies, significantly enhancing the search performance by adapting their search behavior to the fitness landscape of the problem to be solved.

The standard evolution strategy (ES) can be described as follows:

$$\mathbf{x}(t) = \mathbf{x}(t-1) + \tilde{\mathbf{z}} \tag{1.5}$$

$$\sigma_i(t) = \sigma_i(t-1)\exp(\tau' z)\ \exp(\tau z_i), \quad \forall i \in \{1, \ldots, n\} \tag{1.6}$$

$$\mathbf{x}, \tilde{\mathbf{z}} \in \mathbb{R}^n; \qquad z_i, z \sim N(0, 1); \qquad \tilde{\mathbf{z}} \sim N(\mathbf{0}, \boldsymbol{\sigma}(t)^2) \tag{1.7}$$

where $\mathbf{x}$ is the decision vector to be optimized, and τ, τ', and σ_i are the strategy parameters. The values for τ and τ' are specified as follows:

$$\tau = \left(\sqrt{2\sqrt{n}}\right)^{-1}; \qquad \tau' = \left(\sqrt{2n}\right)^{-1}, \tag{1.8}$$

where z_i, z, and $\tilde{\mathbf{z}}$ are normally distributed random numbers and a random vector, respectively, which together characterize the mutation applied on the decision vector and strategy parameters. σ_i are also called step-sizes, similar to the step-size in the gradient-based methods.

Adaptation of the step-sizes is the key trick in evolution strategies that distinguish different variants of evolution strategies. The earliest version in the standard evolution strategy is called the *1/5 rule* [368], which tunes the strategy parameters in a heuristic way. Self-adaptation methods for adjusting mutation steps include the standard adaptation, where each decision variable has a separate step-size, the rotation matrix adaptation method that controls not only the step-size but also the direction of the mutation [368], the generating set adaptation approach which adapts normal mutation distributions independent of the chosen coordinate system [155], and the most popular de-randomized covariance matrix adaptation evolution strategy (CMA-ES) [153, 316]. CMA-ES represents the most efficient evolution strategy for continuous optimization, which will be used in many algorithms presented in this book.

CMA-ES [153] differs from the standard evolution strategy mainly in the following three respects:

- In order to reduce the stochastic influence on the self-adaptation, only one stochastic source is used adapting both decision variables and strategy parameters. In the de-randomized variant [316], the actual step length in the decision space is used to adapt the strategy parameters. Therefore, the self-adaptation of the strategy parameters depends more directly on the local topology of the fitness landscape. The strategy parameters and decision variables are adapted in the follow form:

$$\sigma(t) = \sigma(t-1)\exp\left(\frac{1}{d}\left(|\mathbf{z}| - E[|N(\mathbf{0},\mathbf{1})|]\right)\right) \tag{1.9}$$

$$\mathbf{x}(t) = \mathbf{x}(t-1) + \sigma(t)\,\mathbf{z} \tag{1.10}$$

$$\mathbf{z} \sim N(\mathbf{0},\mathbf{1}).$$

  In CMA-ES, the actual step length in the decision space is used to adapt the strategy parameters. As a result, if the mutation was larger than expected ($|\mathbf{z}| > E[|N(\mathbf{0},\mathbf{1})|]$), the strategy parameters are increased. This ensures that if a larger mutation is successful, a larger mutation will occur again in the next generation. The same argumentation holds if ($|\mathbf{z}| < E[|N(\mathbf{0},\mathbf{1})|]$). Parameter d is used to regulate the adaptation.
- An additional strategy parameter called the cumulative step-size is introduced and adapted. While the standard evolution strategy extracts useful information for adaptation of the strategy parameters from the population, the cumulative step-size adaptation relies on information collected during successive generations, which can be seen as a time-averaged approach, making the adaptation work with a much smaller population size. This is why CMA-ES typically adopts a much smaller population size than that of the standard evolution strategy, making it more practical for expensive optimization problems where the evaluation of the fitness involved computationally intensive simulations or costly experiments. The cumulation of steps is termed the evolution path [153] and is adapted as follows:

$$\mathbf{s}(t) = (1-c)\mathbf{s}(t-1) + c_u\mathbf{z}. \tag{1.11}$$

  The factor c determines the length of the evolution path and $c_u = \sqrt{c(2-c)}$ is introduced for normalization, which calculates the variance of $\mathbf{s}(t)$ in the limit $t \to \infty$.

- The covariance matrix **C** of the probability density function of CMA-ES

$$f(\mathbf{z}) = \frac{\sqrt{\det(\mathbf{C}^{-1})}}{(2\pi)^{n/2}} \exp\left(-\frac{1}{2}(\mathbf{z}^T\,\mathbf{C}^{-1}\,\mathbf{z})\right) \tag{1.12}$$

is adapted for mutation of the decision vector. According to [155], if the matrix B satisfies $\mathbf{C} = \mathbf{B}\mathbf{B}^T$ and $z_i \sim N(0, 1)$, then $\mathbf{B}\,\mathbf{z} \sim N(0, \mathbf{C})$. The adaptation of the decision vector is expressed by

$$\mathbf{x}(t) = \mathbf{x}(t-1) + \delta(t-1)\,\mathbf{B}(t-1)\,\mathbf{z}, \quad z_i \sim N(0, 1), \tag{1.13}$$

where δ is the overall step-size. The covariance matrix is adapted in two steps using the cumulative step-size ($c_{cov} \in (0, 1)$ and $c \in (0, 1)$ that determine the influence of the past during cumulative adaptation:

$$\mathbf{s}(t) = (1-c)\,\mathbf{s}(t-1) + c_u\,\mathbf{B}(t-1)\,\mathbf{z} \tag{1.14}$$

$$\mathbf{C}(t) = (1-c_{cov})\,\mathbf{C}(t-1) + c_{cov}\,\mathbf{s}(t)\mathbf{s}^T(t). \tag{1.15}$$

The next step is to determine matrix **B** based on **C**. Since it is insufficient to derive **B** according to $\mathbf{C} = \mathbf{B}\mathbf{B}^T$, the eigenvectors of **C** are chosen as column vectors of **B**. This lies in the fact that the overall step-size δ must be adapted in the end. For this purpose, the expected length of the cumulative vector $\mathbf{s}_\delta$ needs to be known. Therefore, the variation vector, which depends on **B**, should be $N(0, \mathbf{1})$ distributed. This can be achieved by normalizing the column vectors of **B** by the square root of the corresponding eigenvalues, yielding $\mathbf{B}_\delta$, if the column vectors of **B** are the eigenvectors of **C**. At last, $\mathbf{s}_\delta$ and δ are adapted as follows:

$$\mathbf{s}_\delta(t) = (1-c)\,\mathbf{s}_\delta(t-1) + c_u\,\mathbf{B}_\delta(t-1)\,\mathbf{z} \tag{1.16}$$

$$\delta(t) = \delta(t-1)\cdot\exp\left(\beta\left(||\mathbf{s}_\delta(t)|| - \hat{\chi}_n\right)\right), \tag{1.17}$$

$\mathbf{B}_\delta$ equals **B** with normalized columns in order for $\mathbf{B}_\delta(t-1)\,\mathbf{z}$ to be $N(0, \mathbf{1})$ distributed. $\hat{\chi}_n$ denotes the expectation of the χ_n distribution, which is the distribution of the length of an $N(0, \mathbf{1})$ distributed random vector.

1.1.3 Direct and Indirect Representations

Most evolutionary algorithms for optimization use a direct encoding method, no matter whether it is binary, integer, real-valued, or a tree structure [230]. However, it has been well recognized that biological chromosomes are highly redundant [68]. Research in artificial evolutionary systems has also found that redundant representations suffer less from epistasis, making the evolutionary algorithms more robust, particularly in changing environments [76, 146]. In genetic algorithms, epistasis is defined to be the influence of the genotype at one locus on the effect of a mutation at another locus [444].

In artificial life, one new type of representation has become popular, which is known as embryogeny or developmental coding. As its name suggests, the genotype in developmental coding encodes a set of growing or development rules that define how the phenotype will develop [32], while in direct encoding all details in the phenotype must be encoded. Some indirect coding schemes in genetic algorithms, such as cellular encoding and Lindenmayer systems [66] can also be regarded as embroyogenies. Developmental coding is essential in the computational modeling of biological development, which will be elaborated on in the book in different chapters. In the following, we provide a summary of different encoding (representation) methods.

1.1.3.1 Direct Encoding

In a direct encoding method, all decision variables are directly encoded using a sequence of binary, integer, real values, or symbols. For all direct encoding methods, however, the chromosome length (the dimension of the search space) increases linearly with the number of decision variables, making the EA hard to be scaled to large-scale problems.

1.1.3.2 Indirect Encoding

Indirect encoding methods typically use a rule-based method to encode the information indirectly, so the chromosome length does not depend directly on the number of decision variables. Hence scalability is far better. Some of the indirect encoding methods are described as follows:

- Grammatical encoding uses a recursive rewriting grammar [142, 222]. For instance, one can define rules to transform 1×1 matrices into 2×2 matrices. Then all elements of the new matrix are again transformed into 2×2 matrices and so on.
- Cellular encoding uses a graph grammar method for encoding similar structures such as artificial neural networks [142]. In [142], a grammar tree is adopted with ordered branches. The nodes are labeled with the names of program symbols, which represent instructions for cell processes. Comparable to a Turing machine, a cell executes its instructions step by step.
- Super coupled maps [223]. The model is a continuous-valued, but discrete time model that is based on metabolic reactions. Rules depending on concentrations of several chemicals are encoded in the chromosomes.

It should be pointed out that for a direct encoding, the genotype-phenotype map is stationary, which implies that the developmental process is not considered. By contrast, an indirect encoding method presents a dynamic map between the genotype and phenotype, simulating the biological development. Consequently, most evolutionary neural or morphological developmental systems introduced in this book use indirect encoding, which is typically represented by a gene regulatory network, even if the gene regulatory network itself is directly encoded in the evolution. This means that in evolutionary developmental systems, the gene regulatory network is part of the genotype-phenotype map.

1.1.4 Multi-objective Evolutionary Algorithms

Most real-world optimization and decision-making problems in science, engineering, or daily life involve multiple objectives, and the objectives are usually conflicting with each other. For example, an investment problem aims to maximize the profit and minimize the risk; or when one purchases a mobile phone, one intends to minimize the cost and maximize the performance. Likewise, in engineering design such as an airfoil, one must consider both maximization of the lift and minimization of the drag. In solving all these problems, a set of solutions that present different trade-off relationships between different objectives is sought, which are called Pareto optimal solutions. This differs from single-objective optimization, where one global optimum is sought. Before introducing the definition of Pareto optimality, a definition of Pareto dominance will be provided first, which is the most important concept in multi-objective optimization that enables us to compare the quality of different solutions.

Pareto Dominance: For minimization problems, an objective vector $\mathbf{u} = (u_1, ..., u_m)$ is said to dominate another vector
$\mathbf{v} = (v_1, ..., v_m)$ if and only if $\forall i \in \{1, ..., m\}$, $u_i \le v_i$ and $\exists i \in \{1, ..., m\}$, so that $u_i < v_i$, where m is the number of objectives.

Based on the definition of Pareto dominance, the Pareto optimality can be defined as follows:

Pareto Optimality: A solution $\mathbf{x}^\star$ is said to be Pareto optimal if and only if there is no feasible solution $\mathbf{x}$ for which $\mathbf{v} = \mathbf{F}(\mathbf{x}) = (f_1(\mathbf{x}), ..., f_n(\mathbf{x}))$ dominates $\mathbf{u} = \mathbf{F}(\mathbf{x}^\star) = (f_1(\mathbf{x}^\star), ..., f_n(\mathbf{x}^\star))$, where $\mathbf{F}(\mathbf{x})$ is an m-dimensional vector of objective functions, and $\mathbf{x}$ is an n-dimensional decision vector.

The collection of all Pareto optimal solutions are called the Pareto optimal sets, and the corresponding set of objective vectors ($\mathscr{P}$) is denoted as the Pareto optimal front, Pareto front for short.

The shape of the Pareto front presents an additional dimension of hardness compared to single-objective optimization. A Pareto front is called regular if it spans over the entire objective space, whereas it is called irregular if not, either because the Pareto front is discrete or degenerate [174]. For both regular or irregular Pareto fronts, they can be further grouped into convex, concave or non-concave, discontinuous, and degenerate; refer to Fig. 1.3. A Pareto front is said to be degenerate if its dimension is smaller than $m - 1$, where m is the number of objective functions.

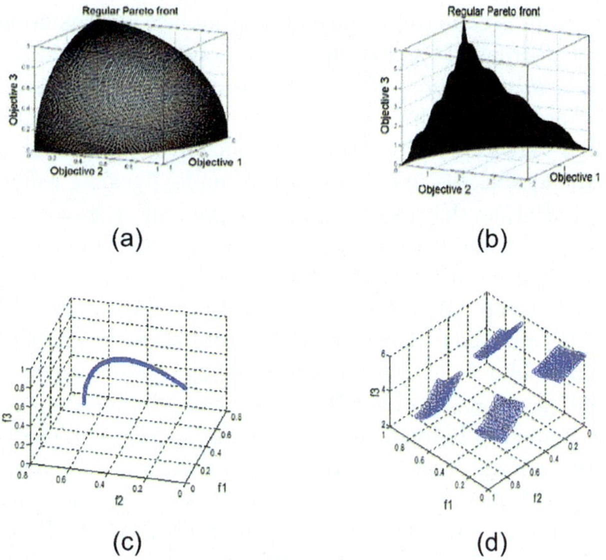

Fig. 1.3 Examples of different types of Pareto fronts. **a** Concave; **b** Non-convex; **c** Degenerate; and **d** Discontinuous

A Pareto front is said to be convex if and only if $\forall \mathbf{u}, \mathbf{v} \in \mathscr{P}, \forall \gamma \in (0, 1), \exists \mathbf{w} \in \mathscr{P} : ||\mathbf{w}|| \leq \gamma||\mathbf{u}|| + (1 - \gamma)||\mathbf{v}||$. Otherwise, the Pareto front is said to be concave or non-convex.

A Pareto front can be neither purely convex nor purely concave, i.e., it is possible that parts of the Pareto front are concave and parts are convex.

Since the early work on evolutionary multi-objective optimization [356], a large body of research on evolutionary multi-objective optimization has been reported; refer to [67]. Over the past decade, more attention has been paid to multi-objective optimization problems that have more than three objectives, often known as many-objective optimization [57].

1.1.4.1 Main Methodologies

Roughly speaking, evolutionary multi-objective optimization algorithms can be categorized into four main approaches:

- *Aggregation-based approach* In this approach, multiple objectives are aggregated into one single objective. The advantage of this approach is that it is easy to apply all existing single-objective evolutionary algorithms to solve multi-objective optimization problems. However, a priori knowledge is needed to determine the weight for each objective and only one Pareto solution can be obtained in one run of optimization. In addition, different weights may result in the same Pareto solution if the Pareto front is concave, and evenly partitioned weights do not necessarily lead to evenly distributed Pareto optimal solutions.

- *Decomposition-based approach* Early ideas in this approach are similar to the aggregation-based approach, where a set of randomly generated or evenly distributed weights are used to convert a multi-objective optimization problem into a number of single-objective problems. For example, dynamic weight aggregation is introduced [194] so that together with an external archive, a set of solutions can be obtained for multi-objective optimization with both convex and concave Pareto fronts by using a set of dynamic changing weights. In [301], a cellular multi-objective genetic algorithm is proposed, in which a set of N evenly distributed weights are used to convert a multi-objective optimization problem into N single-objective sub-problems using weighted aggregation, where each individual in the population solves one of the N sub-problems. To speed up the convergence, crossover is limited to individuals within the same neighborhood. An enhanced version of the above algorithm, called multi-objective evolutionary algorithm by decomposition (MOEA/D) is proposed in [471], where the individuals in a neighborhood work more closely in solving the sub-problems, and the linear weighted aggregation is changed to the Tchebycheff approach or the penalty-based boundary intersection function, making it be more suited to problems with a concave Pareto front.

  The decomposition-based approach also includes another class of multi-objective evolutionary algorithms that use a set of evenly distributed reference vectors in the objective space to decompose multi-objective evolutionary algorithms into a number of single-objective sub-problems, called a reference vector guided evolutionary algorithm for many-objective optimization (RVEA) [57].

  A large number of decomposition-based multi-objective evolutionary algorithms have been proposed, focusing on designing more elaborated reference vectors, e.g., adapting the distribution of the reference vectors to the shape of the Pareto front, instead of using a set of predefined evenly distributed reference vectors, which becomes essential when the Pareto front is irregular.

- *Pareto-based approach* In this approach, the Pareto dominance relationship is applied to sort and rank the solutions in the population. The early Pareto-based approaches assign a scalar fitness value to each individual according to the rank of each individual and the crowdedness of the solution in its neighborhood, thereby making it possible to apply the environmental selection methods proposed for single-objective optimization to multi-objective optimization [114, 169]. In this approach, it is essential to take measures to maintain the diversity of the population, e.g., by taking into account the crowdedness of the solutions in the neighborhood. In addition, it has been found that elitism can significantly enhance the performance Pareto-based multi-objective evolutionary algorithms [480].

  The most popular Pareto-based evolutionary algorithm for multi-objective optimization is the elitist non-dominated sorting genetic algorithm, often known as NSGA-II [83]. NSGA-II uses Pareto dominance to sort the individuals in a combined population that merges the parent and offspring populations into a number of non-dominated fronts. Individuals on each non-dominated front are ranked according to a crowding distance in a decreasing order, where the crowding distance of each individual is calculated based on the Euclidean distance between the individual's left and right neighbors in the objective space. The environmental selection picks the best half of the sorted combined population as the parents of the next generation. NSGA-II combined the parent and offspring population before selection, therefore, it can be considered as an elitism. Note, however, NSGA-II may still lose obtained non-dominated solutions, since it does not have an external archive. Overall, NSGA-II works robustly on multi-objective optimization problems with two or three objectives, and various variants have been proposed to reduce the computational complexity of non-dominated sorting [423, 474].

- *Performance indicator-based approach* Performance indicators are scalar quantitative measures that assess the convergence (closeness to the theoretic Pareto front) and diversity of a non-dominated solution set obtained by an evolutionary algorithm [313]. The two most popular performance indicators are the hypervolume that measures the area/volume or hypervolume (HV) covered by non-dominated solutions surface with respect to a reference point in the objective space [179], and the inverse generational distance (IGD) that calculates the average shortest Euclidean distance between each solution in a reference set and those in the obtained solution set [180].
  Interestingly, since these performance indicators convert the quality of any multi-objective optimization solution set to a scalar value, they have also been made use of for environmental selection. The central idea is to adapt the performance indicators accounting for the quality of a set to a fitness measure of individual solutions [33, 424].

1.1.4.2 Many-Objective Optimization

The previously discussed methodologies have been very successful in solving a wide range of problems with two or three objectives. However, they will face various challenges when handling optimization problems having more than three objectives (often called many-objective optimization problems) for the following reasons. First, the selection pressure drops quickly for dominance-based approaches for a population with a limited size. Second, the computational complexity dramatically increases for performance indicator-based algorithms when the number of objectives is large. Finally, decomposition-based algorithms that have an evenly distributed weight vector or reference vector will also experience difficulties since it is no longer possible to use a limited number of reference vectors to represent the theoretical Pareto front in a high-dimensional space. This becomes even worse when the Pareto front is irregular, i.e., either discrete or degenerate. In essence, the idea of achieving a limited number (say 100 or 200) of evenly distributed solutions to represent the theoretical Pareto front, which works perfectly well for two- or three-objective problems, will clearly fail for many-objective optimization.

To address the above challenges, various ideas have been suggested, ranging from reducing the objective number to modifying the definition of dominance or adding additional criteria in Pareto-based approaches, to reducing the computational costs in performance indicator-based approaches, and to adapting the reference vectors (instead of using predefined, evenly distributed reference vectors) in decomposition-based approaches [252]. Despite all these efforts, they do not really solve the problem since the philosophy of approximating the Pareto front using a limited subset of solutions no longer works. Consequently, decision-making according to user preferences in multi-objective optimization comes into play.

1.1.4.3 User Preferences in Multi-objective Optimization

Although the main research effort in evolutionary multi-objective optimization focuses on obtaining a representative subset of the Pareto optimal solutions, only a small number of solutions will be selected for implementation in practice according to the user's preferences. On the other hand, for many-objective optimization problems, it becomes impossible even to achieve a representative subset of Pareto optimal solutions without limiting the search space based on the user's preferences due to the large objective space.

Usually, a decision can be made based on the user's preferences, if such preferences are available. In general, preferences can be incorporated into optimization either before, during, or after the optimization process takes place [445].

- **Before optimization** If user preferences are available in terms of weights (importance) for different objectives, then different objectives can be combined into a single objective using the weighted aggregation method. Alternatively, one can also incorporate user preferences into Pareto ranking in optimization [74] if the preferences are articulated in terms of the relative importance of different objectives. This approach is also known as the a priori approach.
- **During optimization** This can usually be realized by an interactive process via a user interface, in which the user interacts with the optimizer during optimization. One of the earliest efforts on using interactive approaches in multi-objective optimization can be traced back to [126, 413]. Recently, interactive evolutionary multi-objective optimization has attracted increasing attention to handling many-objective optimization problems with irregular Pareto fronts [174], or for other purposes where the computational resource is a concern [290]. This approach is often called the interactive approach.
  One main challenge in preference incorporation in multi-objective optimization is that user preferences are not easily available in many cases, mainly because the user may not have much a priori knowledge about the problem to be optimized. To address this challenge, one may turn to more natural and less specific preferences such as the knee solutions [466], or design a model that iteratively learns the user's preference.
- **After optimization**. Most evolutionary multi-objective optimization algorithms belong to this a posteriori category. That is, the user is presented with a subset of

Pareto optimal solutions, and needs to choose one or a small number of solutions based on the user's preferences [465].

1.2 Computational Models of Gene Regulatory Networks

1.2.1 Introduction

Generally speaking, neural and morphological development of biological systems is governed by the progressive installation of transcriptional regulatory states that are transiently positioned in embryonic space, cell-cell interaction, and the interaction with the environment [251]. The underlying genetic mechanism is the spatio- and temporal expression of genes encoding sequence-specific transcription factors and other functional proteins.

According to the central dogma of molecular biology, during the expression process, all information encoded in the DNA is first transcribed either to messenger RNA (mRNA) that will then be translated to proteins, or to non-coding RNA such as small RNA (sRNA) that will serve other functions [71, 128], as illustrated in Fig. 1.4. Note that in addition to the two pathways shown in the figure, other changes in gene expression have also been observed. These changes are referred to as epigenetic regulations, in which heritable changes in gene expression are caused by mechanisms other than changes in the underlying DNA sequence, including DNA methylation, histone modifications, and RNA-related silencing as well.

DNA segments consist of *cis*-regulatory elements, including promoters where transcription is initiated, enhancers, repressors (also known as silencers), introns

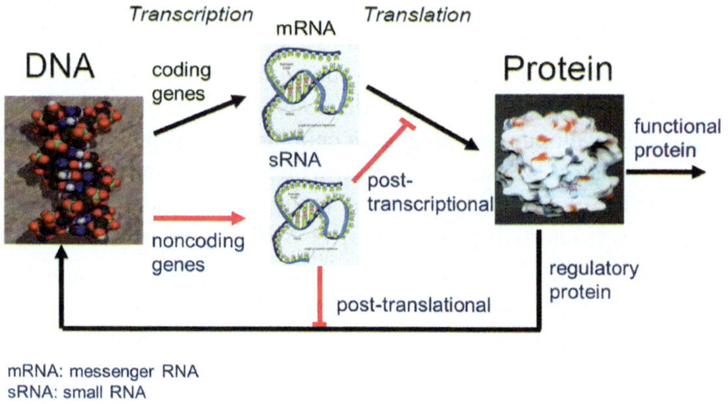

Fig. 1.4 A diagram for central dogma of molecular biology showing two pathways, one for the coding DNA that will first be transcribed into mRNA and then into proteins, the other into non-coding RNA that play a role of silencing in gene expression

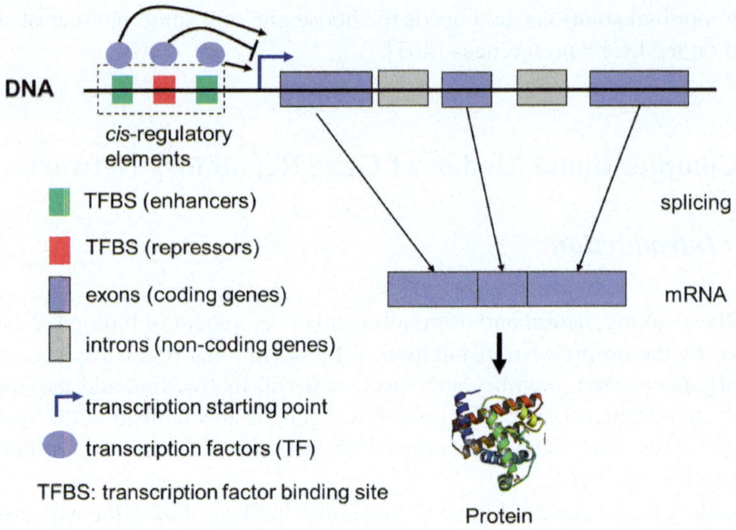

Fig. 1.5 An illustration of a DNA segment of biological systems

(non-coding genes), and exons (coding genes). The enhancers and silencers in front of coding genes will together determine whether the coding genes will be transcribed into mRNA, as shown in Fig. 1.5.

The gene expression process can be described by a gene regulatory network (GRN), in which a group of genes influence each other's level of expression. GRNs are the most important organization level within a cell [72], and computational modeling of GRNs has been the focus of a large body of research in systems biology [148]. The building blocks of gene networks remain elusive [333, 380], however, the role of each gene can be better understood by investigating their regulatory interactions within the GRN [145]. Systems biology can deepen our understanding of the GRNs responsible for the basic biological functions and their robustness, as well as the causes of their breakdowns leading to diseases [217]. It remains a grand challenge to understand the developmental process of cellular systems through the interactions between genes, proteins and small molecules, and between the cells [56].

In the following, a few widely used computational models for analyzing and synthesizing the regulatory dynamics of GRNs will be provided.

1.2.2 Computational Models of GRNs

There are several categories of mathematical models for GRNs, including discrete, continuous, and hybrid models. In addition, some models are deterministic, and

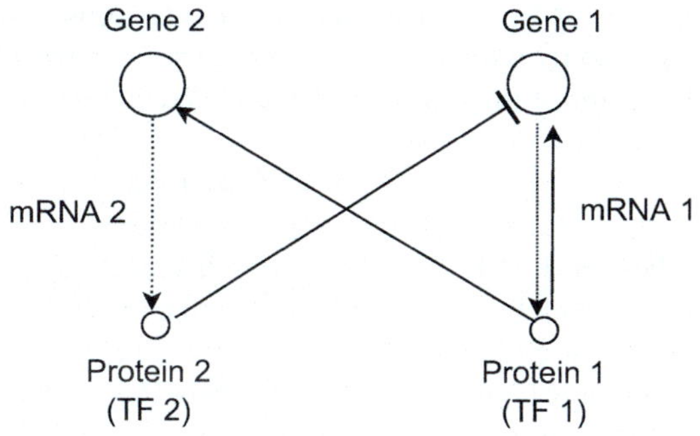

Fig. 1.6 A gene regulatory network consisting of two genes, producing two proteins which are transcription factors (TFs). TF1 activates the expression of both gene 1 and gene 2, while TF2 represses the expression of gene 1

others are stochastic. Below, the two most common GRN models, the Boolean networks and differential equation models [145], will be introduced.

1.2.2.1 Boolean Networks

The most fundamental logic models are Boolean networks [282], which are widely used for modeling complex networks, including social networks and biological networks [282]. For a Boolean network model consisting of two genes, the expression status of each gene can be either active (1) or inactive (0) and interactions can be modeled using *IF* statements. For the GRN shown in Fig. 1.6, a Boolean network will model the regulatory status as follows:

$$
\begin{aligned}
g_1 &= \begin{cases} 0 & \text{if } g_2 \text{ is } 1 \\ 1 & \text{if } g_2 \text{ is } 0 \end{cases} \quad \text{Repressor} \\
g_2 &= \begin{cases} 0 & \text{if } g_1 \text{ is } 0 \\ 1 & \text{if } g_1 \text{ is } 1 \end{cases} \quad \text{Activator,}
\end{aligned}
\tag{1.18}
$$

where g_1 activates itself and g_2, while g_2 represses g_1. As a result, each of the genes will flip from the inactive (0) state to the active (1) state due to the repression of $g1$ by $g2$. It has been demonstrated in [216] that Boolean networks are important and effective tools in understanding the steady states and robustness of GRNs.

1.2.2.2 Ordinary Differential Equations

In Boolean networks, the gene expression states can be either active or inactive, which is oversimplified in describing the gene expression level. A class of more detailed models is ordinary differential equations (ODEs), which are able to model the regulatory dynamics of each gene in the GRN, [55, 350, 464]. Since the states of ODE models can take an arbitrary real value, ODE models can more precisely describe the change of the expression level of each gene over time, although the computational complexity will become much higher. In general for an N gene network, the dynamic behavior of the GRN can be modeled as follows:

$$\dot{x_i} = f_i(x_1, x_2, \ldots, x_N), \tag{1.19}$$

where $f_i, i = 1, 2, \ldots, N$ represents the regulatory relationship between the network genes, which is commonly modeled as a Hill function [418, 456]. Hill functions are nonlinear equations that are derived from Michaelis-Menten enzymatic kinetics [9]. A possible ODE model of the GRN in Fig. 1.6 using Hill functions and the summation logic to combine the autoregulation of g_1 with the repression from g_2 will take the following form:

$$\begin{aligned} \dot{g_1} &= \frac{w_1}{2}\left(\frac{\beta_1}{1+(\phi_1/g_2)^n} + \frac{\beta_1 g_1^n}{\phi_1^n + g_1^n}\right) - \gamma_1 g_1, && \text{Repressor} \\ \dot{g_2} &= w_2 \frac{\beta_2 g_1^n}{\phi_2^n + g_2^n} - \gamma_2 g_2, && \text{Activator} \end{aligned} \tag{1.20}$$

where w_N is the interaction weight, β_N is the maximum activation, ϕ_N is the threshold for the interaction, γ_N is the degradation of the protein from gene N, and n is the Hill coefficient. Nonlinear models are favored since interactions in nature, such as gene regulation, often have nonlinear characteristics in their behavior [256, 441].

1.2.2.3 Other GRN Models

In addition to Boolean network models and ODEs, models based on Bayes' Theorem [58, 355] have also been suggested, which are directed graph models that use an edge to represent a causal relationship between two nodes. There are also models based on Pearson's correlation [92] or mutual information [275, 475] that use gene expression profiles to determine the likelihood of a connection between two genes. Other models such as Petri Nets [216] and state charts [109] are based on the state of a system.

1.2.3 Reconstruction of GRNs

It is non-trivial to understand gene regulatory dynamics in biological systems [411]. The reconstruction of biological GRNs based on expression data is one of the most complex tasks in bioinformatics and systems biology [244, 282]. Current high-throughput experiments provide abundant genome-wide gene expression measurements, making it possible to infer gene networks from the gene expression data [148]. Despite experimental advances in data collection techniques such as next generation sequencing, collecting fine-grained time series gene expression data remains highly costly, resulting in a huge amount of sparse data. That is, the total amount of data is huge due to the huge number of genes involved in a biological GRN; the data points describing the temporal change of gene expression level for each gene are still very limited. For example, for microarray time series data for 3 replicates each with 25 time points costs in the region of £30,000 (over $45,000) [322]. Furthermore, much of the collected gene expression data are heterogeneous and cannot be easily used together, making it particularly challenging to statistically reconstruct large networks [216].

As biological networks are often large [148], particularly for more complex organisms, effective and efficient techniques are required for reconstructing biological GRNs. Typically, machine learning and optimization algorithms are adopted to reconstruct the biological networks based on data, which are often noisy and incomplete [13, 148]. Optimization algorithms rely on evaluations of potential solutions without requiring a prior knowledge, making them more suited for modeling complex biological networks [451]. In addition, optimization algorithms are better suited for determining the topology of the GRN, while machine learning techniques are commonly used to estimate the parameters for a given topology.

Metaheuristics such as evolutionary algorithms (EAs) can be used to solve difficult optimization problems with little or no prior knowledge. It has been shown that metaheuristics are able to solve under-determined problems [296, 393] such as the reverse-engineering of GRNs. Although the role of elitist selection in optimization is a debated issue due to its ability to aide algorithm convergence [199, 418], it has been revealed that elitism is helpful for reconstruction algorithms since it can aide the evolutionary search and reduce the computation time [176, 248, 282]. This can be attributed to the fact that reverse-engineering biological networks typically has a large search space as a result of the system complexity, and even an overly simplified known genome can contain 200,000 interactions [148].

In reverse-engineering biological networks, a common objective is simply to minimize the error between the model output and the gene expression data [307, 460], which can be applied to both Boolean networks and ODEs [148]. This works well for small networks, but, however, can lead to over-fitting, resulting in many false positive connections if there is no constraint on the number of connections between nodes. This becomes a problem of larger networks, not only for biological relevance but it also increases computation time dramatically. Thus, in the reconstruction of GRNs, it makes sense to consider other objective functions such as sparsity and robustness to reflect biological plausibility, or information criteria [248] and penalty terms that alleviate over-fitting [59, 307]. Prior biological information can also be incorporated in optimization to assist the reconstruction process [125, 256]. In these cases, multi-objective evolutionary algorithms can be adopted. Note that several combinations of these objectives can be used, for example, error between the model and the data, sparsity, robustness, number of regulators, maximum connectivity of each node, regulation density, modularity, regulation chain length, and biological plausibility [92]. One additional benefit of multi-objective optimization is that it can provide several Pareto optimal solutions, from which the user can select one or more solutions based on his/her preference.

Finally, it has been shown that multi-objectivization [151, 263] is believed to be helpful in reconstructing GRNs in that it can speed up the convergence or increase the probability to obtain the global optimum. By multi-objectivization, we mean the process of either decomposing a single objective into multiple objectives with similar goals [45, 227], or adding 'helper' objectives [45, 175, 189], which may provide more non-dominated solutions without extra functional evaluations [152, 260]. Nevertheless, the performance of multi-objectivization may be problem-dependent; see, for example, [45, 152] or [422]. More investigations are needed to understand if multi-objectivization can accelerate convergence in reconstructing large biological networks [392].

1.3 A Gene Regulatory Model for Multi-cellular Growth

1.3.1 Motivation

Generally speaking, biological development is the progressive changes in size, shape, and function during the life of a biological system. In this process, the genotype of the biological system is translated into its genotype, i.e., its functioning mature systems. Morphogenesis, part of biological development, describes the embryonic development in which a tissue or organ develops its shape by changing the spatial distribution of cells governed by gene regulations and cell-cell interactions. Thus, it is of great importance to simulate biological developmental processes in computer systems.

For this purpose, computational models of simplified biological development must be developed. Simulated evolutionary development combines evolutionary computation with simplified models of developmental biology. To this end, the computational model must contain at least three main components, a model for cellular growth, a GRN model that controls cellular growth, and an evolutionary algorithm that evolves the GRN to achieve the desired cellular growth behavior. In the following, a model for multi-cellular development [363, 406] will be presented in detail.

1.3.2 A Model for Multi-cellular Development

Computational modeling for simulating biological development has been very popular in the late 1990s and early 2000s [79]. Most of the work focuses on designing particular phenotypic features to be achieved by the developmental process, e.g., growth of two-dimensional patterns like flags [105, 291] and tesselating tiles [31], self-repair [41], or growth of complex shapes and structures [96, 123]. However, a fundamental understanding of the critical properties in biological development, such as stabilized growth and self-repair, still lacks.

It is of critical importance to computationally model the developmental process at an appropriate abstraction level, which influences the behavior and the result of the simulation. For this purpose, a good balance between biological plausibility and computational complexity is essential. While an overly high abstraction level may limit the capability of the model of understanding biological systems, an overly detailed model makes it computationally intractable. In the following, a multi-cellular growth model regulated by a GRN will be introduced, where the developmental behavior and the final shape of the developed morphology are controlled by the GRN.

The cellular model [360, 403] to be presented below consists of multiple artificial virtual cells, each containing a virtual DNA. In addition, the physical interactions between cells are also taken into account. In the following, the genotype of the cellular model will be given, followed by a description of the cellular representation.

1.3.2.1 Genetics

In this model, cellular growth is controlled by a genome stored in a virtual DNA (vDNA). All cells contain the same vDNA, and an identical copy will be made when a cell divides. Inspired by the structure of the biological DNA, the artificial genome consists of regulatory units (RUs) and structural units (SUs), simulating the *cis*-regulatory elements and coding genes, respectively, in biological DNA segments (cf. Fig. 1.5). These units are initially generated in random order. Thus, not all SUs in the vDNA are functional. A functional unit is called a gene, which shall be composed of a group of RUs followed by a set of SUs. The SUs encode various actions once they are activated, while the RUs together, including activating (RU^+) and inhibitory RUs (RU^-), determine whether the following SUs will be expressed or not like the enhancers and silencers in biology. Here, the action encoded by an SU will be performed if it is expressed. Typical actions encoded in the vDNA include the production of transcription factors, cell division, and cell apoptosis. Both RUs and SUs are represented by a set of double precision values.

An illustrative example of a vDNA with two genes is given in Fig. 1.7. Note that the RUs behind the last SU and the SUs in front of the first RU are not functional, and thus are not taken into account in the developmental process.

- **Structural units:** An SU encodes the action to be performed, containing the parameters that specify the corresponding cellular action. Possible actions include cell division, production of the transcription factor (TF) for cell-cell signaling, and production of adhesion molecules on the cell surface, which determine cell-cell adhesion forces.

  More specifically, an SU consists of a five-dimensional vector $\mathbf{x}$ with $x_i \in [0, 1]$, $i = 1, \ldots, 5$. x_1 determines the type t of action encoded by the SU:

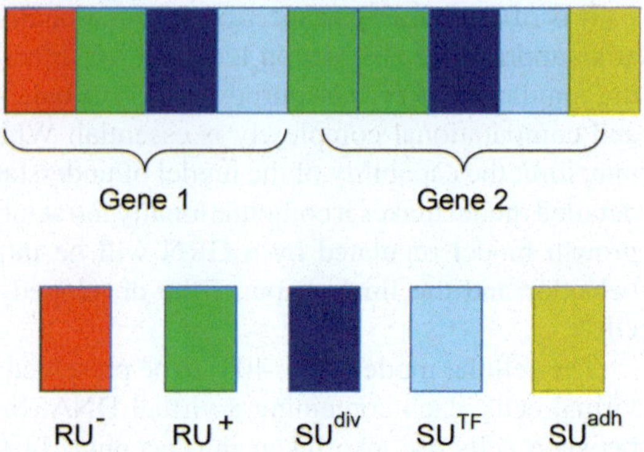

Fig. 1.7 An illustrative vDNA with two genes, each consisting of one or more regulatory units (RUs) and structural units (SUs). A RU^+ means activating, and a RU^- means repressive. An SU encoding the production of a transcription factor (TF) is denoted by SU^{TF}, an SU encoding cell division by SU^{div}, and an SU encoding cell adhesion by SU^{adh}

$$t = \begin{cases} 1 \ \forall \mathbf{x} : 0 \leq x_1 < \frac{1}{3} \\ 2 \ \forall \mathbf{x} : \frac{1}{3} \leq x_1 < \frac{2}{3} \\ 3 \ \forall \mathbf{x} : \frac{2}{3} \leq x_1 < 1. \end{cases}$$

If $t = 1$, a cell division is encoded. x_2 determines the division angle and x_3 to x_5 are not defined. If $t = 2$, a TF will be produced. There, x_2 encodes an affinity label assigned to TF (aff_{TF}), x_3 denotes the amount of TF to be released, x_4 is a diffusion constant, and x_5 represents the decay rate. In case $t = 3$, adhesion molecules are produced, whose type is specified by x_2. In this model, cells containing the same type of adhesion will adhere to each other.

- **Regulatory units:** Two types of RUs are introduced in the cellular model, which either activate or inhibit the expression of a gene. RUs can sense the presence of certain types of TFs in the neighborhood of the cell. If the label of a TF is affine to the label associated with the RU, and if the concentration of the TF is higher than a predefined threshold, an activity value is determined for each RU. The overall activity of the gene is determined by the activating and inhibiting values of all RUs of the same gene.

  In the following, an RU will be specified in detail. Each RU consists of a three-dimensional vector $\mathbf{y}$ with entries $y_i \in [0, 1]$, where $i = 1, \ldots, 3$. y_1 codes for an affinity parameter that determines the affinity between that RU and the surrounding TFs. If the affinity variable γ calculated by

$$\gamma = 0.2 - |\text{aff}_{TF} - y_1|$$

  is positive, then the TF and the RU are affine to each other.

  Let M be the number of RUs of the same gene, L_j the number of TFs that are affine to RU_j, and y_i^j the i-th entry of the vector $\mathbf{y}$ of RU_j. Then the indices k_j, which denote the TFs whose concentration c is greater than y_2^j, is determined by

$$k_j = \{k \in \{1, .., L_j\} : c_k > y_2^j\}.$$

Here, y_2^j can be seen as a threshold of RU_j. The partial activity a_{RUj} for that RU is expressed by

$$a_{RUj} = \sum_{k_j} c_{k_j} - y_2^j.$$

A sum of all partial activities, scaled by y_3, results in

$$\alpha = \sum_{j=1}^{M} a_{RUj} \cdot (2 \cdot y_3^j - 1).$$

y_3^j can be seen as the sign of RU_j, since the term in brackets is negative for $y_3 < 0.5$. The overall activity A of the gene is finally given by

$$A = \frac{2}{1 + \exp(-20 \cdot f \cdot \alpha)} - 1,$$

where f is a value encoded in the vDNA for every gene, representing the slope of the nonlinear function. If A is larger than zero, then the gene is expressed.

Note that the above definitions of SUs and RUs are designed for a particular purpose. The details can be adapted according to the developmental system to be models, e.g., neural development, or morphological development [203, 360, 401, 403]. All parameters can be evolved if the developmental model is embedded in an evolutionary system.

1.3.2.2 Cellular Interaction

To simulate the development, the size of the body plan needs to be defined, typically by grids. The TF concentration inside a grid is considered to be the same. The virtual cells are modeled as spheres with a radius of 1, which communicate with each other by reading and releasing TFs. In addition, cells also interact with each other through rigid body interactions coupled with adhesion forces in cell motion [321]. The nonlinear force function can be modified to prevent cell deformations [403]. However, it is possible to allow for a small overlap between the neighboring cells. Each cell may read TF concentrations from four neighboring grids and interpolate them, since cells are not centered and fixed in a grid. Similarly, each cell may also release TFs to four neighboring grids by increasing the corresponding TF concentration. Diffusion of the TFs is simulated by using the Euler method:

$$\frac{\mathrm{d}u}{\mathrm{d}t} = \mathrm{D}_f \nabla^2 u - \mathrm{D}_c u + s(t),$$

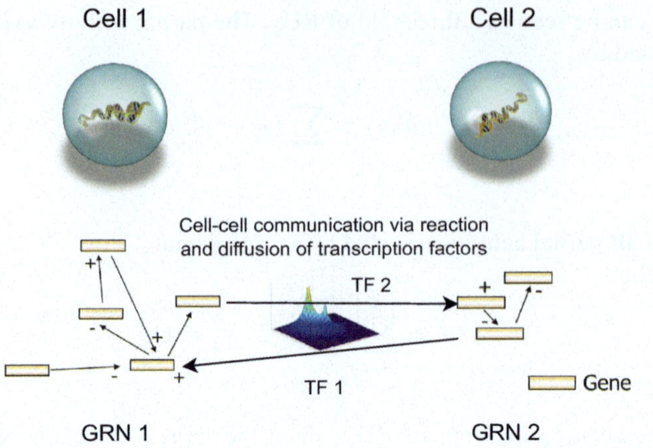

Fig. 1.8 An illustrative example of interaction between two cells via reaction and diffusion of external transcription factors (TFs). Each cell has a gene regulatory network (GRN), producing an external TF. The external TF of Cell 1 diffuses into Cell 2 and activates the expression of the corresponding gene in GRN 1, while the external TF of Cell 2 diffuses into Cell 1 and activates the expression of a gene in Cell 2. Here, physical interaction of the cells is not included

where u is the concentration of the TF, D_f is a diffusion constant, and D_c a decay constant. $s(t)$ is a source term, which signifies the increase in concentration resulting from TF generation. Note that the diffusion of each TF is simulated separately. An illustration of cell-cell communication by means of the reaction and diffusion of transcription factors is given in Fig. 1.8.

The mechanism for cell adhesion implemented in this model is fairly simple, i.e., two cells will adhere to each other if they contain the same type of adhesion molecules, indicating that they are generated by the same gene.

Several models simulating cellular growth and cell-cell interactions have been proposed. In many cases, however, cells are modeled as pixels fixed in a grid [430]. Plant cells and their physical behaviors are simulated by a spring-mass-damper system in [342]. Which cell growth model should be employed depends on the purpose of the model, and the evolved gene regulatory system controlling the cellular model may also differ significantly.

1.3.3 Cellular Developmental Process

The developmental process starts with a single cell containing the vDNA sitting in the center of the simulation area. When the growth process begins, a maternal TF will be released, which maintains a constant concentration in the whole area over the entire developmental process. In this cellular growth model, contrary to most existing models, the maternal TF does not signal any position information, different from that

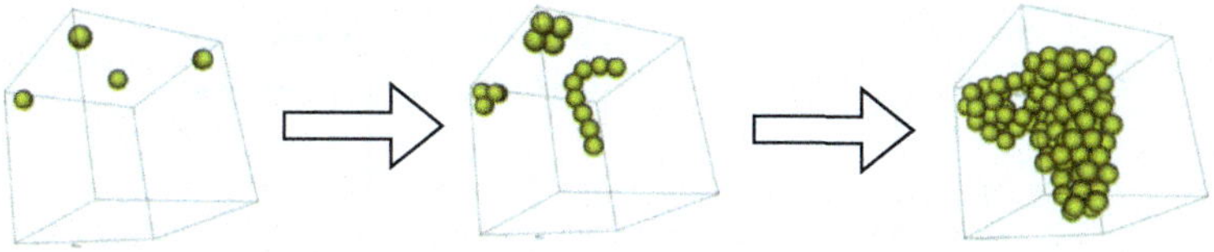

Fig. 1.9 An illustrative example of multi-cellular development

in [41, 104]. In each developmental step, the virtual DNA is translated in all cells at first. If the TFs in the vicinity of the cell activate a gene, the action encoded by that gene will then be performed. Finally, the position of all cells is updated, and the diffusion of the existing TFs will be simulated using the Euler method. An example of the cell growth process is shown in Fig. 1.9.

1.4 Modeling of Morphological and Neural Development

1.4.1 Modeling Morphological Development

In principle, the cellular model presented in the previous section can be used for modeling both morphological and early neural development, depending on the actions encoded in the structural units. For simulating basic morphological development, cellular behaviors such as division, differentiation, movement, death, as well generation of transcription factors will be adequate [255]. However, it is non-trivial to replicate a biologically plausible morphological development process for the following two reasons. First, the growth process must be self-stabilized, i.e., the cell division and cell death must reach an equilibrium state so that the growth process stops at a certain point, like in all biological processes. Second, similar to biological morphogenesis, the growth process should be capable of self-healing in the presence of limited damage to the body plane. Finally, the development must be able to generate a desired morphology, e.g., an elongated shape, a ball-like shape, or even a more complex body plane.

It is non-trivial to achieve autonomous and self-stabilized growth in computational systems, nor a self-repairing capability [362]. It is even more challenging to create a desired morphology in a self-organizing way without a hard limit of the cell growth [450]. To achieve the above properties, an effective evolutionary algorithm must be designed to evolve a gene regulatory network that governs cellular development, together with the evolution of cell-cell interactions and cell-environment interactions.

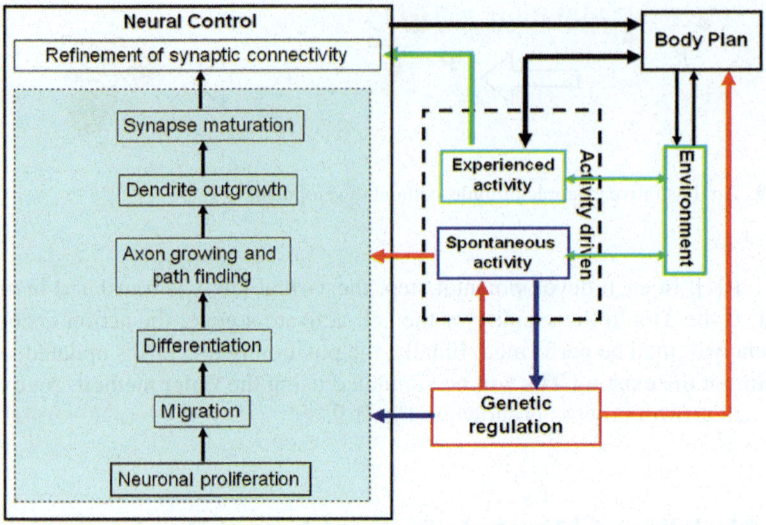

Fig. 1.10 The main procedures of neural development, in which the early development is mainly driven by genetic regulation, while the maturation and refinement of the nervous systems are mainly activity-dependent

1.4.2 Modeling of Early Neural Development

Neural development can be roughly categorized into early genetic-driven development, which lays down the blueprint of the nervous system, and activity-dependent neural plasticity that refines and matures the nervous system; refer to Fig. 1.10. Early neural development includes neuronal proliferation, migration, differentiation, axon growth, and dendrite growth. Although it is also heavily influenced by spontaneous neural activities, early neural development is mainly driven by the gene regulatory network. Therefore, the cellular growth model presented in the previous section can also be applied for modeling early neural development [203], if the structural units encode these neuronal behaviors, and a correct temporal order of these behaviors can be ensured. The hardness here in modeling early neural development is the autonomous genetic control of the developmental process.

1.5 Activity-Dependent Neural Plasticity

1.5.1 Types of Neural Plasticity

Various types of neural plasticity have been observed in experiments and thus also investigated in computational modeling. Generally speaking, neural plasticity can be divided into the following three types.

- *Synaptic plasticity.* Synaptic plasticity means the strengthening or weakening of the synapses that connect neurons and facilitate the transmission of electro-chemical signals [64]. The change in synaptic strength can result from a change in the abundance of neuro-transmitter molecules at the synapse, or due to an increase or decrease in the conduction of post-synaptic receptors. In general, synaptic plasticity refers to any models that directly change the strength of the connections between neurons, including Hebbian, anti-Hebbian rules, and Oja's rule.
- *Intrinsic Plasticity.* Intrinsic plasticity denotes modification of a neuron's intrinsic ability to generate or propagate action potentials [46]. This process is triggered by the neuron itself rather than the synapses between neurons. Intrinsic plasticity can often be seen as the self-regulation of a neuron's activity, which is a homeostatic mechanism to keep the activity within a practical range.
- *Homeostatic plasticity.* Although plasticity based on the correlations between pre- and post-synaptic activity is the most popular and successful, it has been widely shown that weight changes are driven by synaptic plasticity, since increases or decreases in synaptic weights may result in a positive feedback loop by activating or inhibiting the neuronal activities. Hence, homeostatic processes are essential for the stable learning of neural networks by regulating the total or averaging firing rate of a neuron.
  Two main homeostatic mechanisms are synaptic scaling and meta-plasticity. The former refers to a uniform, cell-wide modulation of synaptic efficacies [381] whereas the latter is a uniform, cell-wide modulation of the modification threshold for synaptic plasticity, each controlled by a long-term average of pre- or post-synaptic firing rate [3].

1.5.2 Hebbian Rules

An important landmark in the basis of associative learning in the brain came from Hebb [157] in the form of a highly influential postulate. In essence, it states that if neurons are active at the same time, connections between them will become stronger to facilitate further co-activation. In this way, the coincident neural activity becomes an associated activity and provides a mechanistic basis for the idea of associative learning that, until Hebb's postulation, had existed largely as an abstract notion.

There are numerous quotes from [157] that describe Hebb's postulate. In essence, it says that the connections between two neurons will be strengthened if they are repeatedly active at the same time. On the contrary, the connections will become weaker if one is active and the other is not. This can be described mathematically as follows:

$$\Delta w_i = \eta x_i y, \tag{1.21}$$

where Δw_i is the change in synaptic strength, which increases proportionally to the product of the pre- (x_i) and post- (y) synaptic activity, and $\eta > 0$ is a learning rate. From Eq. (1.21), we can see that the connection strength can only increase, since there is no term that will reduce the synaptic strength, leading to instability in learning. Although an upper bound can be set for the synaptic strength, an inappropriate threshold will lead to saturation that prevents the network from further learning.

1.5.2.1 Anti-Hebbian Learning

An inverted form of Hebbian learning is called anti-Hebbian [21], which is observed to be active in decorrelating sensory inputs in taste and vision between laterally connected ganglion cells. Anti-Hebbian learning can ensure that the output signals of these cells represent changes in the input signal in the most efficient way.

The initial outputs may be correlated because of overlaps in the cell sensitivities to the inputs. After a period of anti-Hebbian training on the lateral connections between the cells, the variables are decorrelated. This is shown to produce a larger spread of output values for correlated inputs and therefore increases the sensitivity for small changes in the input which leads to a more efficient representation of the input variables.

The equation for anti-Hebbian adaptation takes the same form as Eq. 1.21, except that the learning rate must be negative.

1.5.2.2 Oja's Rule

Oja's rule [312] is a modification of the plain Hebbian interpretation that aims to address the stability problem of an exclusively potentiating mechanism. The rule is given as follows:

$$\Delta w_i = \eta(x_i y - y^2 w_i). \tag{1.22}$$

This adaptation mechanism is similar to the Hebbian rule in Eq. (1.21), except that it includes a depressive term, $-y^2 w_i$. The strength in synaptic decrease is proportional to both the current weight of the connection strength and the square of the post-synaptic activity. The larger the weight and the resultant post-synaptic activity is, the greater the depression will be. The depression term in Oja's rule represents self-regulation in plasticity-based learning to make Hebbian-type learning stable.

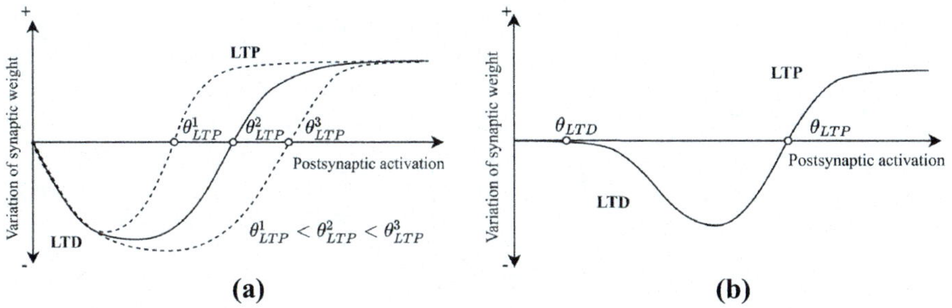

Fig. 1.11 Illustration of **a** the classic BCM rule and **b** a dual-threshold BCM rule

It is shown [312] that when the rule is applied to the incoming synapses of a single neuron, the post-synaptic response extracts the principle component of the presynaptic input vector. This is observed after the rule has been applied long enough to the synapses that have converged to a stable state where the average weight change across the inputs is zero.

1.5.3 BCM Plasticity Rules

The BCM rule [35] is a rate-based Hebbian rule based on a temporal average of pre- and post-synaptic activity. It also regulates the post-neuron firing rate to a desired level as shown in Eq. 1.23:

$$\frac{dw_i}{dt} = y(y - \theta_M)x_i - \epsilon w_i, \tag{1.23}$$

where Θ_M is the regulating parameter, a dynamic threshold changing according to $E[y/y_0]$, where $E[\cdot]$ denotes an average over time, y is the post-synaptic activity, and y_0 is the desired level. A decay parameter ϵw_i is included in the BCM rule for additional stability, which slowly reduces the connection strength and provides a mechanism for uniform weight decay, regardless of the level of activity or correlation. A plot of the BCM weight change is presented in Fig. 1.11a.

It has been argued that existing BCM learning rules can only select a set of fixed rule parameters, which is biologically implausible and practically inflexible to learn the structural information of input signals. To address this, a dual-threshold BCM rule is evolved to regulate the reservoir internal connection weights of the echo-state-network [446], which can contribute to alleviating information loss and enhancing learning performance by introducing different optimal long-term depression thresholds for different post-synaptic neurons. The basic update rule is nearly the same as in Eq. (eq:bcm), which, however, has two thresholds:

$$\frac{dw_i}{dt} = y(y - \theta_{LTP})x_i - \epsilon w_i, \tag{1.24}$$

where w_i is defined as follows:

$$w_i = \begin{cases} w_i + \frac{dw_i}{dt}; & \text{If } y > \theta_{LTP} \\ w_i; & \text{If } y \leq \theta_{LTD}. \end{cases} \tag{1.25}$$

It is shown [446] that the modified BCM rule with two thresholds being optimized by an evolutionary algorithm can result in the synergistic learning of different plasticity rules, effectively improving the learning performance of an echo state network in comparison with other neural plasticity rules on three widely used benchmark tasks.

1.5.4 Homeostatic Regulation

In early simulations incorporating Hebbian learning, it became apparent that a mechanism to reduce synaptic weight is needed to allow stable learning. Von der Malsburg [273] proposed a synaptic scaling technique to keep the sum of the synaptic weights connected to a post-synaptic neuron constant. As a result, any increases in synaptic weight will be automatically balanced out with a decrease in the others, thereby enabling all synapses to change further according to the neural activity.

According to [273], the synapses are normalized in the following way:

$$w_{ik} = w'_{ik} \cdot N \cdot \frac{w_{\text{avg}}}{w'_k}, \tag{1.26}$$

where w'_{ik} is the weight before normalization connecting from neuron i to neuron k, w_{ik} is the normalized synaptic weight, and w'_k is the sum of all weights fanning into neuron k. Finally, w_{avg} is the average weight value of all synapses, where N is the number of driving inputs.

1.5.5 Spike Timing-Dependent Plasticity Rules

All plasticity mechanisms introduced above are based on rate-based neuron activation models, in which the activity level of a neuron is assumed to be its instantaneous rate of generating action potentials. Consequently, these plasticity rules are not directly applicable to spiking neural network models that account for the precise timing of action potentials rather than the average rate of firing.

Experimental studies in several studies [34, 108, 458] have demonstrated that the increase or decrease of the synaptic strength and the amplitude of the change depend

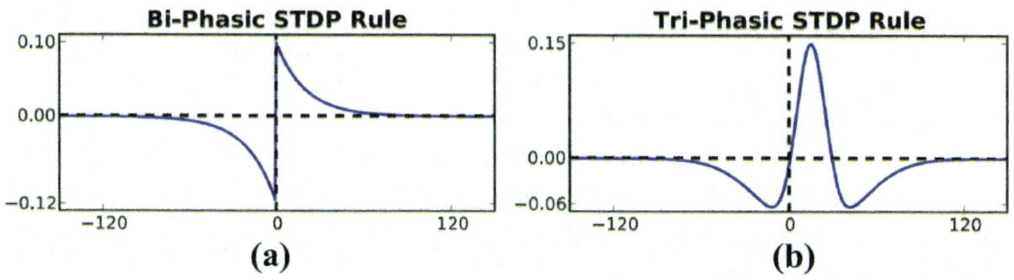

Fig. 1.12 Two main STDP plasticity rules. **a** Bi-phasic STDP rule, and **b** Tri-phasic STDP rule

on both the order of pre- and post-synaptic spikes and the time delay between them. Experimental findings in this research indicate the observed 'learning windows' of spike timing-dependent plasticity (STDP). According to STDP, the synapse between two neurons will be strengthened if the timing of the input spike is slightly earlier than the spike of the output neuron; vice versa, the synapse between two neurons will be weakened if the timing of the input spike is slightly later than the spike of the output neuron. Theoretically, these learning windows present a temporal interpretation of Hebbian learning in which causal co-activation is reinforced and anti-causal co-activation is diminished with respect to the timing of the spikes. Although no explicit self-regulation is included in the standard STDP, the stability of the learning can be achieved due to the depressive regions in the learning window. In the following, we discuss two main computational models that can approximate the observed biological data, namely the Bi-phasic STDP rule and Tri-phasic STDP rule found in neocortical and hippocampal regions of the brain, respectively, as plotted in Fig. 1.12.

1.5.5.1 Bi-phasic STDP Rule

The STDP rule depends on the temporal correlation between pre- and post-synaptic spikes. The synaptic weight change is computed based on the delay between the firing times of the pre- and post-neuron. This is described in a fixed 'learning window' in which the y-axis is the level of weight change, and the x-axis is the time delay between a pre- and post-synaptic spike occurrence. The bi-phasic STDP rule consists of two decaying exponential curves [396], a positive one to potentiate in-order spikes, and a negative one to depress out-of-order spikes. This rule was derived from experimental work carried out on populations of neurons in vitro [34, 276]. Bi-phasic STDP is given in Eq. 1.27:

$$\Delta w(t) = \begin{cases} A_{+} \cdot \exp\left(\frac{-\Delta t}{\tau_{+}}\right) & \text{if } t > 0 \\ A_{-} \cdot \exp\left(\frac{\Delta t}{\tau_{-}}\right) & \text{if } t \leq 0. \end{cases} \tag{1.27}$$

1.5.5.2 Tri-phasic STDP Rule

A tri-phasic STDP learning window consists of a narrow potentiating region for closely correlated activity but depressing regions on either side: for recently uncorrelated activity, and for correlated but late activity. This learning window has been observed in vitro, most notably in the hippocampi, between areas CA3 and CA1 [458]. The tri-phasic STDP is given in Eq. 1.28:

$$\Delta w(t) = A_+ \cdot \exp\left(\frac{-(x-20)^2}{200}\right) - A_- \cdot \exp\left(\frac{-(x-20)^2}{2000}\right). \tag{1.28}$$

Although the two STDP rules described above can well approximate the biological findings, subsequent empirical studies have revealed a more complex relationship between spike-timing, firing rate, stimulus duration, and post-synaptic bursting in controlling the conductance of hippocampal excitatory synapses. In [48], a more detailed biophysical model of single dendritic spines on a CA1 pyramidal neuron was proposed. The model is a parsimonious model of Calcium-driven kinase and phosphotase dynamics that determine the transitions between binary synaptic weight states and are able to describe the NMDAr-dependent Calcium influx generated by different stimulation protocols.

1.6 Summary

Evolution, learning, and development are three adaptation mechanisms in biology. Building computational models for these mechanisms is indispensable both for better understanding these processes on the one hand, and for borrowing ideas from them for solving real-world problems. Note, however, that all the models described in this chapter are at a very high abstraction level, because many biological details are left out. While the computational models of gene regulatory networks lay the foundations for the genetic-driven neural and morphological development, the neural plasticity rules focus on activity-dependent developmental plasticity. Note also that activity-dependent plasticity can be seen as unsupervised learning, and therefore, two other fundamental learning paradigms, supervised learning [344] and reinforcement learning [410], are not touched upon in this chapter.

Chapter 2
Analysis of Gene Regulatory Networks

2.1 Robustness and Evolvability

It is extremely challenging and of paramount importance to understand how natural evolution has shaped the huge biological diversity and complexity of simple particles and molecules. In addition to environmental changes, it is believed that two important principles, i.e., robustness and evolvability, may have played a central role in evolving biological diversity and complexity.

2.1.1 Phenotypic Robustness

Biological robustness means the organisms' ability to maintain relatively their functionality in the presence of a certain degree of internal and external perturbations. Since DNA is subject to constant mutations, insensitivity of the most important biological functions to genetic mutations is indispensable for biological systems to survive. Another important biological property in contrast to robustness is evolvability, which refers to the organisms' ability to evolve heritable novel phenotypic functionalities that help the organism survive and reproduce in a changing environment. Over the past decades, research on robustness and evolvability has become one of the main research topics in systems biology [5, 220, 433, 442].

However, research on understanding biological robustness and evolvability is still in its infancy, and many questions remain open [5, 63, 249]. A well-established quantitative definition for biological robustness and evolvability is still missing, and the biological origin of various biological mechanisms for robustness and evolvability remains elusive.

It is conceivable that robustness and evolvability are two closely related properties, simply because without robustness, evolutionary tinkering will most likely lead to lethal consequences, preventing evolution from creating new functionalities. A

Y. Jin, *Computational Evolution of Neural and Morphological Development*, Natural Computing Series, https://doi.org/10.1007/978-981-99-1854-6_2

is investigated. Innovation was defined as the average local variability (the probability of encountering a new phenotype in the presence of a genotypic mutation). The connection matrix of the mapping was then evolved using a multi-objective evolutionary algorithm. It was shown that there is an inherent trade-off between innovation and robustness.

2.1.3 Trade-Off Between Robustness and Evolvability

Natural evolution can never be single-objective, including the capability to survive and reproduce. Trade-offs between different evolutionary targets have accompanied the history of evolution, mainly due to the limited amount of available food, energy, and time to adapt. A number of trade-off examples in biological systems is listed below

- A trade-off between energy efficiency and functionality in the evolution of the brain size and organization of nervous systems. Many biological findings indicate that the intelligence level of organisms is approximately proportional to their relative brain size, e.g., the ratio between brain weight to body weight. However, a bigger brain is evolved at several costs [8]. First, a big brain often means more energy consumption, and therefore more food consumption, which is always limited. Second, animals with a bigger brain usually have a longer life span, implying that they are likely to experience more extreme environmental changes and thus harder survival environments. For example, in the evolution of nervous systems, it has been shown that energy efficiency has been the main constraint for evolving their functionality. Interestingly, results from computational brain-body co-evolution reveal that such kind of trade-off may lead to the emergence of biologically plausible nervous and body plan structures [208].
- Reproduction and survival (longevity) have been revealed to be two conflicting objectives [400], where evidence has been presented showing that male fruit flies supplied with virgin females have lower longevity than those kept without access to females [299]. This trade-off relationship has also been shown to be important in evolving an optimal lifetime in an artificial evolutionary system [319].
- There is a trade-off between the number of offspring and their size [53].
- Trade-offs have been found in many research topics of bioinformatics and computational biology. In protein sequence alignment, for example, the maximization of the number of matching bases and the minimization of the number of gaps may be two conflicting objectives. In addition, conflicting objectives must also be considered in constructing phylogenetic trees and reverse engineering biological gene regulatory networks [150].

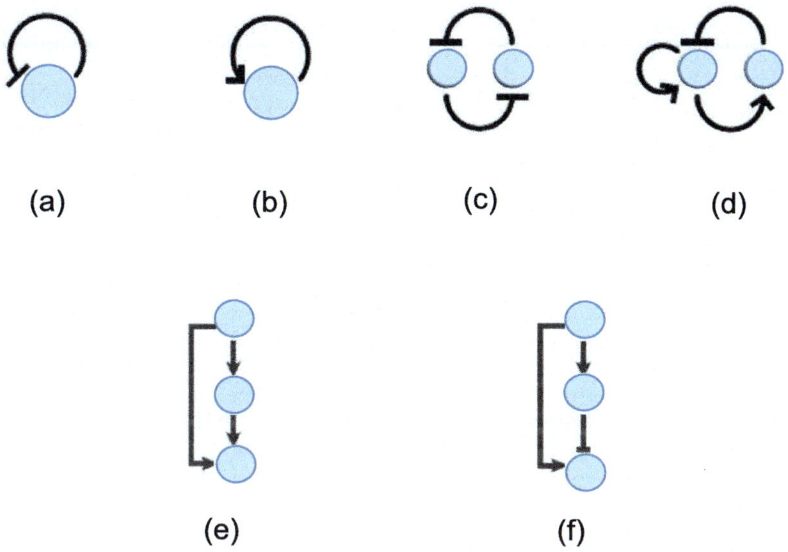

Fig. 2.2 Six typical network motifs: **a** Negative autoregulation; **b** Positive autoregulation; **c** toggle switch; **d** relaxation oscillator; **e** coherent feedforward loops; and **f** incoherent feedfoward loops (adapted from [9])

2.2 Network Motifs and Connectivity

2.2.1 Typical Network Motifs and Their Properties

Research findings in computational systems biology indicate that computational models for development are helpful for us to understand the genetic and cellular mechanisms underlying biological development. In particular, analysis of small, frequently occurring network structures, also known as network motifs, has received much interest [10] and achieved impressive success. Analysis of biological data revealed that such motifs can widely be identified in bacteria and yeast; see, e.g., [17]. It has been found that some motifs may have played an essential role in evolution. For instance, the role of feedback loops is analyzed in [237] and found that more positive feedback loops and less negative feedback loops contribute to the robustness of gene regulatory systems.

Network motifs are sub-networks that occur more often in biological gene regulatory networks than expected at random. In the following, we analyze the occurrence of different types of regulatory motifs, such as autoregulation and feedforward loops, as shown in Fig. 2.2. In the following, we elaborate on the function of a few typical network motifs [10].

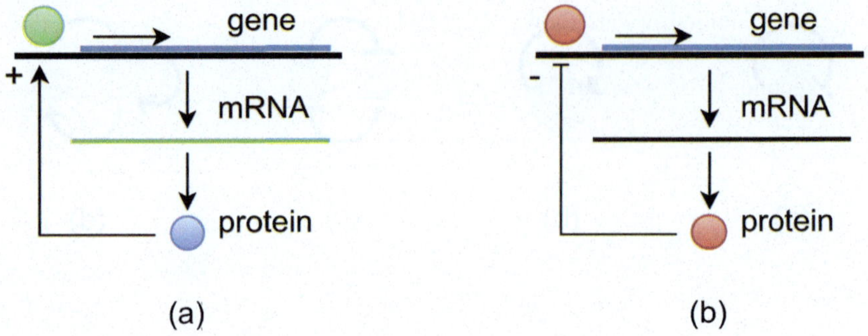

Fig. 2.3 Two most widely seen genetic motifs: **a** Positive autoregulation; **b** negative autoregulation

- The positive autoregulation (PAR) slows down the response time and can result in bi-stability; see Fig. 2.3a. A PAR can be described by the following ordinary differential equation (ODE):

$$\dot{x} = \alpha_1 f(x) - \alpha_2 x, \tag{2.1}$$

$$f(x) = \frac{\beta x^n}{\theta^n + x^n}, \tag{2.2}$$

where x is the concentration of the protein product, $\alpha_1 > 0, \alpha_2 > 0$ are production rate and decay rate, respectively, $f(x)$ is the increasing Hill function, and θ is the dissociation constant.
- Negative autoregulation (NAR) refers to a gene whose regulatory protein (transcription factor) directly represses its own expression, as shown in Fig. 2.3b. Such motifs can accelerate the response time compared to a gene without NAR with the same steady state. It leads to steady states with a rapid rise and a sudden saturation. NAR also promotes robustness. An NAR can be described by the following ODEs:

$$\dot{x} = \alpha_1 f(x) - \alpha_2 x, \tag{2.3}$$

$$f(x) = \frac{\beta \theta^n}{\theta^n + x^n}, \tag{2.4}$$

where $f(x)$ is the decreasing Hill function.
- Toggle switches consist of the cross-repression between the two genes and result in a definite choice between two states. Therefore, from the dynamics systems point of view, toggle switches have two stable states (bi-stability). Depending on an external signal, the toggle switch is able to produce a sharp transition between two steady states of two expression levels.
- A relaxation oscillator involves two genes, in which one gene activates the expression of the other, while the latter represses the expression of the former. Therefore, the relaxation oscillator consists of one positive and one negative feedback loops.

- The coherent feedforward loop 1 (C1-FFL) results in a fast convergence to a steady state but a slow decrease of the concentration.
- The incoherent feedforward loop 1 (I1-FFL) can act as a pulse generator. It can turn a concentration very fast on with an overshoot, and then it converges to its steady state.

There are a lot of different FFLs, among which C1-FFL and I1-FFL are the most frequent ones in E. coli and yeast.

The functional analysis described above is performed on isolated motifs, and therefore their behavior in a whole network can vary depending on other genetic parts of the network. Therefore, we can distinguish static motifs from dynamic motifs. All possible connections of a network define the static network and static network motifs are those in a static network. By contrast, a network that is really used during the development is called a dynamic network, consisting of connections that have been activated (the expression level is above the threshold) in at least one cell during the development. By definition, there are more static motifs than dynamic ones.

2.2.2 Connectivity of Genetic Networks

Many complex networks, including biological networks such as brain networks and gene regulatory networks, show typical types of network connectivity such as small-world networks [43] and scale-free networks.

2.2.2.1 Random Networks

In a random network, connections (edges) between any nodes (vertices) are generated with a connection probability of following a uniform distribution. A random network can be obtained by starting with a number of isolated vertices (nodes), and then successively adding edges (connections) between the nodes at a given probability. In a gene regulatory network, a node represents a gene, and a directed connection/edge represents a regulatory relationship.

2.2.2.2 Small-World Networks

Another widely investigated network connectivity defined by the distance between vertices is called the small-world network, in which all vertices are more likely connected to their neighbors. The small-world network has a small average path length and a high clustering coefficient, and the average path length between any two vertices is proportional to the logarithm of the number of vertices. A small-world network model connects two vertices according to a connection probability and their distance. As a result, a pair of vertices will be more likely connected if their distance is short. In addition, there will be no connections between two vertices if the distance between them is larger than a threshold. The pseudo code for generating a small-world network is given in Algorithm 1, where p_f and p_b represents the connection probability of v_i to v_j and v_j to v_i, respectively. The proposed small-world network model connects two vertices according to the connection probability and their distance. As a result, a pair of vertices will be more likely connected if their distance is short. Additionally, no connections can be established when the distance between two vertices exceeds the distance threshold θ.

Algorithm 1: Generation of Small World Networks

Input: The number of vertices N, the connection probability p_f and p_b, the distance threshold θ.
Output: The adjacent matrix A.

Initialize $A = \{0\}_{N\times N}$;
Add N initial vertices; # Add connections with probabilities p_f, p_b and the distance threshold θ. #
for *Each two vertices* v_i *and* v_j **do**

$$decay = \begin{cases} 0 & |i-j| \geq N \times \theta \\ 1 - \frac{|i-j|}{N\times\theta} & |i-j| \leq N \times \theta \end{cases}$$

  if *With the probability* $p_f \times decay$ **then**
    Add an edge $e_{i,j}$, set $a_{i,j} = 1$;
  end
  if *With the probability* $p_b \times decay$ **then**
    Add an edge $e_{j,i}$, set $a_{j,i} = 1$;
  end
end
return A;

2.2.2.3 Scale-Free Networks

Scale-free networks are the most widely seen network connectivity in many real-world networks. In a scale-free network, some nodes have more connections than the rest in the network structure. The degree distribution $p(d)$ of a scale-free network asymptotically follows the power law, i.e., $p(d) \sim k^{-\gamma}$, where d is the degree of the vertex and γ is a parameter generally with the range $2 \leq \gamma \leq 3$. The Barabási-Albert model is the most widely known generative model of the scale-free network for an undirected graph with $\gamma \approx 3$ [6].

The pseudo code for generating a scale-free network is provided in Algorithm 2, where p_α, p_β, and p_γ are three parameters for controlling the connectivity of the resulting network.

Algorithm 2: Generation of Scale-free Networks

Input: The number of vertices N, the connection probability p_α, p_β and p_γ.
Output: The adjacent matrix A.

Initialize $A = \{0\}_{N \times N}$;
Add two initial vertices v_0, v_1;
The current number of nodes $n = 2$;
Add an edge $e_{0,1}$, set $a_{0,1} = 1$;
while $n < N$ **do**
    Calculate the in-degree d_{in} and out-degree d_{out}; # Add connections with probabilities p_α, p_β and p_γ. #
    if *With the probability* $\frac{p_\alpha}{p_\alpha+p_\beta+p_\gamma}$ **then**
        Select an input vertex v_j with the probability $\frac{d_{in}(v_j)}{m+n}$;
        Add new vertex v_{n+1} and the edge $e_{n+1,j}$;
        Set $a_{n+1,j} = 1, n = n + 1$;
    else if *With the probability* $\frac{p_\beta}{p_\alpha+p_\beta+p_\gamma}$ **then**
        Select an input vertex v_j with the probability $\frac{d_{in}(v_j)}{m+n}$;
        Select an output vertex v_i with the probability $\frac{d_{out}(v_i)}{m+n}$;
        if $e_{i,j}$ *is not exist* **then**
            Add new edge $e_{i,j}$;
            Set $a_{i,j} = 1$;
        end
    else if *With the probability* $\frac{p_\gamma}{p_\alpha+p_\beta+p_\gamma}$ **then**
        Select an output vertex v_i with the probability $\frac{d_{out}(v_i)}{m+n}$;
        Add new new vertex v_{n+1} and edge $e_{i,n+1}$;
        Set $a_{i,n+1} = 1, n = n + 1$;
    end
end
return A;

2.3 Robustness-Evolvability Trade-Off in a Boolean Genotype-Phenotype Map

In this section, we examine the relationship between the robustness and evolvability of a gene regulatory network modeled by a Boolean regulatory network [196]. This is investigated by simultaneously maximizing the robustness and evolvability of a network by optimizing the genotype-phenotype map. To this end, we first present a Boolean network model, whose connectivity is subject to evolution for optimization. This is followed by a quantitative definition of robustness and evolvability based on the model. Finally, the non-dominated solutions that present trade-offs between robustness and evolvability will be analyzed and discussed.

2.3.1 A Boolean Model

Consider the following genotype-phenotype mapping from n-dimensional genotype space $\mathscr{G} \in \{0, 1\}^n$ to m-dimensional phenotype space $\mathscr{P} \in \{0, 1\}^m$:

$$p_i = f_i^{k_i}(c_{i1}(g_1), c_{i2}(g_2), \ldots, c_{in}(g_n)), \tag{2.5}$$

where p_i, $i = 1, 2, \ldots, m$ is the i-th phenotype trait, g_j, $j = 1, 2, \ldots, n$ is the j-th genotype, and $\mathbf{C} = (c_{ij})_{m\times m}$, where

$$c_{ij} = 1, \iff \text{phenotype trait } p_i \text{ is affected by gene } g_j \tag{2.6}$$

$$c_{ij} = 0, \iff \text{phenotype trait } p_i \text{ is independent of gene } g_j. \tag{2.7}$$

$f_i^{k_i}(.)$ is a Boolean function with k_i inputs, where

$$k_i = \sum_{j=1}^{n} c_{ij} \tag{2.8}$$

is also known as the arity of the Boolean function.

If $k_i > 1$, that is, if phenotype p_i is influenced by more than one genotype, it is called polygeny. By contrast, the number of phenotype traits affected by gene g_j is given by the sum of the elements in the column:

$$l_j = \sum_{i=1}^{m} c_{ij}. \tag{2.9}$$

If $l_j > 1$, then genotype g_j is said to be pleiotropy. Without loss of generality, it is assumed that $\mathbf{C}$ is chosen such that $l_j \geq 1$ for all $i = 1, \ldots, n$. This means that each gene affects at least one phenotype trait.

The mapping can be defined by the dependencies between genes and phenotype traits, and a set of Boolean functions that determine the values of the phenotype traits. Since the connection between the genotype and phenotype determines the number of inputs for a certain phenotype trait and thus for the corresponding Boolean function, the dependencies between genes and phenotype traits are determined at first. Once the connection matrix $\mathbf{C}$ has been fixed, the Boolean functions $f_i^{k_i} : \{0, 1\}^{k_i} \rightarrow \{0, 1\}$, $i = 1 \ldots m$ can then be defined.

An example of genotype-phenotype mapping with eight genes and four phenotype traits is given in Fig. 2.4, where the arity of the Boolean function for the four phenotypes is three, six, six, and two, respectively. The connection matrix in this example is as follows:

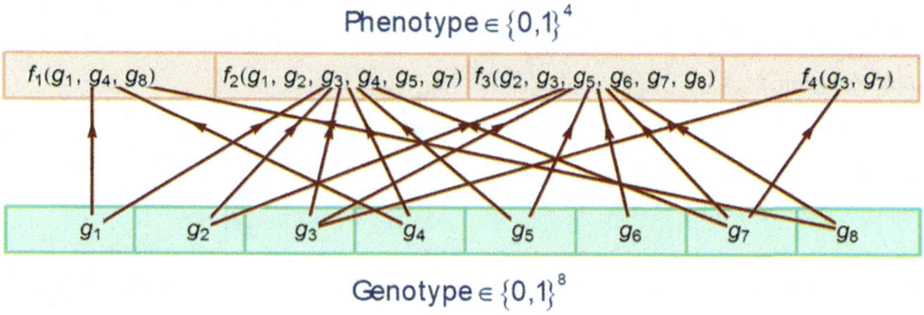

Fig. 2.4 An example of genotype-phenotype mapping, where the number of genes is eight, and the number of phenotype traits is four

$$\mathbf{C} = \begin{pmatrix} 1\,0\,0\,1\,0\,0\,0\,1 \\ 1\,1\,1\,1\,1\,0\,1\,0 \\ 0\,1\,1\,0\,1\,1\,1\,1 \\ 0\,0\,1\,0\,0\,0\,1\,0 \end{pmatrix}. \quad (2.10)$$

In the following, the connection matrix **C** will be evolved together with the Boolean functions to maximize the robustness and evolvability using the NSGA-II [81], which is one of the most popular evolutionary multi-objective optimization algorithms.

2.3.2 Encoding of the Boolean Model

The first stage of the multi-objective evolutionary approach is to determine the scheme for representing the genotype-phenotype mapping model. Here, an encoding with a fixed coding length and relatively compact is adopted.

The encoding of the Boolean model consists of two parts: the encoding of the connection matrix C and the encoding of the Boolean functions f_i, $i = 1, \ldots, m$. The encoding of C is trivial, since each entry c_{ij} can be written in a binary string, leading to a fixed encoding length mn, which is independent of the actual value of C.

It is more subtle to find the encoding for f_i. The connection matrix C determines the polygeny $k_i = \sum_{j=1}^{n} c_{ij}$ of each phenotype trait p_i, i.e., the number of inputs of the Boolean function $f_i^{k_i}$. Since $f_i^{k_i}$ is a Boolean function $f_i^{k_i} : \{0, 1\}^{k_i} \rightarrow \{0, 1\}$, it is fully defined when the corresponding outputs for each 2^{k_i} inputs are determined. It is clear that such a canonical encoding is dependent on the actual value of the connection matrix **C**. During the evolutionary search, the size of the encoding is varying and a set of *ad hoc* search operators must be designed to ensure the feasibility of the newly generated solutions.

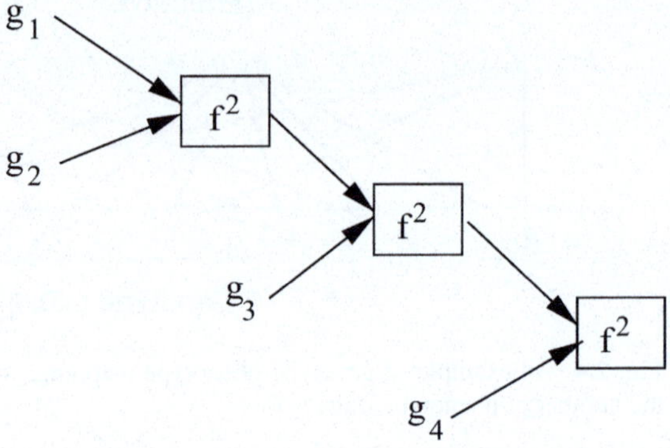

Fig. 2.5 An example of 4-ary extension of a 2-ary Boolean function $f^4 \uparrow_{f^2}$

A simpler option is to define an encoding for $f_i^{k_i}()$ independent of its arity k_i. For this purpose, it is assumed that $k_i = n$ in that a phenotype trait cannot depend on more than n genes. However, this will result in an encoding with a length of $nm + 2^n m$, making the length of the chromosome become intractably large even for a medium size of genotype and phenotype spaces, e.g., $m = 8$ and $n = 16$. To resolve this problem, additional restrictions on the Boolean model can be made to achieve a reasonably small encoding length. Since the encoding of the Boolean functions $f_i^{k_i}()$ and its dependency on k_i is the most important, we can define a class of Boolean functions that are independent of the polygeny k_i.

Let $f^{k_e} \in \mathscr{F}^{k_e}$ denote an arbitrary Boolean function with an arity of k_e. Then the k-ary extension of f^{k_e}, $f^k \uparrow_{f^{k_e}} : \{0, 1\}^k \to \{0, 1\}$, is recursively defined by

$$f^k \uparrow_{f^{k_e}} (g_1, \ldots, g_k) = \begin{cases} f^{k_e}(g_1, \ldots, g_{k_e}), & \text{if } k = k_e; \\ f^{k_e}(f^{k-1} \uparrow_{f^{k_e}} (g_1, \ldots, g_{k-1}), g_k), & \text{if } k > k_e; \end{cases} \tag{2.11}$$

for $k \geq k_e$. If $k < k_e$, then,

$$f^k \uparrow_{f^{k_e}} (g_1, \ldots, g_k) = f^{k_e}(\underbrace{1, \ldots, 1}_{k_e - k}, g_1, \ldots, g_k), \tag{2.12}$$

where the first $(k_e - k)$ positions are set to 1. f^{k_e} is called the elementary Boolean function and k_e its elementary arity. Figure 2.5 shows an example of a 4-ary extension of a 2-ary Boolean function f^2.

Having defined the k-ary extensions $f^k \uparrow_{f^{k_e}}$ of Boolean functions, we consider the following three different approaches to encoding the Boolean functions, each resulting in a reasonably compact and fixed encoding length.

2.3.2.1 Multiple Boolean Functions

All $f^k \uparrow_{f^{k_e}}$ have the same encoding size 2^{k_e}, independent of their actual arity k, since only the elementary Boolean function f^{k_e} must be encoded. As a result, a restricted version of the Boolean model can be defined by determining a fixed elementary arity k_e for the model and choosing every Boolean function $f_i = f^k \uparrow_{f_i^{k_e}}$, $i = 1, \ldots m$, where k_i is the polygeny of the corresponding phenotype trait p_i determined by the connection matrix **C**. Note that each phenotype function, f_i is the k-ary extension of different elementary Boolean functions $f_i^{k_e}$. This model restriction will result in an encoding size of $nm + m2^{k_e}$ bit, which is reasonably small for $k_e = 2, 3, 4$. The structure of the encoding is schematically described below

$$\overbrace{\boxed{c_{11}}\boxed{c_{12}}\boxed{\cdots}\boxed{c_{mn}}}^{\text{connection matrix}} \quad \overbrace{\boxed{f_1^{k_e}}\boxed{f_2^{k_e}}\boxed{\cdots}\boxed{f_m^{k_e}}}^{\text{elementary Boolean functions}} \tag{2.13}$$

2.3.2.2 Single Boolean Function Encoding

An even simpler encoding can be realized when the Boolean model is restricted to a single elementary Boolean function. To this end, a fixed elementary arity k_e for the model is determined and the local Boolean functions are k_i-ary extensions of the same elementary Boolean function: $f_i := f^k \uparrow_{f^{k_e}}$, $i = 1, \ldots m$. This restriction will result in an encoding size of $nm + 2^{k_e}$ bits. The structure of the encoding is visualized below

$$\overbrace{\boxed{c_{11}}\boxed{c_{12}}\boxed{\cdots}\boxed{c_{mn}}}^{\text{connection matrix}} \quad \boxed{f^{k_e}} \tag{2.14}$$

2.3.2.3 Majority Rule Encoding

To further simplify the Boolean model, we get rid of an explicit definition of the Boolean functions f_i. Instead, we determine the phenotype traits using the majority rule. In case of a tie, the first input will be taken as the output of the majority rule, without the loss of generality:

$$f_i(g_{j_1}, g_{j_2}, \ldots, g_{j_{k_i}}) = \begin{cases} 1, & \text{if } \sum_{l=1}^{k_i} g_{j_l} > k_i/2; \\ 0, & \text{if } \sum_{l=1}^{k_i} g_{j_l} < k_i/2; \\ g_{j_1(i)}, & \text{otherwise.} \end{cases} \tag{2.15}$$

Hence, only the connection matrix needs to be encoded:

$$\boxed{c_{11}}\boxed{c_{12}}\boxed{\cdots}\boxed{c_{mn}} \tag{2.16}$$

The above different forms of encoding strongly restrict the original model. As a result, each of them presents a different model on its own.

2.3.3 *Qualitative Measures for Robustness and Innovation*

2.3.3.1 Local Neutrality Versus Local Variability

There is no well-established quantitative definition for robustness and evolvability for biological systems. To rigorously analyze the robustness and evolvability trade-off relationship, such definitions are indispensable. Here, we define a robustness measure as the local neutrality degree and an evolvability measure to be the maximum local innovation for the gene regulatory network modeled by a Boolean network.

Given a genotype-phenotype mapping (the Boolean network) $\phi\colon \mathcal{G} \rightarrow \mathcal{P}$ and a neighborhood relation N over the set of genotypes $\mathcal{G}$, the local neutrality degree $\nu_\phi(g)$ of mapping ϕ is defined by [246]

$$\nu_\phi(g) := \frac{|\{g' \in \mathcal{G} : \phi(g) = \phi(g') \wedge N(g, g')\}|}{|\{g' \in \mathcal{G} : N(g, g')\}|} \tag{2.17}$$

Similarly, given the genotype-phenotype mapping $\phi\colon \mathcal{G} \rightarrow \mathcal{P}$ and a neighborhood relation N over the set of genotypes $\mathcal{G}$, the local variability $\delta_\phi(g)$ of mapping ϕ at genotype g is defined by

$$\delta_\phi(g) := \frac{|\{g' \in \mathcal{G} : \phi(g) \neq \phi(g') \wedge N(g, g')\}|}{|\{g' \in \mathcal{G} : N(g, g')\}|} \tag{2.18}$$

The variability δ defined above captures the fraction of unique phenotypes in the non-neutral neighborhood, which is accomplished by the set notation. Accordingly, $\forall g \in \mathcal{G}$, $\nu_\phi(g) + \delta_\phi(g) \leq 1$ holds by definition. Two illustrative examples of how to calculate the local neutrality degree and local variability are given in Fig. 2.6.

The definition of local neutrality degree has been discussed in a number of studies. However, the local neutrality degree measure is largely investigated with respect to single genotypes or neutral set mapping to the same phenotype. Since we aim to evaluate different characteristics of neutrality with respect to entire genotype spaces, we adopt here the local neutrality degree to calculate the mean neutrality degree of a genotype-phenotype mapping, which can be accomplished by randomly sampling genotypes and averaging their local neutrality degree. A high mean neutrality degree means that there are many genotypes with a high local neutrality degree, and therefore the mean neutrality degree indicates how many neutral mutations or how much neutrality is provided by the mapping. In the genotype space where the neutrality degree is high, mutations are most likely neutral rather than deleterious.

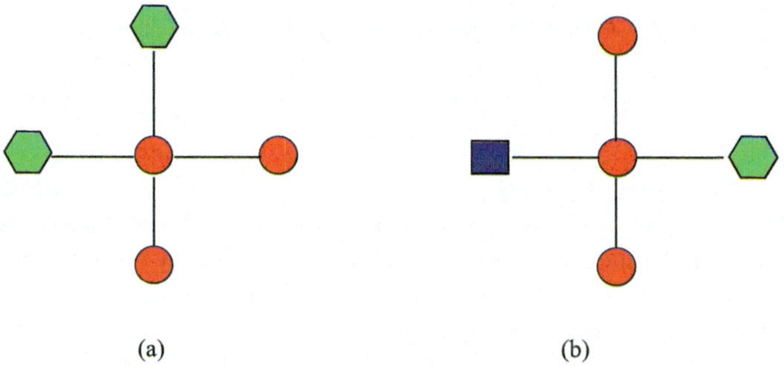

Fig. 2.6 Two illustrative examples of calculating local neutrality degree and local variability. In the figure, filled circles, squares, and pentagons denote different phenotypes, while the lines represent local neighborhood in the genotype. **a** $\nu_\phi(g) = 0.5$, $\delta_\phi(g) = 0.25$, and **b** $\nu_\phi(g) = 0.5$, $\delta_\phi(g) = 0.5$

Consequently, populations on neutral networks tend to drift toward the regions with a high neutrality degree to evolve stronger robustness. Therefore, the mean neutrality degree can be employed to represent the mutational robustness.

2.3.3.2 Neutrality Versus and Uniformity Entropy

To gain a deeper insight into the relationship between neutrality degree and entropy, we investigate here another measure for innovation by examining the Shannon evenness of the reachable phenotypes. In calculating the entropy, the genotypes are sampled randomly from the genotype space, and mapped to their associated phenotypes. The occurrence of each phenotype is counted and normalized to determine the probability (p_i) of locating the i-th phenotype by randomly choosing a genotype. The Shannon evenness (uniformity entropy) can then be calculated as follows [272]:

$$E = \frac{H}{H_{\max}}, \tag{2.19}$$

$$H = -\sum_{i=1}^{S} p_i \ln p_i, \tag{2.20}$$

$$H_{\max} = \ln S, \tag{2.21}$$

where H is known as Shannon index of diversity, $H_{\max}$ is the maximum value of H, and S is the number of different phenotypes.

High uniformity entropy means that the estimated phenotype distribution is highly uneven, and therefore the success rate of the evolutionary search to discover a new phenotype is low. On the contrary, if the uniformity entropy is low, the phenotype distribution is flat and all phenotypes can be located by randomly sampling the genotypes with nearly the same probability.

2.3.4 *Multi-objective Optimization of the GPM*

The most significant difference of Pareto-based multi-objective optimization compared to single-objective optimization is the selection method. Here, we adopt the fast elitist non-dominated sorting genetic algorithm (NSGA-II) [82] for the environmental selection, which consists of four major steps. First, the parent and offspring populations are combined, therefore, NSGA-II is an elitism. Second, the combined population is sorted according to the non-dominated ranks. During the ranking, non-dominated solutions in the combined population are assigned a rank 1, which belongs to the first non-dominated front. These individuals are removed temporally from the population and the non-dominated individuals in the rest of the population are identified, which consists of the second non-dominated front of the population and are assigned a rank 2. This procedure repeats until all individuals in the combined population are assigned with a rank from 1 to R, assuming that R non-dominated fronts can be identified in total. Here, the smaller the rank, the more preferred a solution is. Third, a crowding distance reflecting the crowdedness in the neighborhood of a particular solution is calculated. The crowding distance of the i-th solution s_i^j in the j-th ($j = 1, ..., R$) non-dominated front is the Euclidean distance between the two neighbors of solution s_i^j in the objective space:

$$d_i^j = \sum_{k=1}^{m} |f_k(s_{i-1}^j) - f_k(s_{i+1}^j)|, \tag{2.22}$$

where m is the number of objectives of the multi-objective optimization problem, and solutions s_{i-1}^j and s_{i+1}^j are the two neighboring solutions of solution s_i^j. A large distance is assigned to the two extreme solutions on each non-dominated front. Here, the larger the crowding distance, the less crowded it is around solution s_i^j, and the more preferred the solution is in the selection. Therefore, solutions on the same non-dominated front will be sorted in a descending order according to the crowding distance. By far, the combined population is completely sorted. Finally, the top half of the sorted combined solution are selected as the parents of the next generation.

2.3.5 *Experimental Results*

2.3.5.1 Trade-Off Between Neutrality and Variability

In the experiment, different encoding methods for the Boolean model are examined with respect to the mean variability and mean neutrality degree. The non-dominated solutions for a population containing 50 individuals that is run for 100 generations are summarized in Fig. 2.7.

It can be seen that the obtained results heavily depend on the encoding method. The random initial population with the multiple Boolean function encoding is unevenly

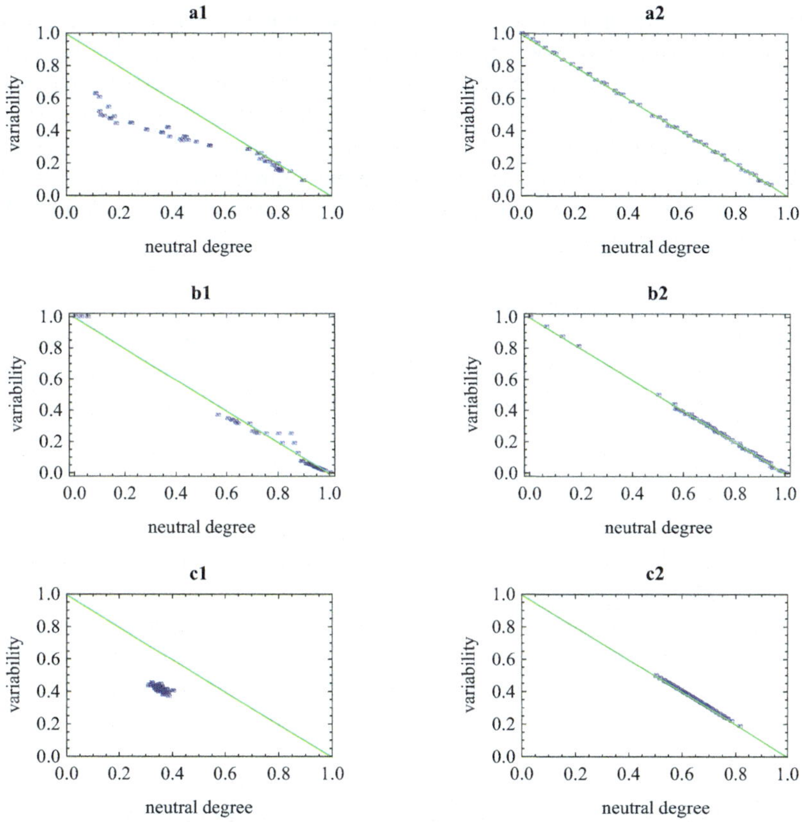

Fig. 2.7 Non-dominated solutions of the Boolean model ($n = 16, m = 8$) by simultaneously maximizing the mean neutrality degree $\bar{\nu}_\phi$ and maximizing the mean variability $\bar{\delta}_\phi$. The Boolean model is encoded by different techniques: **a1, a2** Multiple Boolean functions encoding with an elementary arity 2. **b1, b2** Single Boolean function encoding with an elementary arity 2. **c1, c2** Majority rule encoding. Plots in the left panel show the results of 50 randomly initialized genotype-phenotype mappings with $n = 16$ and $m = 8$. Those in the right panel show the approximated Pareto optimal set of the genotype-phenotype map after 100 generations. The dotted line indicates the theoretical Pareto front determined by the definition of the neutrality degree and variability, i.e., $\bar{\nu}_\phi + \bar{\delta}_\phi = 1$. The different encoding methods result in different approximations of a Pareto optimal set

distributed across the objective space. The mappings lie at some distance from the true Pareto front $\bar{\nu}_\phi + \bar{\delta}_\phi = 1$, although the mappings with $\bar{\nu} > 0.6$ lie nearly on the theoretical Pareto front. After 100 generations, the mappings distribute evenly on the theoretical trade-off surface. This is different from the scenario where single Boolean functions are encoded. The randomly initialized population lies nearly on part of the theoretical Pareto front where $\bar{\nu} < 0.5$. After multi-objective optimization is completed, the mappings are more evenly distributed along this segment of the Pareto front. Only a few mappings lie on the Pareto front section with $\bar{\nu} < 0.5$. The solutions seem to have a uniform distance from each other. For $0.2 < \bar{\nu} < 0.5$,

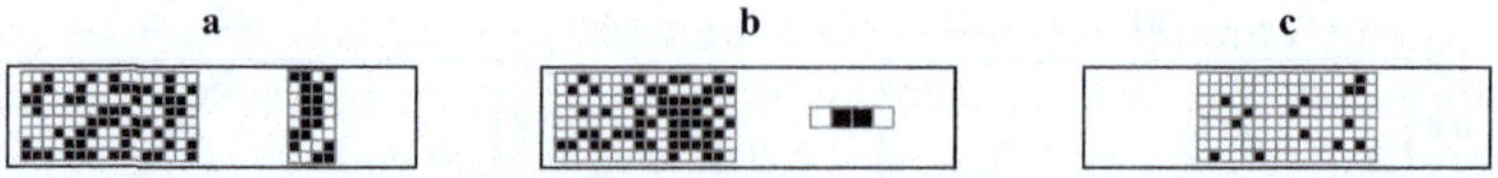

Fig. 2.9 Visualization of parameters in the Boolean model ($n = 16, m = 8$) resulting from different encoding methods: **a** Multiple Boolean function encoding, **b** single Boolean function encoding, and **c** majority encoding

is inevitably the mapping with higher mutational robustness. We think that non-surjective genotype-phenotype mappings must be considered when defining robustness.

2.3.5.3 Visual Analysis of the GPMs

To better understand how the Boolean model is capable of implementing such different mappings, we analyze the underlying parameters of the model when local neutrality and local variability are maximized. These are specified by the connection matrix **C** and the Boolean functions f_i, $i = 1, \ldots, m$. Figure 2.9 visualizes three examples of encoding structures of different encoding methods. However, no significant features or patterns in the encoding structure can be identified from the visualized structures. A profound analysis of the parameters appears hard due to the high degree of freedom in the model. A statistical approach is presented in [143] to interpret the parameter structure of the majority rule encoding.

In Fig. 2.9, each frame corresponds to an implementation of the connectivity matrix **C** and the set of phenotype functions (if necessary) of a genotype-phenotype mapping. The results in Fig. 2.9a are obtained when the multiple Boolean functions encoded with an arity of 2 are adopted. The left array depicts the connection matrix, whereas the right array illustrates the phenotype functions f_i, $i = 1, \ldots, m$. Row i depicts the truth table of the elementary Boolean function f_i^2. The results for the single Boolean functions encoded with an elementary arity 2 are given in Fig. 2.9b, where the left array shows the connection matrix, and the right array depicts the truth table of the elementary Boolean function f^2. When the majority rule encoding is adopted, the array indicates the connection matrix; see Fig. 2.9c. In all descriptions, black array entries correspond to 1, whereas white array entries correspond to 0. An entry $c_{ij} = 1$ in the connection matrix indicates that phenotype trait p_i depends on gene g_j. Therefore, the pleiotropy of gene g_j is given by the sum of column j, whereas the polygeny of phenotype trait p_i is given by the sum of row i.

2.3.6 Discussions

In the above, it is shown that the Pareto optimization approach can be used to analyze the robustness-evolvability trade-off in a class of redundant Boolean representations for simulated evolutionary systems. By introducing quantitative measures for robustness and innovation, both measures are maximized by optimizing the genotype-phenotype mapping. This way, an explicit and quantitative trade-off relationship between robustness and innovation has been revealed. Such trade-off relationships indicate that no single representation can achieve maximum robustness and innovation simultaneously. An optimal representation then depends on the environment. A representation with a higher innovation ability is preferred in a rapidly changing environment to enhance the survival probability, whereas, in a slowly changing environment, stronger robustness may enable the system to reach a higher fitness.

2.4 Evolvability of Evolutionary Developmental Systems

2.4.1 Motivation

In this section, the relationship between innovation and robustness of a Boolean genotype-phenotype mapping is analyzed. This is achieved by simultaneously maximizing the robustness and evolvability using a multi-objective optimization algorithm and then analyzing the obtained non-dominated solutions. It is shown that regardless of the differences in encoding schemes and the Boolean functions, a hard trade-off between robustness and evolvability exists, indicating that the Boolean network is not a biologically plausible model for the genotype-phenotype map. This is not surprising, since the Boolean network is a stationary model that is not able to include developmental plasticity, which is extremely important for biological systems.

In the following, we examine the robustness-evolvability relationship for a multi-cellular growth model like the one introduced in Sect. 1.3 of Chap. 1, and the results were reported in [205]. The cell growth model is a gene regulatory model expressed in terms of a set of continuous nonlinear differential equations, whose phenotypes are described by the states of the equations representing the concentration of regulatory and functional proteins involved in cellular growth. In addition, an evolvability measure independent of any specific fitness function is proposed, which is applicable to gene regulatory models based on differential equations. With the help of the proposed evolvability measure, the change of evolvability of the cell growth model when evolving regulatory dynamics such as limit cycles are examined. In contrast to the stationary Boolean network model, it is shown that there is no longer a hard trade-off between evolvability and robustness for the gene regulatory network model for cell growth, and evolvability steadily increases even without an explicit selection pressure on evolvability, particularly when the mutation step-size is small.

2.4.2 *A Developmental Model for Cell Growth*

2.4.2.1 Physical Environment

In the cell growth model, which is similar to the one presented in Sect. 1.3, each cell is represented by a three-dimensional sphere and contains a virtual DNA. Cells are located in a bounded environment, which is a three-dimensional space in which the cells may divide, migrate, and interact with each other. For simulating the diffusion of transcription factors (TFs), the simulated environment is divided into a number of cubes. In the simulation, the concentration of the TFs is defined on the vertices of the cubes, and the concentration at an arbitrary point within the cube is approximated by a linear approximation of the concentrations at the eight neighboring vertices.

2.4.2.2 Genome Structure

The virtual DNA for cell growth consists of a number of genes, each again composed of a number of structural units (SUs) and regulatory units (RUs). An SU defines a particular cellular function such as cell division, cell death, and production of transcription factors (TFs), while the RUs, either activating or repressive, determine whether the SUs between them and the next RUs in the chromosome will be expressed, i.e., whether the functions they encode will be performed. Whether a TF activates or represses a gene depends on the degree of match between the affinity value of a TF and that of a RU. Each SU has nine loci of real values, specifying the properties of the encoded cellular functions. Each RU consists of eight loci encoding the parameters of a sigmoid activation function, namely θ, σ, and C, and defining the affinity properties of the RU, including the affinity value a^{RU} and affinity range ε. The exact meaning of the parameters will be described in the next subsection. The structure of a genome is plotted in Fig. 2.10a, which contains the loci of a RU and an SU. Likewise, details of an SU for producing TFs are provided in Fig. 2.10b. It should be mentioned that some of the loci both in the SUs and RUs are not used in the current model, and that a valid gene should start with one or more RUs followed by one or more SUs. That is, genes starting with an SU or ending with a RU will not be transcribed, which can be seen as non-coding genes.

2.4.2.3 Evolved Gene Regulatory Networks

Cellular growth of the considered system is under the control of a gene regulatory network (GRN) in a virtual DNA illustrated in Fig. 2.10a. The activation level of an SU is determined by the sum of activity values of all its RUs (β_l^{RU}):

$$\beta_l^{\mathrm{gene}} = \sum_{l=1}^{n_{\mathrm{RU}}} \beta_l^{\mathrm{RU}}, \tag{2.23}$$

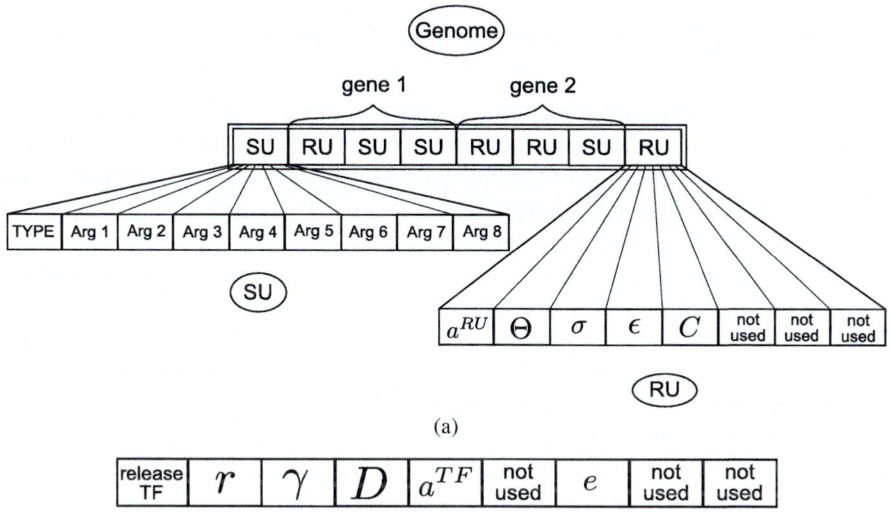

Fig. 2.10 **a** Structure of a genome. **b** The loci in an SU encoding the cellular function of producing a transcription factor

where n_{RU} is the number of RUs of the l-th gene. If this activation level β_l^{gene} is larger than zero, the gene is activated and its cellular functions are expressed (carried out).

The activity level of one RU is determined by the sum of the regulatory activities of all internal TFs (denoted by index j) and external TFs (denoted by index i):

$$\beta^{\mathrm{RU}} = f_{\mathrm{nl}}^{\mathrm{RU}} \left(\sum_{i=1}^{n_{\mathrm{TF}}} f_{\mathrm{aff}}^{\mathrm{RU}}(d_{\mathrm{aff},i}) \cdot c_i^{\mathrm{TF}} + \sum_{j=1}^{n_{\mathrm{TF}}} f_{\mathrm{aff}}^{\mathrm{RU}}(d_{\mathrm{aff},j}) \cdot c_j^{\mathrm{TF}} \right), \tag{2.24}$$

where c_i^{TF} and c_j^{TF} are the concentration of the external and internal TFs, respectively, d_{aff} is the matching degree between the affinity values of the RU and the TF:

$$d_{\mathrm{aff}} = \min(\,\mathrm{abs}(a^{\mathrm{RU}} - a^{\mathrm{TF}}),\ 1 - \mathrm{abs}(a^{\mathrm{RU}} - a^{\mathrm{TF}})\,), \tag{2.25}$$

and $f_{\mathrm{aff}}^{\mathrm{RU}}(d_{\mathrm{aff}})$ is the affinity function:

$$f_{\mathrm{aff}}^{\mathrm{RU}}(d_{\mathrm{aff}}) = e^{-0.5 \cdot \left(d_{\mathrm{aff}}/\varepsilon\right)^2}. \tag{2.26}$$

Finally, $f_{\mathrm{nl}}^{\mathrm{RU}}(z)$ is a nonlinear activation function:

$$f_{\mathrm{nl}}^{\mathrm{RU}}(z) = 2 \cdot (2 \cdot \sigma - 1)\left(\frac{1}{1 + e^{(-30 \cdot C \cdot (z) - \Theta)}} - 1\right). \tag{2.27}$$

The most important dynamics of the GRN for cellular growth is reaction and diffusion of the external TFs in the environment. The TFs are produced by the SUs shown in Fig. 2.10. The type of the produced TF is identified with an affinity value a^{TF} (ID). The produced TF can be partly internal and partly external, which is specified by a ratio $0 \leq e \leq 1$ encoded in the gene. The internal TF remains inside the cell and can regulate the expression of the genes inside the cell only, while the external TF can diffuse out of the cell and into other cells, thereby regulating the activation of genes in these cells. Here, index i denotes external TFs and index j internal TFs. The production of a particular TF depends on the production rate r, the current activity of their corresponding genes β_m^{gene}, and the factor e. The Heaviside step function $H(\beta^{\mathrm{gene}})$ describes the activation function of the genes. If β^{gene} is higher than zero, $H(\beta^{\mathrm{gene}})$ is one and if it is below zero $H(\beta^{\mathrm{gene}})$ is zero. In case two different SUs have exactly the same affinity value (a^{TF}) in the genome, all parameters are averaged over the two SUs and then combined into one TF. If there are $n_{\mathrm{gene},i}$ genes producing the same TF, the production of the TF can then be described by

$$\frac{\partial \mathrm{TF}_i}{\partial t} = \sum_{m=1}^{n_{\mathrm{gene},i}} H(\beta_m^{\mathrm{gene}}) \cdot r_i \cdot \beta_m^{\mathrm{gene}} \cdot e_i \qquad \text{external TFs} \tag{2.28}$$

$$\frac{\partial \mathrm{TF}_j}{\partial t} = \sum_{m=1}^{n_{\mathrm{gene},j}} H(\beta_m^{\mathrm{gene}}) \cdot r_j \cdot \beta_m^{\mathrm{gene}} \cdot (1 - e_j) \qquad \text{internal TFs.} \tag{2.29}$$

Taking into account the decay of the produced proteins with a decay rate γ, the concentration of internal TFs can be described as follows:

$$\frac{\partial \mathrm{TF}_j}{\partial t} = \sum_{m=1}^{n_{\mathrm{gene},j}} H(\beta_m^{\mathrm{gene}}) \cdot r_j \cdot \beta_m^{\mathrm{gene}} \cdot (1 - e_j) - \gamma_j \cdot \mathrm{TF}_j. \tag{2.30}$$

The concentration of external TFs is influenced by both diffusion and decay, which can thus be described as follows:

$$\frac{\partial \mathrm{TF}_i}{\partial t} = \sum_{m=1}^{n_{gene,i}} H(\beta_m^{\mathrm{gene}}) \cdot r_i \cdot \beta_m^{\mathrm{gene}} \cdot e_i - \gamma_i \cdot \mathrm{TF}_i + D_i \cdot \nabla \mathrm{TF}_i. \tag{2.31}$$

Thus, we can see that in this model, the multi-cellular growth is governed by the dynamics of a set of coupled nonlinear differential equations.

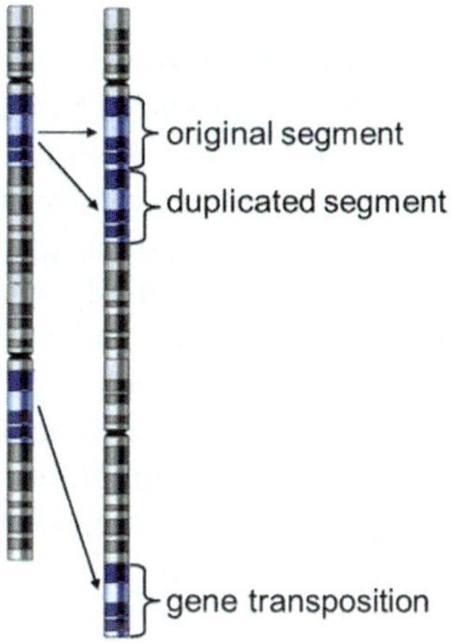

Fig. 2.11 An illustration of gene duplication and gene transposition

2.4.2.4 Genetic Variations

For the GRN to properly control cell growth, a variant of an evolution strategy is adopted to evolve the parameters of the GRN. Like a canonical evolution strategy, the parent population contains μ individuals, which are mutated and recombined to create an offspring population of λ individuals. In addition to mutations and recombination, other biologically plausible genetic variations, including gene duplication, deletion, and transposition, are also introduced. At a predefined probability, one of the genetic operators, i.e., duplication, deletion, or transposition, will be applied to the offspring produced by mutation and recombination. Gene duplication is the main driving force behind many major transitions in evolution [270, 271], which randomly copies a segment of genes and inserts it into the genome at a random position. In gene transposition, a randomly chosen segment of genes is moved from the current position to another randomly chosen position. The deletion operator simply removes a randomly chosen gene segment. An illustrative example of gene duplication and gene transposition is provided in Fig. 2.11.

2.4.3 Fitness-Independent Evolvability Measures

To investigate the relationship between robustness and evolvability of the developmental model described in Sect. 2.4.2, we first need to define quantitative measures for robustness and evolvability independent of a specific evolutionary task. As previously discussed, most existing evolvability measures are more or less fitness-dependent.

Note also that most existing evolvability measures are suggested for discrete direct (stationary) or developmental (dynamic) representations [63, 196]. By contrast, the development model studied here is more complex in that both genotypic and pheno-

where C_1 and C_2 are two chromosomes, M is the length of the target chromosome, d_i^* is the i-th element distance, and V_k^* is the k-th gap penalty on the path. Note that the cost for aligning chromosome C_1 with chromosome C_2 may be different from that for aligning C_2 with C_1, partly because the lengths of the two chromosomes are different. Thus, it makes sense to define the genotype distance D between the two chromosomes as

$$D(C_1, C_2) = E(C_1, C_2) + E(C_2, C_1). \tag{2.34}$$

The characters 'ABC' in the above example represent genes in the virtual DNAs, such as RUs and SUs. Note that the distance defined above applies to two RUs or two SUs, and it should be noted that the distance between a RU and an SU will be very large, which can be set to a big constant. This way, an alignment between a RU and an SU will never be taken. Note that the resulting alignment depends on the penalty value p. If the value is very small, the algorithm tries to match elements which are close in element-wise distance and uses many transposition steps for this. If the penalty value is large, the algorithm matches values with an eventually high element-wise distance to receive long local alignments, which discourage duplication or transpositions. The choice of the gap penalty is in itself a multi-objective optimization problem, because an alignment with more or less transpositions will be created depending on the value of p.

2.4.3.2 Definition of a Phenotype Distance

For the developmental model for cellular growth used here, there are multiple options for defining the phenotype of the system. For instance, the final morphology (shape) of the cellular structure when the growing process is complete can be treated as the phenotype. Another option is the cellular behaviors resulting from the given genome, such as cell division and cell death.

However, cellular behaviors are specific to the tasks, which are fitness-dependent. Here, we propose a generic evolvability measure for developmental systems and therefore use the dynamics of the GRN as the phenotype, which is independent of a specific fitness function. Consequently, two different properties related to the GRN dynamics may be used as the phenotype of the developmental model, i.e., the activity of the genes and the gradients of the TF concentrations. Eventually, we decide to use the concentration of transcription factors as the phenotype, since TFs play a more fundamental role in regulating cellular behaviors.

Nevertheless, it is non-trivial to calculate the phenotype distance when the TF concentrations are employed as the phenotype of the developmental model because there are following different ways of calculating the phenotype distance.

- Since the final steady state of the corresponding TFs of the two genomes may be directly related to the final behaviors of the cellular system, the differences in the steady state of the TFs can be used as the phenotype distance. If the TF

concentrations exhibit oscillatory behaviors, we may compare the two limit cycles in terms of their frequency, amplitude, and mean values.
- While the steady states may be directly related to the cellular behaviors, the whole dynamics before approaching the steady states may also be important. Therefore, we can calculate the Euclidean distance between two trajectories of the TFs as the phenotype distance.
- Use the change in the dynamics of the TF concentrations. That is, we can calculate the difference in the gradients on a number of points on the trajectories of the TF concentrations. Finally, the average over all different vectors of the gradients can be used as the final phenotype distance.

The first option is very interesting, since the final morphology of the cellular structure when the growing process is complete is largely determined by the steady state of the GRN. Unfortunately, it is hard to automatically detect the steady states, in particular when the final states are limit cycles in a high-dimensional space.

The second option seems very easy, but it is only available for comparing TF dynamics at only one point in the environment (one cell), making it unpractical to be used as the phenotype distance for multi-cellular organisms.

Interestingly, the gradient field of a TF is independent of the specific position in the environment. Instead, the gradient depends only on the current state of all TFs and defines the direction in which the state will move in the next time-step. Therefore, we can adopt the difference between the gradients of the TF concentrations for calculating the phenotype distance. This measure is very generic and applicable to all cell growth models controlled by a GRN described by ordinary differential equations.

Since it is infeasible to calculate the gradients in the whole state space, we need to discretize the state variable denoting the TF concentration on equally distributed grids, which needs to be specified by the user. For example, the number of sample points for each state is set to $n_{\text{sample}} = 5$, meaning that we need to compare the difference of the entire gradients on $n = n_{\text{sample}}^{2n_{\text{TF}}}$ points, since there are $2n_{\text{TF}}$ TFs as there are both internal and external TFs. In addition, the range of the sampling points for each TF needs to be determined, which is the bounding box including the trajectories of the growing process of both phenotypes in the state space. If there is more than one cell in the individuals, the bounding box can be set to cover all trajectories of all cells of both phenotypes. The upper and lower bounds of the bounding box can be determined by finding out the minimum and maximum values observed in the whole developmental process.

The algorithm for calculating the phenotype distance that compares the gradients of the TFs in the cellular developmental model can be divided into the following four steps:

- Calculate the corresponding TFs of both phenotypes by comparing their TF on a number of time instances in the growing process.
- Determine the sampling range of all TFs by sorting out the maximum and minimum of their concentrations in all cells of both phenotypes during the whole growing process.

- Divide the range into n_{sample} equal grids for every TF and use these samples as the sampled points for comparing the differences of the phenotypes.
- Calculate the two gradient fields and obtain their differences on the specified grids. The resulting difference vectors are then averaged over all sampling points. The average vector of the differences in gradients is then the phenotype distance. Each value depicts the phenotype difference of one TF.

This above phenotype distance is not only fitness-independent but also independent of the duration of the growing process and the physical interactions between cells. This measure indicates the difference in the capability of genomes to show a specific dynamical behavior. As the GRN dynamics control the whole development process of the cellular system, this measure is very general yet able to describe the fundamental cellular behaviors of the genome.

2.4.3.3 A Fitness-Independent Evolvability Measure

Based on the genotype and phenotype distances defined above, we can now define a fitness-independent evolvability measure. This evolvability measure is essentially a statistical dispersion degree calculated based on a set of mutants, that is, new individuals generated by random mutations. Then, the phenotype and genotype distances of all generated mutants to the original individual are calculated. We can then define the evolvability measure according to the distribution of these distances. Figure 2.13 gives an illustrative example of three typical types of dispersion, representing different scenarios of evolvability-robustness trade-off.

Most evolvability properties can be analyzed using the defined evolvability. For instance, Fig. 2.13a shows an example of a robust individual, since the majority of its mutants are distributed in the lower part of the dispersion distribution, where only small changes in the phenotype are observed regardless of whether the genotypic changes are small or large. By contrast, the example in Fig. 2.13b shows that all mutants are located quite evenly in the distribution of evolvability, implying that this individual is highly evolvable since both robust and innovative offspring can

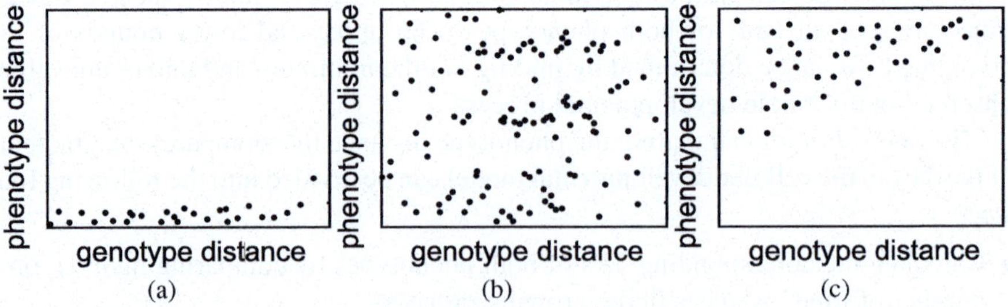

Fig. 2.13 Examples of **a** a very robust individual, **b** an evolvable individual having a good degree of both robustness and innovation, and **c** a highly innovative individual

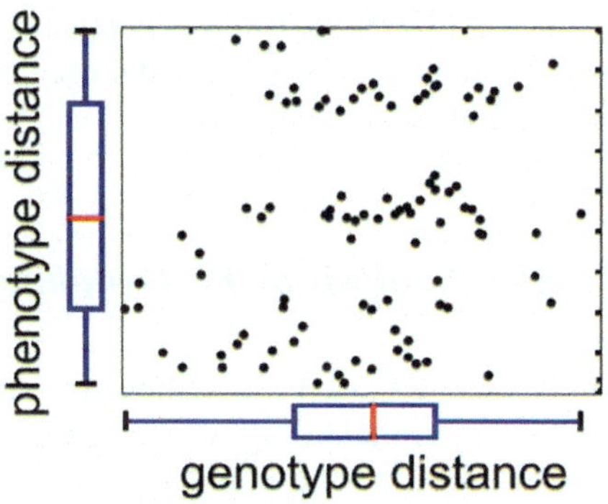

Fig. 2.14 Evolvability with the box-and-whisker plot of all genotype and phenotype distances

be generated from the individual. Finally in Fig. 2.13c, most of the mutants of the individual are located in the upper-left part of the dispersion plot, indicating both small and large changes in the genotype will cause large changes in the phenotype. Therefore, this individual can be regarded to be very innovative.

From the above examples, we know that the evolvability of an individual cannot be determined simply by calculating the ratio between the phenotype and genotype distances. To denote the evolvability of an individual more accurately, it is necessary to compare the distribution of its mutants using a quantitative measure. To this end, the interquartile range of the five-number summary from the descriptive statistics [164] can be adopted. The five-number summary is often visualized in a common box-and-whisker plot. The interquartile range (IQR) is the distance between the first and third quartiles, which is visualized as a box in the box-and-whisker plot. The first quartile is a line under which 25% of all points are arranged. The third quartile is then the line under which 75% of all data points lie. This means that in the IQR 50% of all data points are situated. The final evolvability measure E can be defined as the multiplication of the IQR of the phenotype distances (IQR_p) and the IQR of the genotype distances (IQR_g):

$$E = \mathrm{IQR}_p \cdot \mathrm{IQR}_g. \tag{2.35}$$

This multiplication represents an area in which an individual's mutant can be distributed upon a mutation; see Fig. 2.14. As mentioned above, one graph is only for one transcription factor. So the evolvability measure should be a vector of the same size as the phenotype distance vector. The mean of this vector can be used as an index for measuring evolvability.

2.4.3.4 An Evolvability Measure Based on Diversity

While statistical dispersion provides an effective measure for evolvability, it is natural to use a diversity index to define evolvability, again using the genotype-phenotype distances. Here, we use two popular diversity measures, i.e., the Simpson diversity and Shannon diversity for analyzing changes in the evolvability of the developmental in the course of evolution.

2.4.4 Analysis of TF Dynamics

In this section, we illustrate how to quantitatively evaluate the evolvability of a given individual. In addition, we examine, with the proposed evolvability measure, how the evolvability of a population changes in evolving limit cycles. Interestingly, we show that in evolving limit circles, where evolvability is not directly under selection pressure, evolvability can also evolve, confirming the findings in [95, 184, 192].

2.4.4.1 Experimental Settings

In the simulation, the developmental model is evolved using the evolutionary algorithm described in Sect. 2.4.2.4 to achieve a limit cycle. In the simulations, the parent population size is 40 and the offspring population size is 100. The fitness function is to compare the time series of the TF concentrations with a sine function, which defines the desired frequency of the limit cycle, similar to [196]. Since only the frequency matters and the amplitude is less significant in evolving limit cycles, we normalize the TF concentrations to be zero mean and between -1 and 1:

$$X_k^* = \frac{1}{X\mathrm{max}} \cdot (X_k - X\mathrm{mean}), \tag{2.36}$$

$$S_k = \sin\left(6\pi \frac{k}{N}\right). \tag{2.37}$$

Finally, we use the following mean squared error as the fitness function:

$$E = \frac{1}{N} \cdot \sum_{k=0}^{N-1} (X_k^* - S_k)^2 - 1.0. \tag{2.38}$$

Thus, the minimum fitness value is -1.

The Runge-Kutta method has been used for solving the differential equations of the gene regulatory network, in which the step-size is set to 0.25. The developmental time lasts for 300 iterations for each individual. In other words, $N = 300$ in Eq. 2.38.

The evolution was run for 500 generations, and 50 independent runs were performed for each setup. We report here three sets of evolutionary runs for comparison. In the first set of simulations, we limited the maximum number of RUs to six and the maximum number of SUs to three. In this setup, only mutations and gene transcriptions are used for genetic variations. That is to say, the genome length will be constant

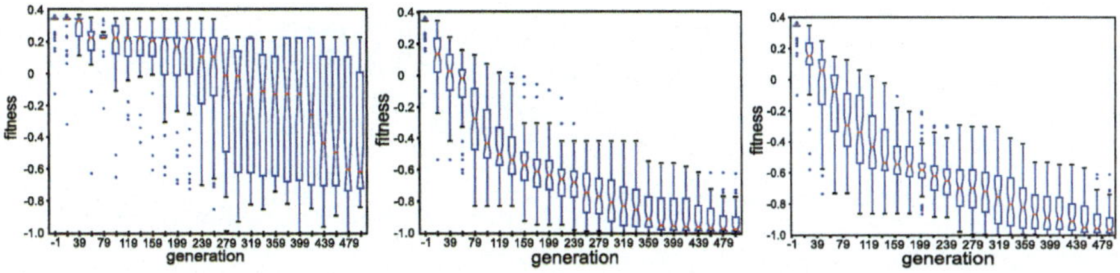

(a) (b) (c)

Fig. 2.15 Box plots of the best fitness from 50 independent runs. **a** No elitism, fixed genome length, **b** elitism, fixed genome length, and **c** elitism, variable genome length

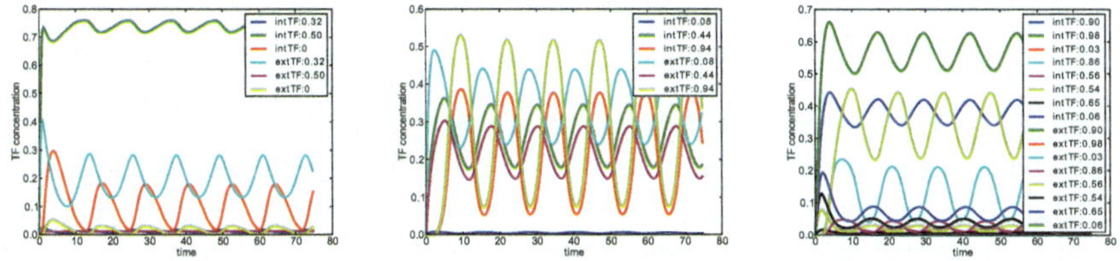

(a) (b) (c)

Fig. 2.16 Evolved TF dynamics of the best individual from the 50 independent runs. **a** No elitism, fixed genome length, **b** elitism, fixed genome length, and **c** elitism, variable genome length

in these evolutionary runs. As indicated in [193], GRNs with this complexity should be adequate for generating sustained oscillation. No elitism strategy was adopted in the first set of evolutionary runs. In the second set of runs, all the parameter setups are the same as in the first set, and the only difference is that three elitists are always passed to the parent population of the next generation. In the third set of runs, gene duplication and deletion are also used for genetic variations in addition to mutations and gene transpositions. However, the allowed maximum number of RUs and SUs is limited to 20 to reduce the total amount of simulation time.

The best fitness in every 20 generations of 50 runs for the three setups are presented in Fig. 2.15a, b, c, respectively. We can see that the elitism strategy greatly improved the convergence of the evolution of limit cycles, as also found in [193]. However, the use of more RUs and SUs did not help improve the performance much, although the best individual from the runs with a variable genome length does have better fitness than that in the runs with a fixed genome length. The dynamics of the best individuals from the three runs is presented in Fig. 2.16a, b, c, respectively.

2.4.4.2 State-Space Analysis of TF Trajectories

As discussed in Sect. 2.4.3, we use the gradients of the TFs as the phenotype for investigating the evolvability of the developmental model. At first, we examine the state-space trajectories of the best individuals found in the evolution (with three SUs

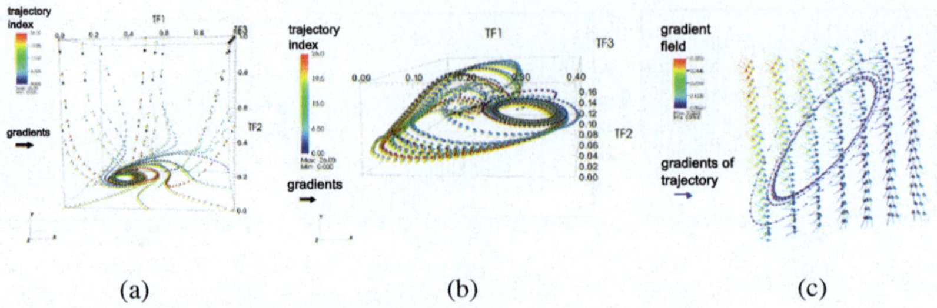

Fig. 2.17 **a** State-space trajectories of three internal TFs with different initial states. **b** State-space trajectories of three external TFs for different initial states. **c** Comparison of the gradient field of the sampled state space and the real trajectory of the growth process

and six RUs). We are interested in visualizing the gradient fields of the TFs in the whole space since we use the gradient fields for calculating evolvability. An easy way to get the gradient field is to change the initial states of the internal TFs and let the individual develop from all these initial states. The initial values of the external TFs were set constantly to zero. All the resulting trajectories would then give us a clue about the gradient field of the GRN dynamics. The gradient field of the above-mentioned best individual is plotted in Fig. 2.17a. Every point (little colored dots) is the state of all internal TFs at one growing (developmental) time-step. Points in the same color belong to the same trajectory independent of the initial states. All the trajectories converge finally to the same limit cycle. The black arrows represent the gradients at each growing time-step.

In Fig. 2.17b, the trajectories of the external TFs are also shown for the same initial states in Fig. 2.17a. Of course, all the trajectories start in the origin since the external TFs are always initialized to zero. Therefore, the different trajectories of the external TFs are produced from the different initial states of the internal TFs. Similar to the internal TFs, all trajectories of the external TFs end up in the same limit cycle.

The phenotype distance defined in Sect. 2.4.3.2 depends on the gradients at different sampling points in the state space. By choosing specific initial states, most of the trajectories will take similar paths. Hence, the gradient field in a large part of the state space has not been shown in Fig. 2.17a,b. However, it is of interest to visualize the whole state space as it is used in the phenotype distance. The gradient is described by Eqs. (2.30) and (2.31). It depends only on the current state of the internal and external TF concentrations. Hence, it is easy to determine the gradient vector (time derivative of the state vector) at each sample point. The resulting 'flow field' of the state space is shown in Fig. 2.17c. In the figure, the large arrows denote the gradient vectors at the sampled point. For comparison, the trajectory of the growing process is also shown denoted by small blue arrows, which is a limit cycle. Note that the sampled gradient field does not agree with the trajectory at first sight. This is due to the dimensionality of the state space. In the chosen genome, there are three concentration values for internal TFs inside the cell and three values for external TFs evaluated at the cell center. The state space is therefore six-dimensional. In the visualization, only three dimensions can be shown and the concentration of the external TFs changes throughout the development. If, for example, the three internal TFs are varied, the concentration of the three external TFs cannot be determined. In this example, all external TFs are set to zero to determine the gradient field. However, the result is only a projection of the whole state space. Thus, it is impossible to show a limit cycle in terms of the gradient field.

2.4.5 *Evolution of Evolvability*

To investigate how the evolvability of the system changes during evolution, we need to show the evolvability graphs of the TFs showing the relationship between the genotype and phenotype distances. Here, we show the evolvability graphs of two individuals, one is randomly initialized, with three SUs and six RUs (generation 1), and the other is the best individual obtained in generation 284 from the evolutionary runs with a fixed genome, whose TF dynamics has been studied above. The evolvability graphs of the three internal and three external TFs of the two individuals are shown in Figs. 2.18 and 2.19, respectively. In the figures, one dot represents one mutant from the respective individual with a mutation step-size of 0.1. Comparing Figs. 2.18 with 2.19, we can observe that the spread in the phenotype of the mutants in generation 284 is much larger than that of the initial individual with respect to a similar spread in the genotype distance. The same conclusion can be drawn from the evolvability graphs for a step-size of 0.001, as shown in Figs. 2.20 and 2.21, respectively. To check the change of evolvability more quantitatively, we show the evolvability graphs of TF3 of the two individuals in Fig. 2.22 again with the boxplot. From the figure, the increase in phenotype distance can be observed clearly, given the same genotype distance. The evolvability measure for the two individuals calculated from TF3 is 3.91×10^{-5} and 1.04×10^{-3}, respectively.

These findings are very interesting, since these may suggest that the evolution of evolvability has been observed without an explicit selection pressure on evolvability. To further verify this hypothesis, the evolvability measures of all TFs in generation 1, 100, 200, and 284 for different mutation step-sizes are calculated and plotted in Fig. 2.23. Note that the evolvability measure is normalized with the used mutation step-size to make them comparable. Although there are a few exceptions, a general trend of increasing evolvability over the generations can be observed for all these setups.

2.4.6 *Discussions*

A quantitative evolvability measure is suggested, which is independent of a specific fitness function. This evolvability measure is generic in the sense that it is applicable to all developmental models in which a set of differential equations are used for

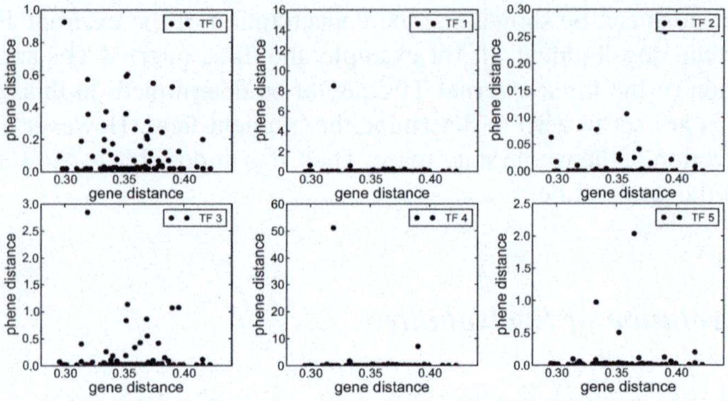

Fig. 2.18 The evolvability graph of the six TFs of the individual from generation 1, step-size $\sigma = 0.1$

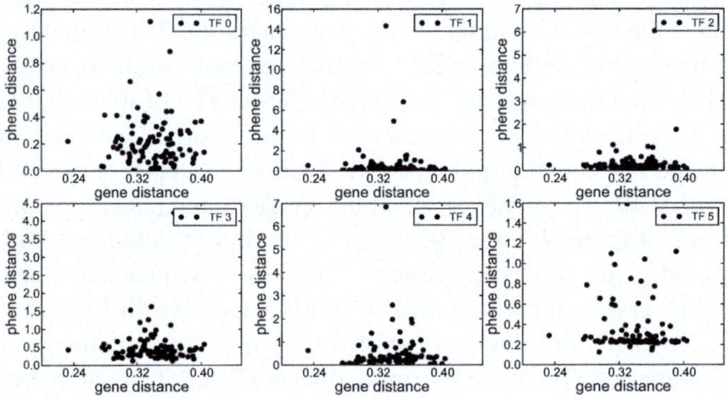

Fig. 2.19 The evolvability graph of the six TFs of the best individual in generation 284, step-size $\sigma = 0.1$

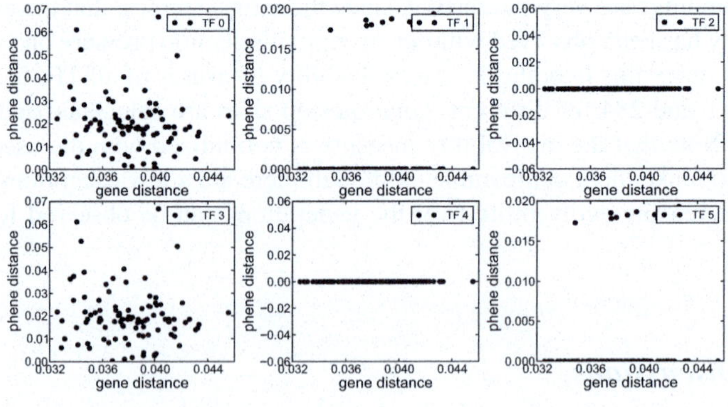

Fig. 2.20 The evolvability graph of the six TFs of the individual in generation 1, step-size $\sigma = 0.001$

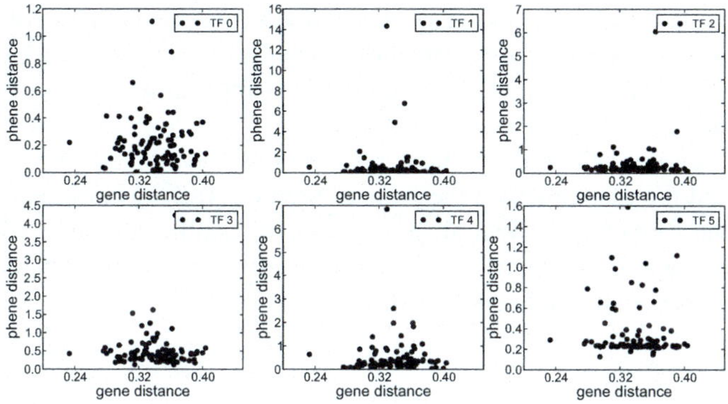

Fig. 2.21 The evolvability graph of the six TFs of the best individual of generation 284, step-size $\sigma = 0.001$

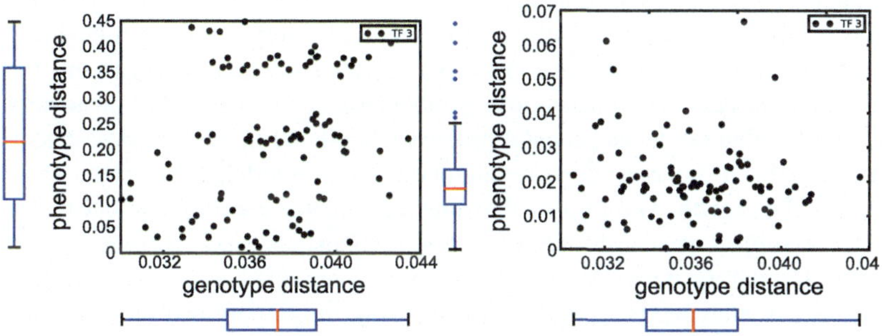

Fig. 2.22 Comparison of evolvability graph of the TF3 at generation 1 and generation 284

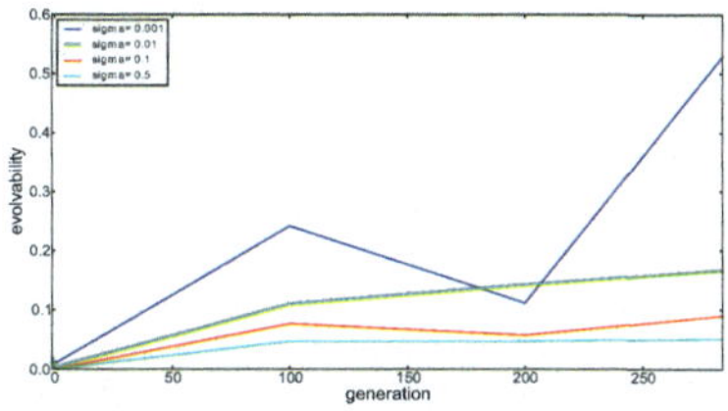

Fig. 2.23 Evolution of evolvability for different mutation step-sizes. The evolvability measure is normalized with the mutation step-size for them to be comparable

modeling the gene regulatory network underlying the individual development. Methods for calculating the genotype and phenotype distances were suggested. Based on the ratios between these distances calculated for a large number of mutants of an individual, the multiplication of interquartile range (IQR) in both genotype and phenotype distances is used as the evolvability measure of the individual. The IQR-based measure can take into account both individual's innovation ability and robustness, both of which are most important for evolvability. With the help of this evolvability measure, we found that evolutionary developmental systems that evolve for limit cycles have also increased their evolvability steadily, which suggests that the evolution of evolvability is selectable even without explicit selection pressure.

One weakness of the IQR-based evolvability measure is that it cannot distinguish between the case in which the mutants are uniformly distributed on the diagonal of the graph and the case in which the mutants are uniformly distributed in the whole graph. In the former case, neutrality may still be insufficient, since robustness is an important factor for evolvability. Other measures such as Simpson's index [272] based on the genotype and phenotype distances defined here can also be considered.

2.5 Summary

This chapter discusses two fundamental concepts of biological gene regulatory networks, robustness and evolvability, and the relationship between robustness and evolvability. Then, we simultaneously evolve robustness and evolvability for two gene regulatory networks, one stationary model represented with a Boolean network, which means that no developmental process is modeled in the genotype-phenotype map, the other a dynamic gene regulatory network modeled by virtual DNA for multi-cellular growth. The empirical results indicate that a stationary gene regulatory network is subject to a hard trade-off between robustness and evolvability, whereas the gene regulatory network for cellular growth modeled by ordinary differential equations can alleviate such trade-offs, implying the importance of developmental plasticity in evolving new functions of biological systems.

Chapter 3
Evolutionary Synthesis of Gene Regulatory Dynamics

3.1 Introduction

Gene regulatory networks play a central role in the evolution and development of biological systems. To understand natural evolution, it is of essence to understand the structure and dynamics of gene regulatory networks. There are two possible ways to achieve this goal. On the one hand, the availability of a huge amount of microarray data for gene expression makes it possible to reconstruct the underlying gene regulatory network for a particular function using mathematical methods [79, 243]. Two main difficulties arise in building mathematical models of gene regulatory networks. First, the microarray data are often very noisy and the data may be collected in different labs under different conditions. Second, the dimension of the model, i.e., the number of genes involved, is extremely high, whereas the number of records is usually very limited (less than 10).

On the other hand, a small number of sub-networks, also known as network motifs, occur very often in complex gene regulatory networks. These network motifs serve as building blocks of regulatory networks, and the dynamics of the whole networks can be analyzed by looking at these motifs. Detection and analysis of the regulatory motifs have been shown to be very helpful in understanding the functions of complex biological networks in systems biology [9, 10].

While most work concentrates on the analysis of dynamics of network motifs, the focus of this chapter is to investigate *in silico* whether it is possible to evolve the desired dynamics given the structure of a network motif, or to evolve more complex regulatory dynamics from simple network motifs. To this end, we employ evolutionary algorithms to evolve the parameters of motifs, the regulatory relationships for a given number of genetic nodes, or the connectivity between different given motifs. Though the evolution of the desired dynamics for a given network motif appears straightforward at first sight, it has been found non-trivial to evolve some particular dynamics, such as undamped oscillations, i.e., limit cycles. In addition, only a limited range of oscillation periods can be evolved, and once undamped oscillation is

Y. Jin, *Computational Evolution of Neural and Morphological Development*, Natural Computing Series, https://doi.org/10.1007/978-981-99-1854-6_3

evolved, the evolved regulatory dynamics will be rather robust against perturbations in the parameters of the network motifs, although it may depend on the initial states.

Various ideas for evolving dynamics for gene regulatory networks *in silico* have been reported in the literature. Both bi-stable switches and oscillators are evolved based on a number of predefined basic biochemical reactions [116], where the fitness function is the difference between the real output and the desired amplitude A_1 at half-integer periods, and the difference between the real output and the desired amplitude A_2 at integer periods. However, it was indicated in [61] that the results reported in [116] are not easily reproducible, which implies that successful evolution of sustained oscillation is sensible to experimental settings. In [61], a correlation-based fitness function has been suggested, although no definite conclusion can be drawn regarding whether this fitness function makes it easier to evolve oscillators. Similarly, two different fitness functions are suggested for evolving oscillation [320], where the first approach is the same as in [116], and the second one is based on the condition for generating Hopf bifurcation. In [225], oscillatory dynamics is evolved using the *Biosys* model developed in [330].

The next step from evolving regulatory dynamics for given motifs is to evolve more complex regulatory structures from individual motifs such as autoregulation, feedforward loops, and feedback loops [10]. At first, the coupling of basic regulatory motifs is substantial for natural evolution to create more complex and robust behaviors. For instance, it has been shown in [73] that positive feedback loops are essential for generating multi-stability. It has also been demonstrated that coherently coupled feedback loops can enhance the robustness to state perturbations in a Boolean network [235]. In addition, it has been found that networks with a larger number of positive feedback loops and a smaller number of negative feedback loops are likely to be more robust [236]. A study in [428] indicated that interlinked positive and negative feedback loops can achieve oscillatory dynamics with a more widely and flexibly tunable frequency with nearly constant amplitudes.

Finally, the cis-regulation logic plays a key role in regulatory dynamics, and the functionality of gene regulatory networks [357] consisting of competitively binding activators and repressors can be controlled more robustly.

In this chapter, we present how evolutionary algorithms can be used to evolve typical genetic regulatory dynamics [199]. We start with evolving the parameters of given genetic motifs to generate the desired regulatory dynamics such as bi-stability and oscillation. Then, we evolve regulatory relations (activation or inhibitory) and regulatory logic (how different regulatory relations are combined) to generate robust oscillatory dynamics. Finally, we examine if evolutionary algorithms can couple the basic motifs so that more complex regulatory dynamics can be evolved, verifying the hypothesis that complex dynamic behaviors can be evolved from tinkering with basic regulatory motifs [294].

3.2 Evolving Parameters of Genetic Switches and Oscillators

3.2.1 *Computational Models*

A simplified view of gene expression consists of two main steps, namely transcription and translation, where in transcription an mRNA copy of the DNA is made, and in translation, proteins are synthesized from mRNA. Whether a structural gene will be expressed (activated) or not (repressed) is decided by a series of interactions within the regulatory network.

If the expression of a gene can be regulated (activated or repressed) by the protein produced by itself, it is termed autoregulation. The mathematical model of gene expression with autoregulation can be described by

$$\frac{d[R]}{dt} = -\gamma_R[R] + \alpha_R H([P]), \tag{3.1}$$

$$\frac{d[P]}{dt} = -\gamma_P[P] + \alpha_P[R], \tag{3.2}$$

where $[R]$ and $[P]$ are the concentration of mRNA and protein, respectively, γ_R and γ_P are the decay rate of the mRNA and protein, α_R and α_P are the synthesis rate of the mRNA and protein, and $H(X)$ is the Hill function. If the autoregulation is a repression, also known as negative autoregulation, the Hill function can be described by

$$H_r(x) = \frac{\beta}{\theta^n + x^n}, \tag{3.3}$$

and if the autoregulation is activation, the Hill function can be written as

$$H_a(x) = \frac{\beta x^n}{\theta^n + x^n}, \tag{3.4}$$

where β is the activation coefficient, θ is the threshold, and n is the Hill coefficient.

Three network motifs will be studied here, including the positive autoregulation (PAR), toggle switch with positive autoregulation, and relaxation oscillation, as shown in Fig. 3.1a, b, c, respectively. For convenience, we provide again the mathematical model of the three motifs as follows. The mathematical models of a PAR can be given by

$$\dot{x}_1 = -a_{11}x_1 + a_{12}H_a(x_2), \tag{3.5}$$

$$\dot{x}_2 = a_{21}x_1 - a_{22}x_2, \tag{3.6}$$

where x_1 is the concentration of mRNA, and x_2 the concentration of protein. There are different ways to combine multiple regulatory inputs to the promoter of a particular

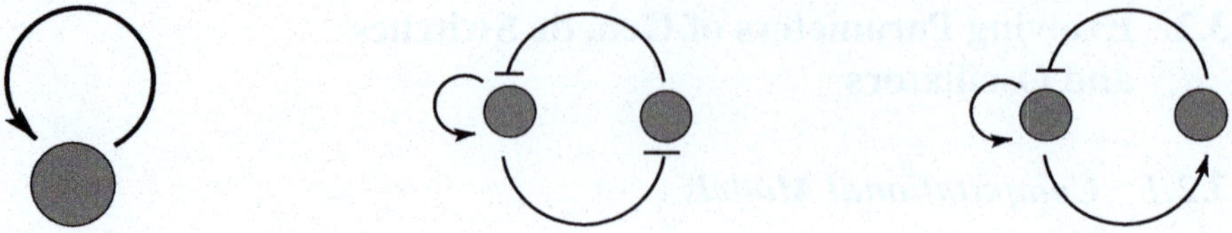

Fig. 3.1 Three widely seen regulatory motifs. **a** PAR, **b** toggle switch with PAR, and **c** relaxation oscillation

gene, leading to slightly different mathematical models. In [10], logic functions such as logic AND or OR are suggested to combine multiple regulatory inputs. Here, the regulatory inputs are supposed to be additive. In this case, the dynamics of the toggle switch with PAR can be described by

$$\dot{x}_1 = -a_{11}x_1 + a_{12}H_a(x_2), \tag{3.7}$$

$$\dot{x}_2 = a_{21}x_1 - a_{22}x_2 + a_{23}H_r(x_3), \tag{3.8}$$

$$\dot{x}_3 = a_{32}H_r(x_2) - a_{33}x_3, \tag{3.9}$$

where x_1 represents the mRNA concentration, and x_2, x_3 the concentration of two proteins.

Finally, the oscillation motif can be described by

$$\dot{x}_1 = -a_{11}x_1 + a_{12}H_a(x_2), \tag{3.10}$$

$$\dot{x}_2 = a_{21}x_1 - a_{22}x_2 + a_{23}H_r(x_3), \tag{3.11}$$

$$\dot{x}_3 = a_{32}H_a(x_2) - a_{33}x_3. \tag{3.12}$$

Similar to the toggle switch, x_1 represents the mRNA concentration, and x_2, x_3 denote the concentration of two proteins.

3.2.2 *Evolutionary Synthesis of Genetic Dynamics*

3.2.2.1 Evolution Strategy for Parameter Optimization

Evolution strategies are very efficient for evolving real-valued parameters. Here, we adopt a canonical evolution strategy for evolving the desired dynamics since the structure of the regulatory motifs is fixed and only the parameters are evolved. In a canonical evolution strategy, the mutation of the decision variables (parameters in the GRN model) is performed by adding an $N(0, \sigma_i^2)$ distributed random number. The step-sizes σ_i are also encoded in the genotype and subject to mutations. The ES used here can be described as follows:

$$\mathbf{x}(t) = \mathbf{x}(t-1) + \tilde{\mathbf{z}} \tag{3.13}$$

$$\sigma_i(t) = \sigma_i(t-1)\exp(\tau' z)\exp(\tau z_i);\ i = 1, ..., n, \tag{3.14}$$

where $\mathbf{x}$ is an n-dimensional decision vector to be evolved, $\tilde{\mathbf{z}}$ is an n-dimensional random number vector with $\tilde{\mathbf{z}} \sim N(\mathbf{0}, \boldsymbol{\sigma}(t)^2)$, and z and z_i are normally distributed random numbers with $z, z_i \sim N(0, 1)$. Parameters τ, τ', and σ_i are called strategy parameters, where σ_i is mutated as in Eq. (3.14) and τ, τ' are constants defined as follows:

$$\tau = \left(\sqrt{2\sqrt{n}}\right)^{-1};\ \tau' = \left(\sqrt{2n}\right)^{-1}. \tag{3.15}$$

In our study, the (μ, λ)-ES selection strategy is adopted. In the evolution, all parameters in the regulatory models, i.e., four decay and synthesis rates in the PAR model and seven in the toggle and oscillation models, together with the activation coefficient (β), the threshold (θ), and Hill coefficient (n) are encoded in the genome and are subject to evolution. A (50, 300)-ES is adopted in the experiments. All parameters to be evolved are randomly initialized between 0 and 4. According to the physical meaning of the parameters, a lower bound is set to 0, but no upper bound is given. The initial step-size is set to 0.1. In all simulations, 1000 generations are run for each case.

3.2.2.2 Fitness Definition

To evolve the bi-stable dynamics for the PAR, the initial state of the regulatory model is randomly set between 0 and 4. The target function of the protein concentration of x_2 for the PAR is as follows:

$$x_2^d(\infty) = \begin{cases} 0, \text{ if } x_2(0) < 2; \\ 4, \text{ if } x_2(0) \geq 2. \end{cases} \tag{3.16}$$

The same initial states are set for the toggle switch motif, and the target of x_3 is defined similar to Eq. (3.16).

The target function for x_3 in evolving oscillation is defined by a sinus function as follows:

$$x_3^d(t) = \sin(2\pi t/\mathrm{T}) + 1.0, \tag{3.17}$$

where t is the time instant, and T is the period of the desired oscillation.

The evolution of oscillation is non-trivial. As indicated in [61], the success rate of achieving the desired dynamics is very low when both the structure and parameters are evolved. We find even in our experiments where the structure of the network motif is predefined, is it not straightforward to evolve limit cycles? This issue will be discussed further in discussing the experimental results.

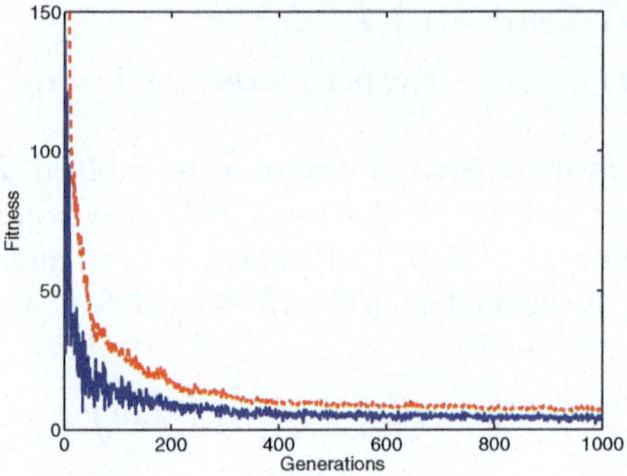

Fig. 3.2 Mean and best fitness in evolving bi-stability for the positive autoregulation motif

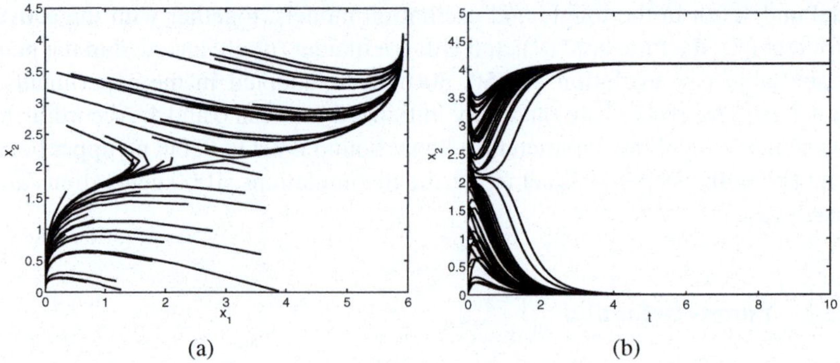

Fig. 3.3 Dynamics of the evolved positive autoregulation motif for different initial states. **a** state-space trajectories and **b** time course

3.2.2.3 Evolving Bi-Stability

The mean fitness (dotted line) and best fitness (solid line) in evolving bi-stability for the PAR are presented in Fig. 3.2. Once the evolutionary optimization is complete, the regulatory dynamics of the genetic system is verified using 50 randomly initialized states. The state-space trajectories of x_1 and x_2 are plotted in Fig. 3.3a, and state x_2 over time is provided in Fig. 3.3b. It can be seen that the evolved bi-stability behavior is insensitive to changes in the initial states. It is also found that the evolution of bi-stability has been successful for all test runs.

Similar results have been obtained in evolving the toggle switch for the positive autoregulation motif. Its state-space trajectories and time course are presented in Fig. 3.4a, b, respectively.

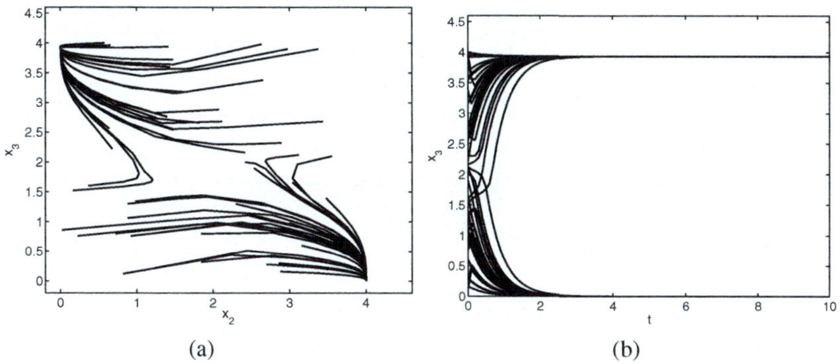

Fig. 3.4 Dynamics of the evolved toggle switch with PAR for different initial states. **a** state-space trajectories and **b** time course

3.2.3 Evolvability in Evolving Sustained Oscillation

In the first runs, the ES was able to achieve a limit cycle in 10 independent runs. We then check the success rate for six different desired oscillation periods T and four different encoding schemes for the Hill coefficient n:

$$n = N \times g_n, \tag{3.18}$$

where g_n is the gene that encodes the Hill coefficient in the chromosome, and $N \geq 1$ is a constant that adjusts the coding range of this gene. Apparently, if $N = \infty$, the Hill function will become a step function, no matter how g_n changes. In our experiments, simulations are conducted for $N = 1, 5, 10$, and ∞. The success rates for different N's are given in Fig. 3.5, where '0' means no success in ten independent runs has been achieved, whereas '1' indicates that in all ten runs it has been successful in evolving stable oscillation. Note that in many unsuccessful runs, the state (x_3) does oscillate, but becomes asymptotically stable; see, e.g., in Fig. 3.6. In this case, we do not count it as a successful run.

Fig. 3.5 The success rate in evolving stable oscillation for different T's and N's

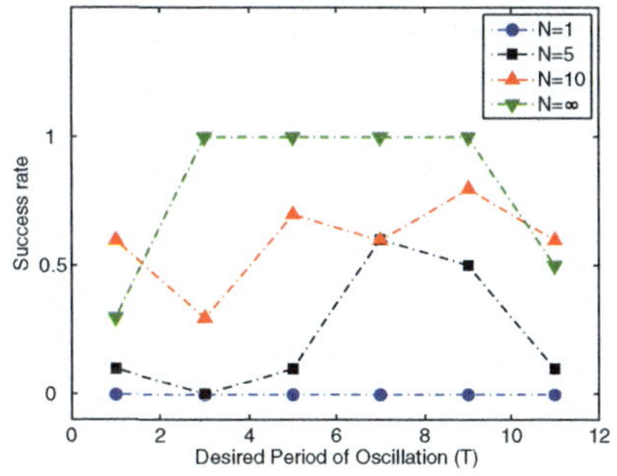

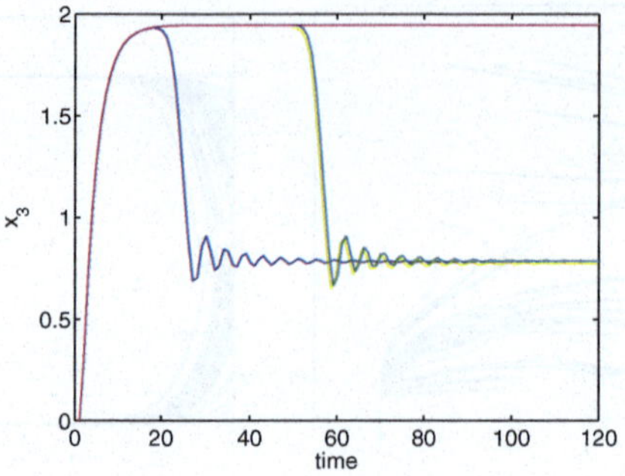

Fig. 3.6 Evolved damped oscillation when $N = 5$ and $T = 7$

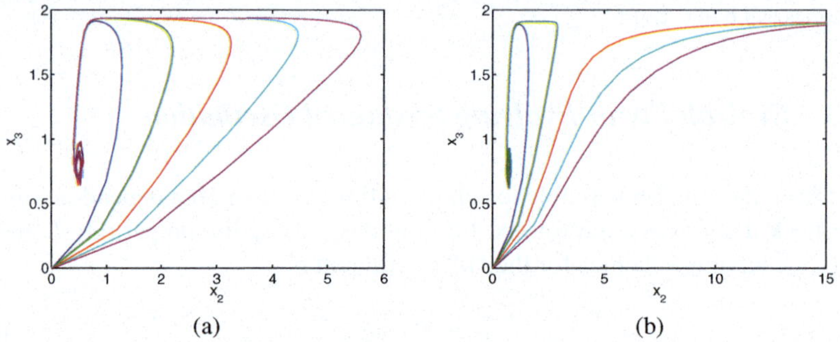

Fig. 3.7 Two oscillators evolved in two independent runs for $T = 5$ and $N = 10$. In **a**, the evolved oscillator shows undamped oscillation for all five initial states, whereas in **b**, the oscillator exhibits a limit cycle for only two of the initial states

From Fig. 3.5, we find that the larger the Hill coefficient, the easier it is to evolve undamped oscillation. When the Hill function becomes a step function ($n = \infty$), the success rate becomes one for the desired oscillation periods ranging from 3 to 9. Meanwhile, we find that it is challenging for the ES to find an evolutionary path toward a larger n that facilitates a sustained oscillation, although no upper bound for the Hill coefficient n is set during the evolution. This problem may be attributed to the fitness function used in the study here. Unlike in evolving the switching behavior, a smaller (better) fitness value does not necessarily lead to a stable oscillation.

For the successfully evolved oscillators, we first verify the robustness of the oscillatory dynamics against different initial states once the evolutionary optimization is completed. Five different initial states are tested for two evolved oscillators evolved in two independent runs, when $N = 10$ for the desired oscillation period of $T = 5$. Although the evolved regulatory dynamics appears to be similar, their behaviors for $x_1(0) = 1.0, 2.0, 3.0, 4.0, 5.0$ are different, as shown in Fig. 3.7. In Fig. 3.7a, all five initial states of x_1 lead to stable oscillation, while in Fig. 3.7b, the evolved motif can produce stable oscillation only for two of the five initial states of x_1.

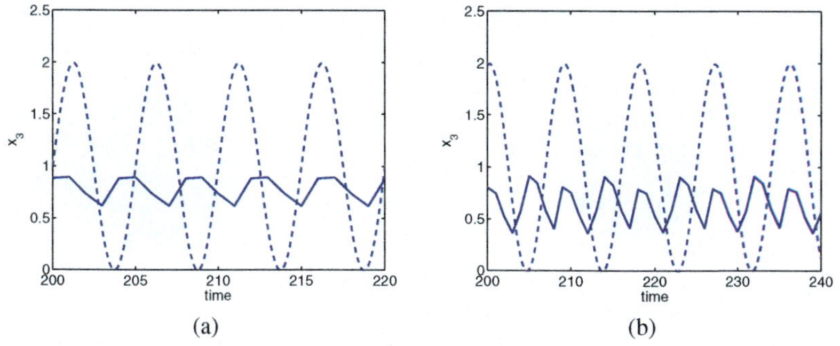

Fig. 3.8 The real and desired states of x_3 of an evolved oscillator. **a** $T = 5$, $N = 10$, **b** $T = 9$, $N = \infty$

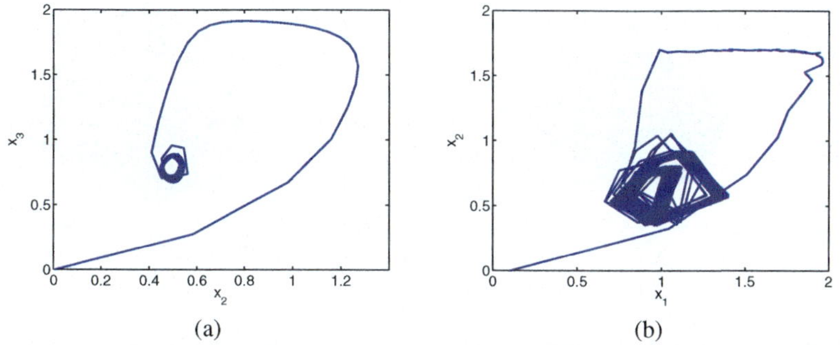

Fig. 3.9 Oscillatory dynamics of the evolved oscillator for **a** $N = 10$, $T = 5$, and **b** $N = \infty$, $T = 9$

In the following, we investigate if the evolved oscillator is able to approximate the desired oscillation period. The time course of x_3 of two evolved oscillators for the desired oscillation period $T = 5$, $N = 10$, and $T = 9$, $N = \infty$ are presented in Fig. 3.8a, b, respectively. From the figures, it can be seen that the evolved oscillations are roughly the same as the desired ones. The state-space trajectories of these two oscillators are plotted in Fig. 3.9a, b, respectively.

3.2.3.1 Robustness of Evolved Oscillators

Here, compare the robustness of the evolved oscillators against perturbations in the parameters of the Hill coefficients. Once the evolution is completed, noise with a level varying from 5% to 20% is added to the parameters of the oscillators, one with a Hill coefficient $n = 10.7$, and the other with $n = \infty$. The state-space dynamics of the two oscillators in the presence of noise in their parameters are plotted in Fig. 3.10 when $n = 10.7$, and Fig. 3.11, whereas the time courses of x_3 are shown in Figs. 3.12 and 3.13, respectively. From these results, we can conclude that the evolved oscillators

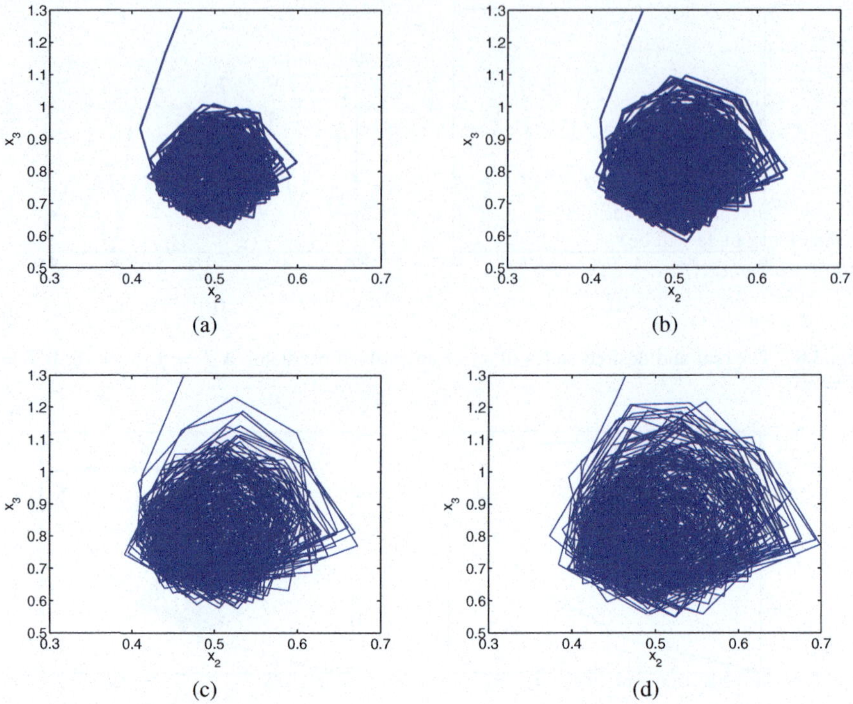

Fig. 3.10 The oscillatory dynamics when the parameters of the evolved oscillator ($n = 10.7$) are perturbed with various noise levels. **a** 5%, **b** 10%, **c** 15%, and **d** 20%

are robust to perturbations in the parameters. However, the oscillator with an infinite Hill coefficient shows more robust oscillatory behavior than the one with a smaller Hill coefficient.

3.3 Evolution of Regulatory Logic

3.3.1 Motivation

In the previous section, we show that evolutionary algorithms are able to shape *in silico* typical regulatory dynamics by evolving the parameters of motifs whose regulatory logic and relationship are predefined. In this section, we are interested in evolving the *cis*-regulation logic, which also plays an important role in the dynamics and functionality of gene regulatory networks. An experimental analysis of *cis*-regulation logic of the genes in sea urchin for the development was conducted in Yuh et al. [467]. A comprehensive investigation of regulatory logic in gene networks

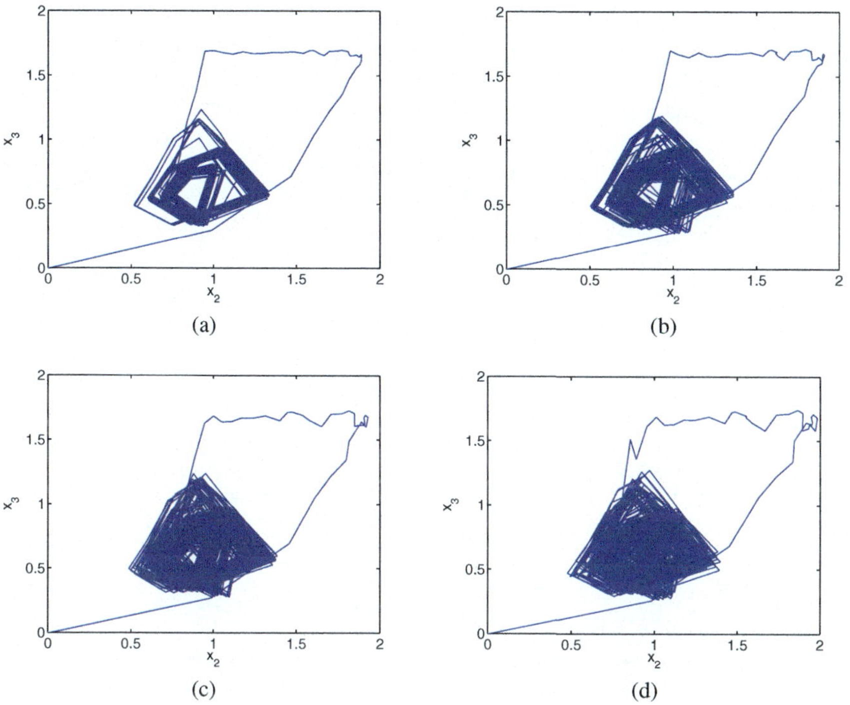

Fig. 3.11 The oscillatory dynamics when the parameters of the evolved oscillator ($n = \infty$) are perturbed with various noise levels. **a** 5%, **b** 10%, **c** 15%, and **d** 20%

was conducted in [357], which concluded that networks consisting of competitively binding activators and repressors can be controlled more robustly.

In addition to analysis of network motifs, *in silico* synthesis of typical regulatory dynamics can offer us insight into how nature has shaped the evolution of regulatory motifs. Francois and Hakim [115] evolved both bi-stable and oscillatory dynamics using a number of predefined basic biochemical reactions, where the difference between the output of the regulatory model and the desired amplitude at the half-integer and integer periods was adopted as the fitness function. Paladugu et al. [320] employed a fitness function based on the condition for generating Hopf bifurcation to make it easier for the evolution to generate sustained oscillation. In [226], a gene regulatory network was used for evolving biological clocks in the presence of periodic environmental stimuli, where both the number and activation type (activating or repressive) of regulatory units of each gene are subject to evolution. The evolved clock has been shown to be robust to perturbations.

In this section, we investigate *in silico* the role of *cis*-regulation logic and the coupling of feedback loops in evolving oscillatory dynamics. The investigation is divided into two stages. In the first stage, we analyzed the robustness of two motifs with a predefined structure to perturbations in parameters and initial states. In the

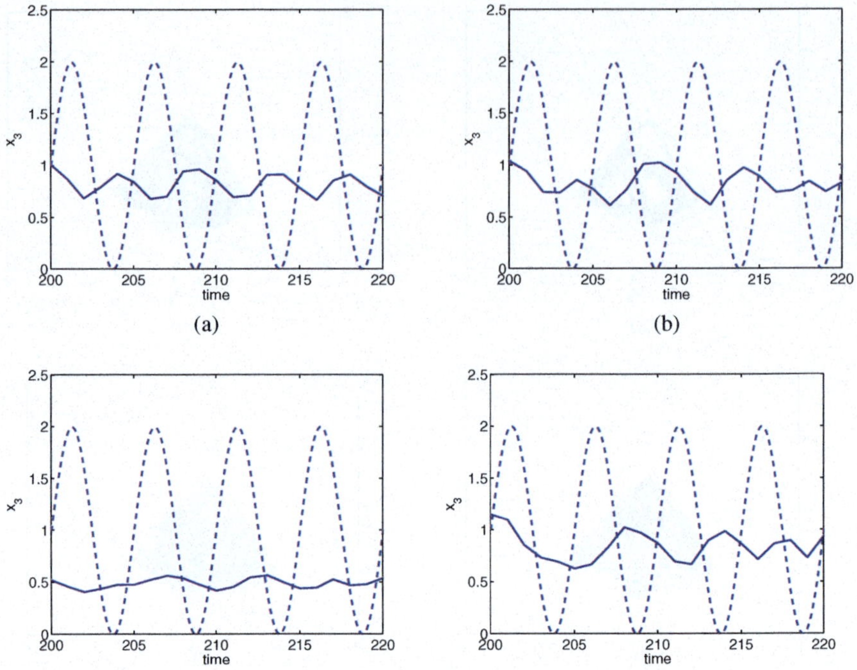

(a)

(b)

(c)

(d)

Fig. 3.12 Time course of x_3 of the oscillator when the parameters of the evolved oscillator ($n = 10.7$) are perturbed with various noise levels. **a** 5%, **b** 10%, **c** 15%, and **d** 20%

second stage, some components of the regulatory motifs and their regulatory logic are evolved. Our results demonstrated that the robust motif emerged from *in silico* evolution without an explicit selective pressure on robustness, which suggests that robustness to perturbations in parameters and regulatory dynamics is an emergent property of evolution, confirming the findings in other research [231].

3.3.2 Gene Regulatory Network Motifs and Regulatory Logic

Here, we mainly consider the gene regulatory motifs consisting of three genes, namely two regulatory genes and one target gene. In the *in silico* evolution, the target gene is expected to generate sustained oscillatory dynamics. There are regulatory interactions, either activating or repressive, between the two regulatory genes and the target gene, although no direct interactions between the two regulatory genes are considered here, thus forming two feedback loops. With this simplification, we can have ten different network structures in total, taking into account the structural symmetry between the two regulatory genes. Each motif has two feedback loops, a

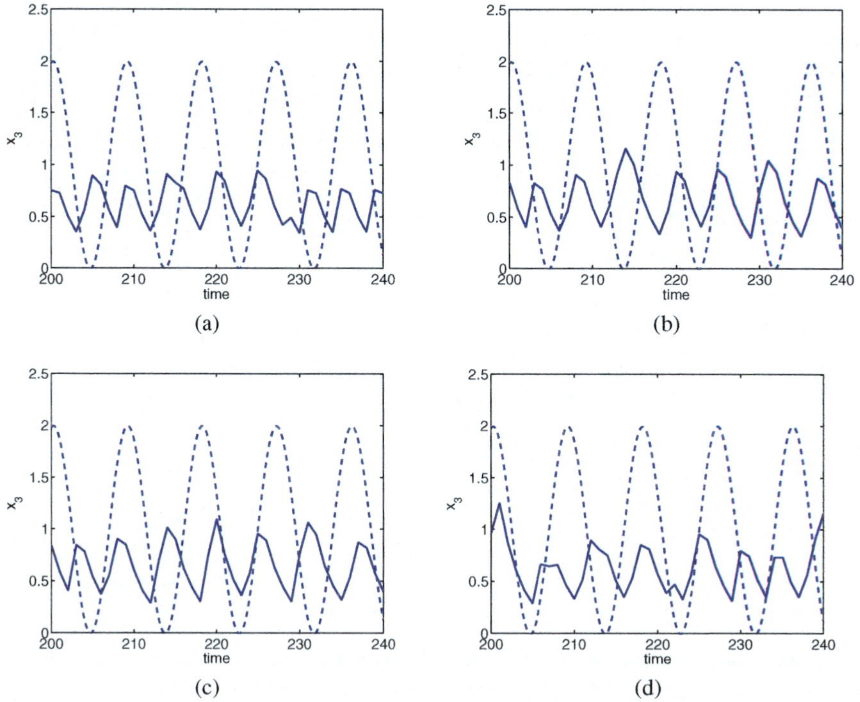

Fig. 3.13 Time course of x_3 of the oscillator when the parameters of the evolved oscillator ($n = \infty$) are perturbed with various noise levels. **a** 5%, **b** 10%, **c** 15%, and **d** 20%

positive feedback loop and a negative feedback loop. A further simplification used here is that the regulatory relationship between one of the regulatory genes and the target gene is fixed. Consequently, we can have four different possible motifs, as shown in Fig. 3.14. In the motifs shown in Fig. 3.14b, d, the target gene is consistently regulated by two activating regulatory genes, which are termed as *consistently regulated motifs* (CRM). By contrast, the target gene is inconsistently regulated in the motifs in Fig. 3.14a, c, which are called *inconsistently regulated motifs* (IRM). Moreover, in motifs in Fig. 3.14a, d, two positive feedback loops are coupled. In contrast to that, a positive feedback loop is interlinked with a negative feedback loop in the motifs in Fig. 3.14b, c.

Here, the dynamics of the regulatory motifs is described using a set of differential equations. For instance, the regulatory dynamics of the CRM in Fig. 3.14b can be described using the following ordinary differential equations (ODEs):

$$\dot{x}_1 = a_{13} H_{13}(x_3) - a_{11} x_1, \tag{3.19}$$

$$\dot{x}_2 = a_{23} H_{23}(x_3) - a_{22} x_2, \tag{3.20}$$

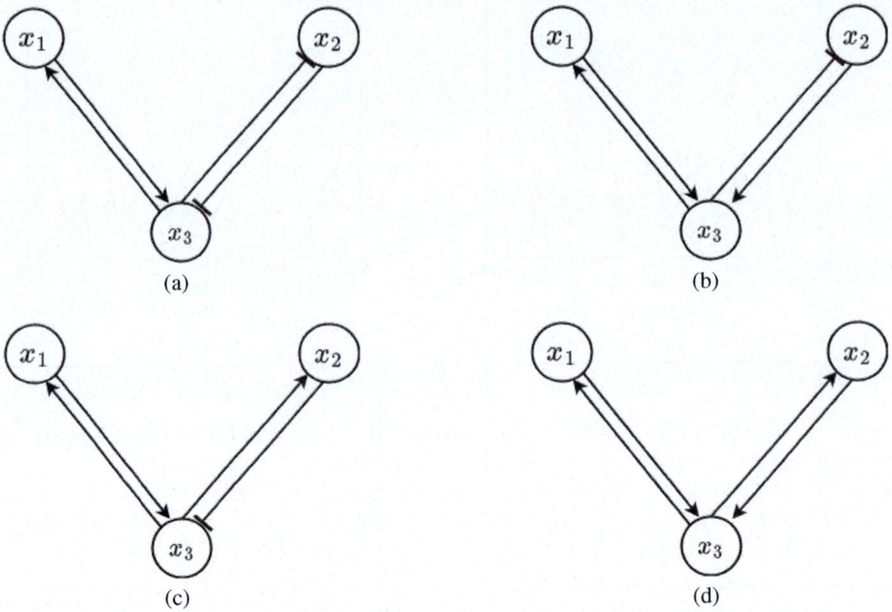

Fig. 3.14 Four gene regulatory motifs studied. The target gene is consistently regulated in motifs **b** and **d**, whereas the target gene is inconsistently regulated in motifs **a** and **c**

$$\dot{x}_3 = a_3 L(H_{31}(x_1), H_{32}(x_2)) - a_{33} x_3, \tag{3.21}$$

where x_1 and x_2 are the concentration of the protein products of the regulatory genes, x_3 is the concentration of the protein of the target gene, a_{11}, a_{22}, and a_{33} are the decay rate of the proteins, and a_{13}, a_{23}, and a_3 are the parameters representing the strength of gene-protein interactions. CRM is composed of three activating regulations and one repressive regulation distinguished by different Hill functions. An activating Hill function, e.g., H_{13} can be described as follows:

$$H_{13}(x_3) = \frac{\beta x_3^n}{\theta_1^n + x_3^n}, \tag{3.22}$$

where n is called the Hill coefficient. By contrast, a repressive Hill function, e.g., H_{23}, can be represented by

$$H_{23}(x_3) = \frac{\beta}{1 + (x_3/\theta_2)^n}. \tag{3.23}$$

In Eq. (3.21), both H_{31} and H_{32} are activating Hill functions similar to H_{13}, and $L(H_{31}(x_1), H_{32}(x_2))$ is a logic function combining the regulatory activity of the two regulatory genes. The dynamics of other regulatory motifs can be described in a similar way by changing the type of the Hill functions according to the coupling structure in the motif. For instance, in IRM in Fig. 3.14c, H_{23} should be activating and H_{32} should be inhibitory.

One assumption is that various transcription factors bind to the binding site of the target gene independently. We considered two situations in the following. First, both transcription factors are needed to activate the expression of the target gene. Second, either of the two regulatory genes will be adequate to activate the expression of the target gene. In the first case, we used fuzzy logic 'AND', which can be described using either the Zadeh 'AND' or the probabilistic 'AND' [191]:

$$\text{Zadeh 'AND'}: \; L(x, y) = \min(x, y), \tag{3.24}$$

$$\text{Probabilistic 'AND'}: \quad L(x, y) = xy. \tag{3.25}$$

For the second case, the following Zadeh 'OR' or probabilistic 'OR' logic operators were adopted:

$$\text{Zadeh 'OR'}: \; L(x, y) = \max(x, y), \tag{3.26}$$

$$\text{Probabilistic 'OR'}: \; L(x, y) = x + y - xy. \tag{3.27}$$

In addition, we also considered the case where a summation operation is performed as the logic function, which can be seen as a combination of the probabilistic 'AND' and probabilistic 'OR':

$$\text{summation}: \; L(x, y) = \frac{1}{2}(x + y). \tag{3.28}$$

3.3.3 *Evolution of Regulatory Motifs*

Like in the previous section, we adopt a canonical evolution strategy (ES) for evolving the parameters in regulatory motifs and a hybrid evolutionary algorithm for evolving the structure, regulatory logic, and the parameters of the motifs.

3.3.3.1 Genetic Representations and Variations

In the first stage of the work, the parameters of two motifs with predefined regulatory logic and regulating structure are evolved using an ES with one global step-size only as the strategy parameter for self-adaptation [18]. In the ES, the genome of each individual consists of two chromosomes. The first chromosome contains 12 real-valued loci, including three parameters representing the strength of protein-protein interactions, three degradation parameters, four regulation thresholds for the four Hill functions, one parameter for the maximum gene expression level, and one for the Hill coefficient. The second chromosome has one locus only, i.e., the strategy parameter.

For evolving the structure and regulatory logic of the motifs in the second stage, each genome has a binary chromosome in addition to the two real-valued ones used in the first stage. The binary chromosome is composed of two loci representing the interactions between regulatory gene 2 and the target gene in the motifs, and two loci encoding the regulatory logic that combines the activity of the two regulatory genes. The representation schemes used in the two stages are illustrated in Fig. 3.15a, b, respectively. In Fig. 3.15b, c, c_{23} represents either an activating ($c_{23} = 1$) or repressive ($c_{23} = 0$) regulatory interaction from the target gene (g_3) to regulatory gene 1 (g_1). Similarly, c_{32} denotes an activating or repressive regulation from the regulatory gene (g_2) to the target gene (g_3). If both c_{23} and c_{32} are activating, or both are inhibitory, then g_2 and g_3 form a positive feedback loop. Otherwise, $g_2$and g_3 form a negative feedback loop.

The two binary bits (l_1, l_2) for encoding the regulation logic can be decoded into an integer number between 0 and 3, which represents the Zadeh 'AND' logic (0), the probabilistic 'AND' (1), the probabilistic 'OR' (2), and the summation logic (3), respectively. The Zadeh 'OR' logic is excluded from this representation since we

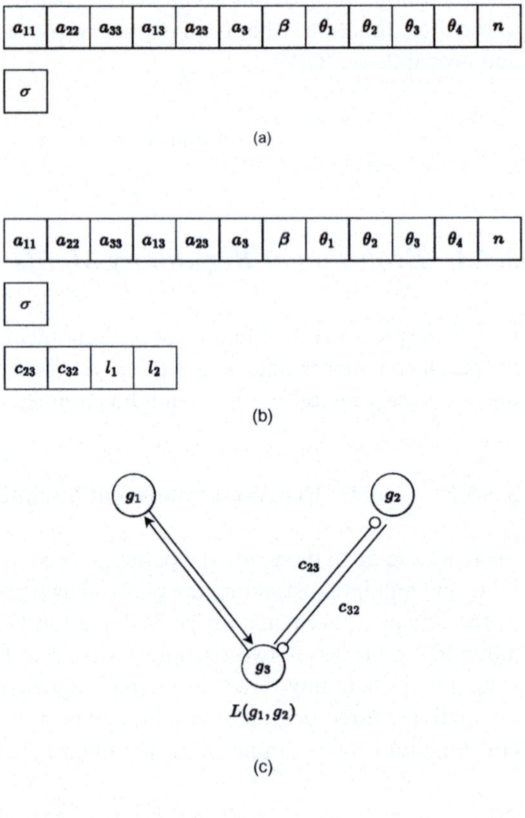

Fig. 3.15 Genetic representations for *in silico* evolution of sustained oscillation. **a** Evolution of the parameters only. **b** Evolution of the regulatory logic, interactions between the genes, and the parameters. **c** A generic regulatory motif whose interactions between regulatory gene 2 and the target gene are determined in evolution

never have succeeded in evolving any sustained oscillation in the first stage with this logic operation.

For evolving the real-valued parameters in the gene regulatory network model in the first stage, Gaussian mutations as well as discrete recombination were employed for genetic variations. Both crossover and bit-flip mutations were applied to the binary chromosome encoding the regulatory logic and the interactions between regulatory gene 2 and the target gene in the second stage.

3.3.3.2 Fitness Function and Selection Strategy

To evolve limit cycle (sustained oscillation), the sinus function is adopted to define the desired state for the target gene:

$$x_3^d(t) = \sin(2\pi t/T) + 1.0, \quad (3.29)$$

where x_3^d is the desired state denoting the protein concentration of the target gene, T is the desired period of the limit cycle, and the amplitude is arbitrarily set to 1.0. The following fitness function is adopted to minimize the difference between the real state of the target gene (x_3) and the desired state:

$$f = \sum_{t=0}^{N} (x_3(t) - x_3^d(t))^2, \quad (3.30)$$

where N is the number of time-steps simulated in solving the dynamics of the regulatory motifs.

For evolution strategies, elitism is usually not recommended to avoid getting stuck in local optimums. It has been found, however, that elitism can effectively help improve the convergence performance in evolving oscillatory dynamics of genetic motifs, and therefore is adopted in all simulations presented in the next section.

3.3.4 *Empirical Results*

3.3.4.1 Experimental Settings

The differential equations are solved using the modified Euler's method with a step-size of 0.1. In the simulations, all regulatory parameters were randomly initialized between 0 and 4, and the initial step-size of the ES was set to 0.5. The parent and offspring population sizes were 60 and 400, respectively. For parameter optimization in the first stage, the Gaussian mutation and discrete recombination are applied with a probability 1, and each evolution is run for 500 generations. For both structural and parameter optimization in the second stage, the parameter settings for changing real-valued chromosomes are the same as in the first stage, and a probability of 0.9 and 0.02 was applied for the crossover and bit-flip mutation, respectively. Each evolution is also run for 500 generations. All the above settings are specified according to the results of pilot studies.

3.3.4.2 Evolvability of Predefined Regulatory Motifs Through Parameter Optimization

At the first stage, two regulatory motifs with a predefined structure are chosen, one is the CRM in Fig. 3.14b, and the other is IRM in Fig. 3.14c, both of which have a positive and a negative feedback loop coupled together. The main reason for choosing these two motifs lies in the findings suggesting that a negative feedback loop coupled with a positive feedback loop is able to generate more robust and flexible oscillation [429]. Five different logic operations described in Eqs. (3.24)–(3.28) are considered for the two motifs, thus resulting in ten different motifs in total.

To evaluate the evolvability of the two different motifs, we attempted to evolve sustained oscillation for different desired periods, namely $T = \{1, \ldots, 11, 12\}$. For each desired period, ten independent runs are conducted. Then, the number of runs in which sustained oscillation has been evolved successfully is countered. The evolution of sustained oscillation is said to be successful if the change in the amplitude of the oscillation is less than 1% within 400 time-steps. The evolution results for the CRM and IRM are presented in Figs. 3.16 and 3.17, respectively.

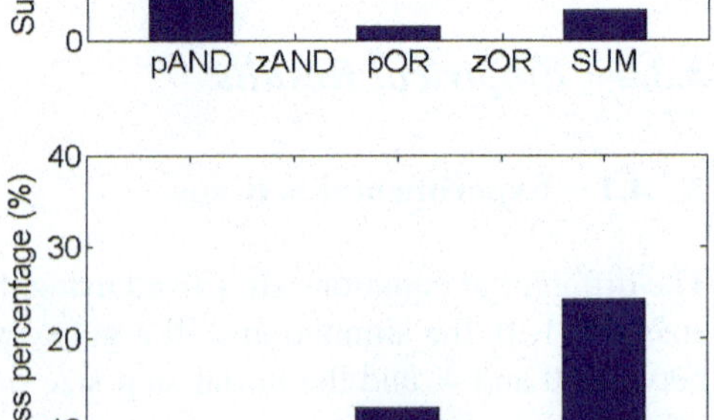

Fig. 3.16 Success ratio in percentage for the CRM with five different logic operations, respectively, in evolving limit cycle (stable oscillation) with the desired period of $T = \{1, 2, \ldots, 12\}$

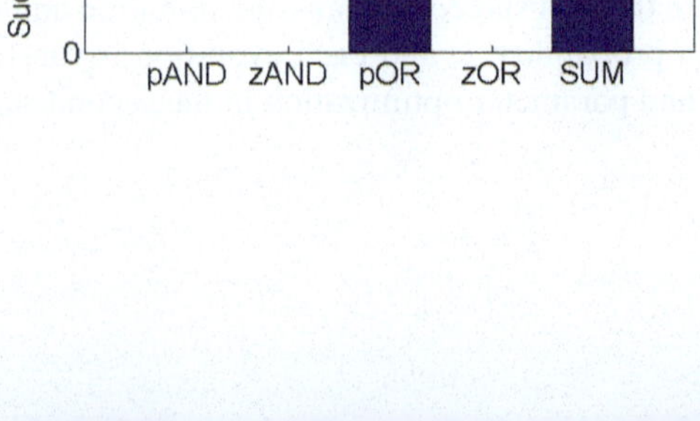

Fig. 3.17 Success ratio in percentage for the IRM with five different logic operations, respectively, in evolving limit cycle with the desired period of $T = \{1, 2, \ldots, 12\}$

From Figs. 3.16 and 3.17, we can see that the success ratios for the CRM and IRM are considerably different. The CRMs using the probabilistic 'AND' logic and summation logic exhibited the highest and second highest success ratios on average among all CRMs. By contrast, the IRMs with the summation and probabilistic 'OR' logic have the highest success percentage among all IRMs. Overall, the success ratio for evolving sustained oscillation is low. However, a high Hill coefficient is required neither for the CRM nor for the IRM in contrast to the relaxation motif studied in the previous section.

From these results, it is observed that at least one run has been successful in 120 independent runs in evolving sustained oscillation for the CRM with probabilistic 'AND', probabilistic 'OR' and summation, and the IRM with probabilistic 'OR' and summation, respectively. Among them, the IRM with the summation logic shows the highest success ratio.

3.3.4.3 Robustness Analysis of the Evolved Motifs

In this subsection, the robustness of the evolved oscillatory dynamics is examined by varying the initial concentrations or by perturbing the parameters of the regulatory motifs. The robustness of regulatory dynamics to perturbations in initial concentrations of the regulatory genes is of interest since the perturbations in initial concentration reflect to a certain degree the influences of genes from outside of the motif. During the evolution, the initial concentrations of the three proteins are set to (1, 1, 0). To study the robustness of the evolved oscillation, the initial concentrations are set between 0 and 4 randomly. It is found that when the initial states of the evolved CRM with the probabilistic 'AND' are randomly initialized, the state of the target gene converges to a point attractor instead of producing a limit cycle for all 16 evolved motifs showing sustained oscillation when their initial state is set to (1, 1, 0). For the CRM with the summation logic, sustained oscillation has been evolved in four runs. In the robustness test, all four motifs are able to maintain their limit cycle when their states are initialized randomly between 0 and 4. From these results, it can be concluded that sustained oscillation of the CRM with the summation logic is less sensitive to perturbations in the initial concentrations than the CRM with the probabilistic 'AND' logic.

For the IRM, the results were quite different. In all 29 successful runs of the IRM with the summation logic and in 15 successful runs for the IRM with the probabilistic 'OR' logic, all randomly initialized states converge to the evolved limit cycle. This indicates that the evolved oscillatory dynamics of IRM is robust to changes in the initial states of the genes.

The different motifs show very different behaviors regarding their sensitivity to perturbations in the parameters. It is found that the CRMs, with either the probabilistic logic or the summation logic, are vulnerable to noise in the parameters. Figure 3.18 shows the 50 state-space trajectories of a CRM with the summation logic under 1% (left panel) and 5% (right panel) noise level. The 50 initial states were generated randomly in the interval of [0, 4]. The motif is still able to generate sustained

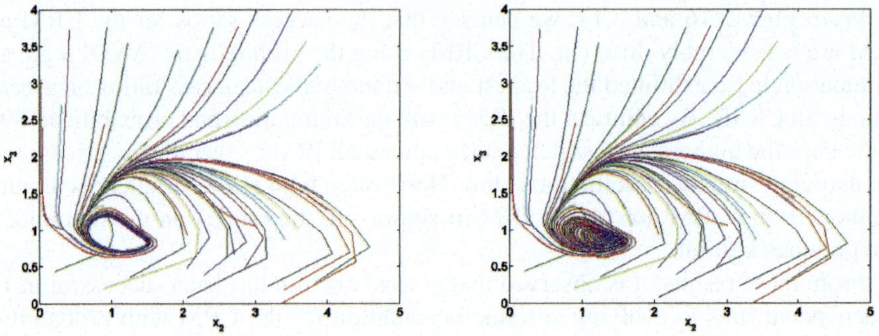

Fig. 3.18 State-space trajectories of a CRM in the presence of perturbations in the parameters, where the x-axis denotes x_2 and the y-axis x_3. Left panel: with 1% noise. Right panel: 5%

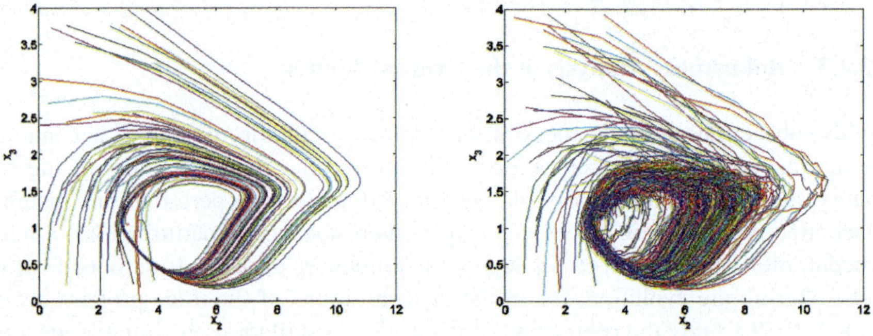

Fig. 3.19 State-space trajectories of an IRM in presence of noise in the parameters, where the x-axis denotes x_2 and the y-axis x_3. Left panel: with 1% noise. Right panel: 5% noise

oscillation for all the initial states when the noise level in the parameters is set to 1%. However, when the noise level increases to 5%, the oscillation became damped, and system states finally converged to a point attractor.

For IRMs with either the probabilistic 'OR' logic or the summation logic, the regulatory system was able to generate sustained oscillation in the presence of 1% or 5% noise level in the parameters. Fifty trajectories of an IRM using the summation logic are presented in Fig. 3.19, where the left panel shows the results with 1% noise and the right panel the results with 5% noise in the parameters. It is evident that in both cases, the regulatory motif is able to maintain stable oscillation.

To summarize, the above results suggest that the IRM using either the summation logic or the probabilistic 'OR' is robust to the initial concentrations and regulatory parameters.

3.3.4.4 Evolutionary Emergence of Robust Motifs

After evolving and analyzing the regulatory motifs with the predefined regulatory logic and structure, it is of interest to see whether robust regulatory motifs can emerge *in silico* evolution when part of the motif structure and the regulatory logic are subject to evolution without an explicit selection pressure on robustness. Two sets of simulations are conducted to answer this question. In the first set of simulations, the structure (interactions between regulatory gene 2 and the target gene) and regulatory logic are evolved, together with the parameters of the regulatory motif, for generating stable oscillation with the desired period $T = \{4, 5, \ldots, 18\}$. For each desired period, twenty independent runs are conducted. The resulting motif structures and logic operators evolved in the successful runs (in which a limit cycle is evolved successfully) are presented in Fig. 3.20a, b, respectively. From 300 runs, the limit cycle was evolved successfully in a total of 65 runs, resulting in a success rate of 21.7%. From the results, we can see clearly that the IRM (structure '10') using summation as the regulation logic, which was shown to be the most robust motif in the previous section, emerged in 63 among the 65 successful runs (Fig. 3.21).

In the second set of simulations, all experimental settings are the same as those in the first set, except that 5% noise is added to the parameters during evolution. The

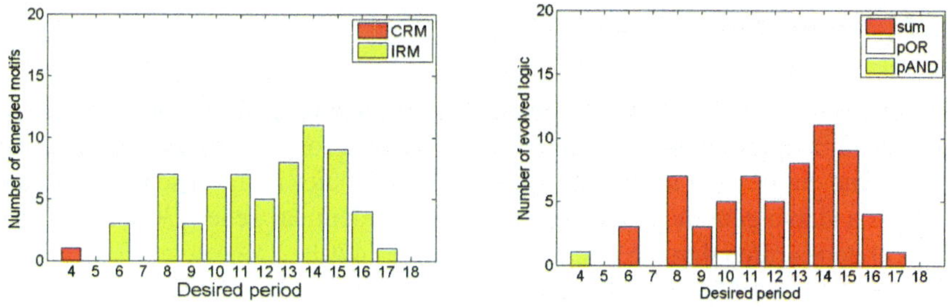

Fig. 3.20 Evolved motif (left) and regulatory logic operator (right) without noise in the successful runs

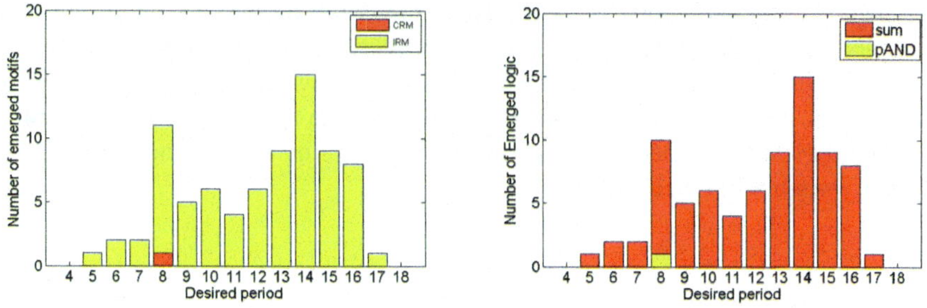

Fig. 3.21 Evolved motif (left) and regulatory logic operator (right) with 5% noise to the parameters added during evolution in the successful runs

purpose is to investigate the influence of noise on the evolvability of the system. The resulting motif structures and regulation logic operators are plotted in Fig. 3.20a, b, respectively. Among a total of 300 runs, 79 runs have been successful in evolving limit cycles, resulting in a success ratio of 26.3%. This indicates that adding a certain level of noise to the parameters during the evolution has enhanced evolvability. The IRM with summation logic has emerged in 78 of the 79 successful runs.

One question that arises regarding the above results is why the IRM using probabilistic 'OR' has evolved only once in all the 600 evolutionary runs, although it has also been shown to be insensitive to changes in the initial states and regulatory parameters. Our hypothesis is that IRM with the probabilistic 'OR' may be less robust compared to the IRM with the summation logic, which can in part be confirmed by the fact that in evolving the parameters with a fixed structure, the IRM using summation has achieved sustained oscillation 29 times in 120 independent runs, while the IRM with probabilistic 'OR' has been successful for 15 times. Nevertheless, a more rigorous conclusion can be made only if a quantitative measure for comparing the robustness of the two motifs is available.

3.4 Evolutionary Coupling of Genetic Motifs

3.4.1 Motivation

It is hypothesized that GRNs evolve from simple structures to more complex networks via gene duplication [470], although the mechanisms for this process remain elusive [351]. Gene duplication leads to functional redundancies in the absence of mutation [470], as a stable genome is unlikely to maintain identical genes [308]. As a result, identical genes often diverge in natural evolution [470].

Generating motifs with multi-stability, such as in a toggle switch, require positive feedback loops [73, 138], whereas a negative feedback circuit is required to produce oscillations in biochemical systems [138]. Biological systems also use negative feedback loops to improve robustness [202] and often consist of interconnected positive and negative feedback circuits [428]. Thus, the combination of simple motifs, for example, the toggle switch and oscillator, is of great interest in systems biology. Therefore, this section goes a step further from evolving the regulatory structure and logic as done in the previous section by investigating the capability of coupling motifs in *in silico* evolution. Specifically, evolutionary coupling of a genetic oscillator and a toggle switch will be considered.

3.4.2 *Genetic Motifs*

3.4.2.1 Genetic Oscillator

An oscillatory GRN can be modeled using ordinary differential equations (ODEs), where the Hill functions are often adopted to describe activating or repressive interactions defined as

$$H_{ij}^{a}(x_j) = \frac{\beta x_j^n}{\theta_i^n + x_j^n}, \tag{3.31}$$

$$H_{ij}^{r}(x_j) = \frac{\beta}{1 + (x_j/\theta_i)^n}, \tag{3.32}$$

where i and j represent the gene interaction pair, x_i and x_j, n is the Hill coefficient, β is the production rate, and θ_i is the threshold for gene x_i.

The logic to combine multiple genes into a single gene can be modeled using logic gates, as discussed in the previous section, which defines the combination of the states of two genes x and y:

$$L(x, y) = \frac{1}{2}(x + y). \tag{3.33}$$

The simplest oscillatory motif consists of three genes, see Fig. 3.22a, which can be described by the following ODEs:

$$\dot{x}_1 = a_{12}H_{12}^{a}(x_2) - a_{11}x_1 , \tag{3.34}$$

$$\dot{x}_2 = a_{23}L\Big(H_{21}^{a}(x_1), H_{23}^{a}(x_3)\Big) - a_{22}x_2 , \tag{3.35}$$

$$\dot{x}_3 = a_{32}H_{32}^{r}(x_2) - a_{33}x_3 . \tag{3.36}$$

In this section, NAR is used in all genes described by the negative second term in each equation, where x_2 is the target gene and the summation logic function is used to combine the interactions from the two regulatory genes x_1 and x_3.

The three-gene network exhibits oscillatory dynamics, as shown in Fig. 3.22a, regardless of the initial state of the system.

3.4.2.2 Toggle Switch

A two-gene toggle switch is illustrated in Fig. 3.22b, which can be modeled using the following equations:

$$\dot{x}_4 = b_{12}H_{45}^{a}(x_4) - b_{11}x_4 , \tag{3.37}$$

$$\dot{x}_5 = b_{21}x_4 - b_{22}x_{5i} . \tag{3.38}$$

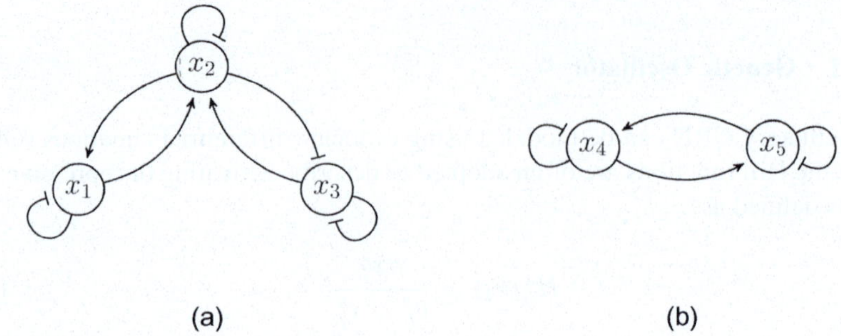

Fig. 3.22 Two genetic motifs. **a** A self-sustained oscillation, and **b** a toggle switch, where arrows and bars represent activating and repressive interactions

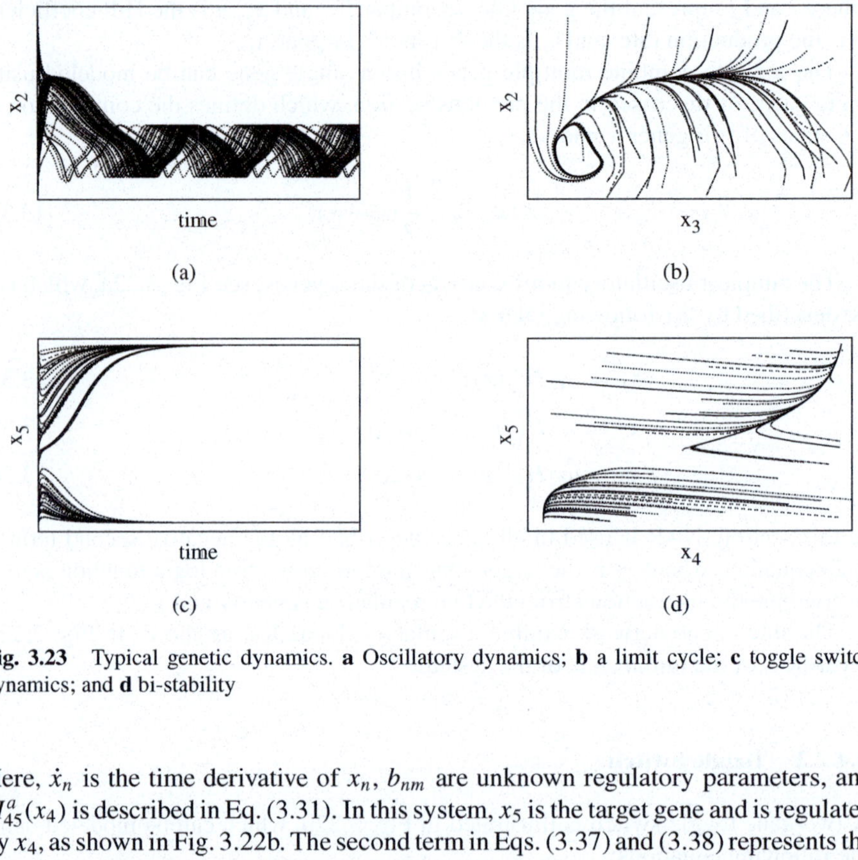

Fig. 3.23 Typical genetic dynamics. **a** Oscillatory dynamics; **b** a limit cycle; **c** toggle switch dynamics; and **d** bi-stability

Here, $\dot{x}_n$ is the time derivative of x_n, b_{nm} are unknown regulatory parameters, and $H^a_{45}(x_4)$ is described in Eq. (3.31). In this system, x_5 is the target gene and is regulated by x_4, as shown in Fig. 3.22b. The second term in Eqs. (3.37) and (3.38) represents the negative autoregulation of the system through parameters b_{11} and b_{22}, respectively, in both the target and regulatory genes.

This two-gene system demonstrates convergence to one of two stable states regardless of the initial conditions of the system, as shown in Fig. 3.23c, however, the initial state will determine to which stable point the state converges. The existence of two stable points leads to the formation of a bi-stability when viewed in the phase plane; refer to Fig. 3.23d. The phase plane clearly shows the trajectory of the states toward one of the stable points depending on the initial state of the genetic motif.

3.4.3 Evolving Regulatory Dynamics of Coupled Motifs

The connection of two network motifs can be achieved in two different master/slave arrangements. In the first instance, the oscillatory motif controls the toggle switch, as shown in Fig. 3.24a. In the second arrangement, the toggle switch controls the oscillator, as shown in Fig. 3.24b. The connection between the two motifs is a simple linear function, since the exact form of the coupling has been demonstrated not to be important, however, the control parameters between motifs need to be appropriately tuned [138]. Also in [138], it has been found difficult to manually define these parameters, making the evolutionary methods favorable in terms of their simplicity by allowing the algorithm to tune these values by optimizing a fitness function. The connection function is given by

$$M_{ij} = c_0 x_j + c_1, \tag{3.39}$$

where M_{ij} is the connection from gene x_j to gene x_i, and c_n represents the connection parameters. M_{ij} is then added to the equation representing $\dot{x}_i$.

The dynamics of the bi-stability motif will become

$$\dot{x}_4 = b_{12} H_{45}(x_4) + b_{11} x_4 + M_{42} \ , \tag{3.40}$$

where

$$M_{42} = c_0 x_2 + c_1 . \tag{3.41}$$

To ensure that a self-sustained oscillation and a bi-stability are to be combined, the two network motifs are evolved separately. Once the dynamics of these two motifs has successfully evolved, we fix the value of a_{nm} and b_{nm} and evolve only the connection parameters between the two motifs, i.e., c_0 and c_1. This way, the master gene directly controls the slave gene, while the original dynamics of the master gene remains unchanged.

3.4.4 Simulation Results

The parameter settings used in all experiments, including those for evolving the toggle switch and oscillator, as well as for coupling the two motifs, are as follows: parent size $\mu = 50$, offspring size $\lambda = 300$, crossover probability $\sigma_{Cross} = 0.9$, flip mutation

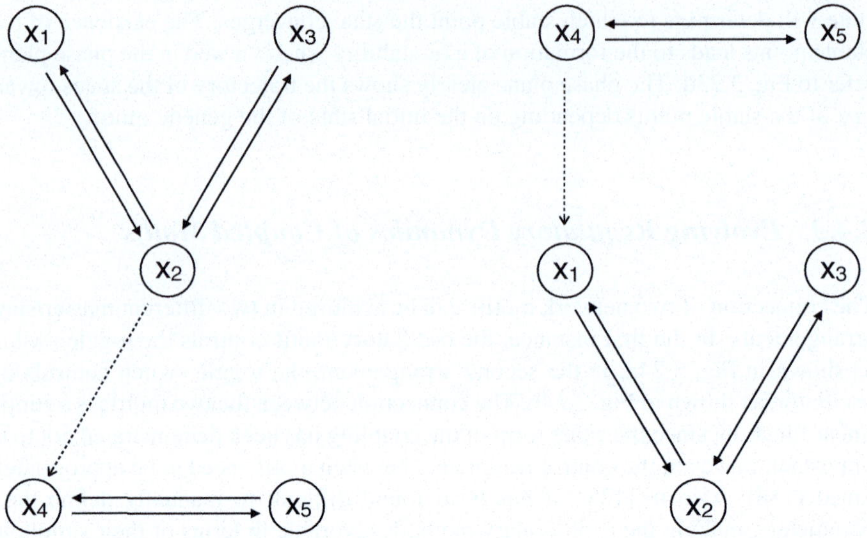

(a) Connection setting one (b) Connection setting two

Fig. 3.24 Connection between the two network motifs. **a** The oscillator controls the toggle switch, and **b** the toggle switch controls the oscillator. Here, the solid arrows represent activating or repressive regulatory connections between genes and the dashed lines represent a possible coupling between the network motifs. Note that the simple linear connection is a one-way interaction

probability $\sigma_{Flip} = 0.02$, Gaussian mutation probability $\sigma_{Gaussian} = 1$, step-size in Euler solver $h_{Euler} = 0.1$, the number of generations is set to 100, and the initial step-size of ES $\sigma = 0.1$. The parameter settings are also used in evolving the connections to produce more complex dynamics from the networks shown in Fig. 3.24a.

3.4.4.1 Oscillator Controls Toggle Switch

As we discussed before, there are two different ways of coupling the oscillator and toggle switch. Let us look at the instance when a CRM in conjunction with summation logic is used to produce a bi-rhythmicity, as shown in Fig. 3.24a. The coupling of the CRM to the oscillator is expected to result in an oscillation around each of the stable points of the toggle switch. In the phase plane, this genetic circuit will be able to generate limit cycles forming around the stable points of the bi-stability, as shown in Fig. 3.25. These dynamics are produced by using the fitness function with the following desired state:

$$x_d = 2\sin(2\pi t/T) + 1. \tag{3.42}$$

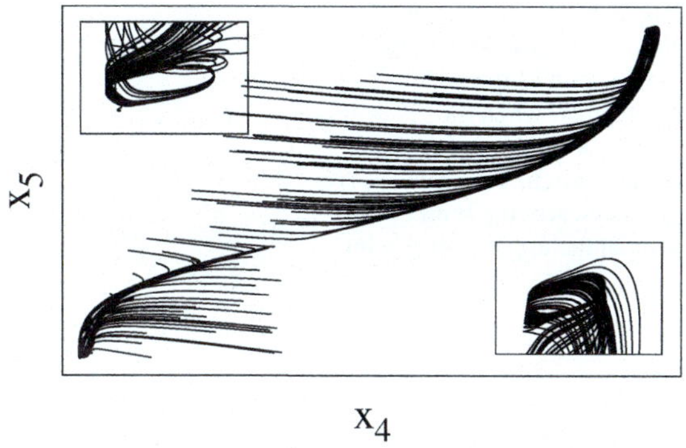

Fig. 3.25 A bi-rhythmicity produced using a CRM with the summation logic. The zoomed sub-figures show the formation of limit cycles around the lower and upper stable points of the bi-stability

Multiple oscillatory dynamics [137], which are functions of the control parameters, are closely related to the appearance of bi-rhythmicity. It is also stated that the control parameter can be perturbed, causing the system to go from one steady state to either of the bi-rhythmical stable point limit cycles, which differ in the period and amplitude. A similar property characterizes the thalamic neurons, which are capable of having different oscillation periods based on the state of the membrane [188, 262]. Phase plane analysis of these neurons reveals multiple excitability thresholds [137], giving rise to the bi-rhythmicity. The different levels of attraction to the stable limit cycles result from the different sensitivities to perturbations in the oscillatory regimes [137]. This gives rise to a larger attraction region for the upper limit cycle compared with the lower limit cycle shown in Fig. 3.25. It has been demonstrated that under the proper conditions, the system can reversibly switch from one limit cycle to the other [137].

In the second coupling instance, an IRM with the summation logic is used to produce a limit cycle around the upper stable state of the toggle switch only. This is shown in Fig. 3.26 and produced using

$$x_d = 5\sin(2\pi t/T) + 1, \tag{3.43}$$

as the desired state of the connected gene in the slave network shown in Fig. 3.24a.

Using this setting with the probabilistic 'OR' logic, the toggle switch stable states split, leading to a doublet of stable states. The two bi-stability lines can be clearly seen by looking at the phase plane, shown in Fig. 3.27. These dynamics were produced using the same desired state and fitness function that produced the bi-rhythmicity. Here, however, a different random number is used in running the code and therefore produced different dynamics.

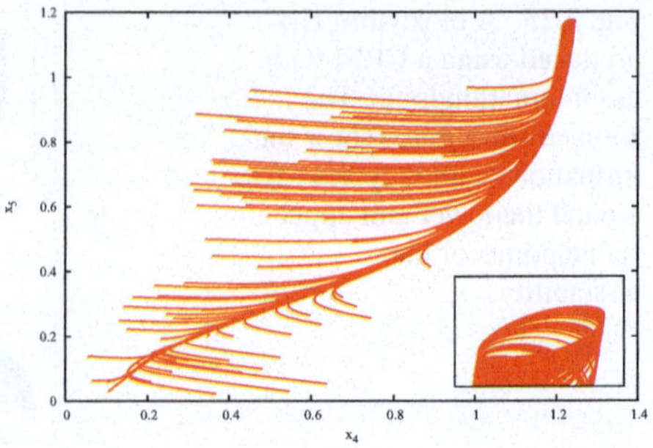

Fig. 3.26 Formation of a limit cycle around the upper stable point in the bi-stability, when the oscillator is coupled with the toggle switch. The inset shows clearly the limit cycle around the upper stable point

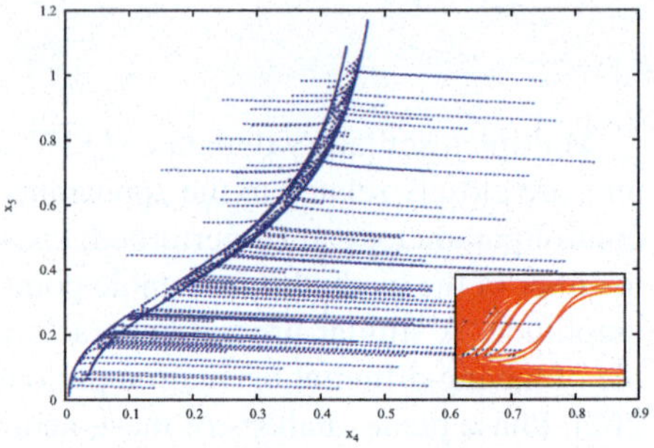

Fig. 3.27 A coupling of the oscillator and the toggle switch leads to a splitting of the stable points and the formation of a doublet bi-stability. The inset shows the splitting of the stable points in the toggle switch

3.4.4.2 Toggle Switch Controls Oscillator

In the second instance, the toggle switch is coupled with the oscillator, as shown in Fig. 3.24b. With the summation logic and using a CRM for the three-gene network, the coupled genetic motif can enable the stable points to determine the outcome of the oscillation. The upper stable point leads to an equilibrium point as the result of a damped oscillation, whereas the lower stable point leads to a self-sustained oscillation, as shown in Fig. 3.28. In the coupled circuit, the toggle switch determines the initial value of the toggle switch, thereby determining whether the oscillation is self-sustained or damped. The desired state in the fitness function used to control whether the oscillation is self-sustained or damped is given as follows:

$$x_d = 2\sin(2\pi t/T). \tag{3.44}$$

The same network setting that produces the dynamics shown in Fig. 3.28 can also produce a permanent change in the oscillation by simply changing the desired state, as described in Eq. (3.45). Instead of determining whether or not a self-sustained oscillation is evolved, this configuration determines the amplitude and period of the oscillation, as shown in Fig. 3.29:

$$x_d = \begin{cases} 2\sin(2\pi t/T) & \text{if } x_{4i} < 0.5 \\ \sin(2\pi t/T) & \text{if } x_{4i} \geq 0.5. \end{cases} \tag{3.45}$$

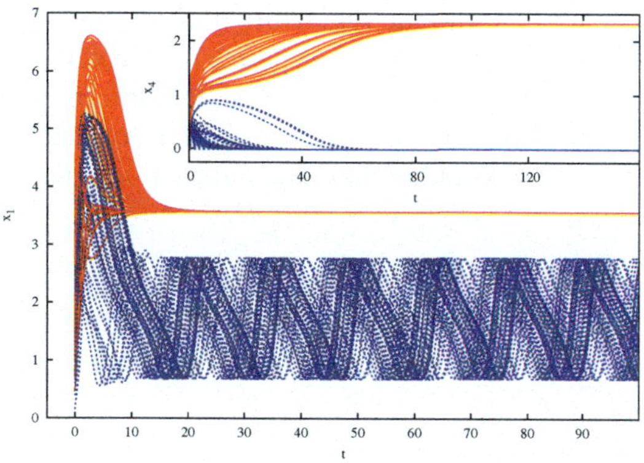

Fig. 3.28 Toggle switch controls whether an oscillator will produce self-sustained or damped oscillation based on the value of the stable point, which determines the initial state of the toggle switch, thereby determining whether or not a self-sustained oscillation is evolved from the coupled motifs

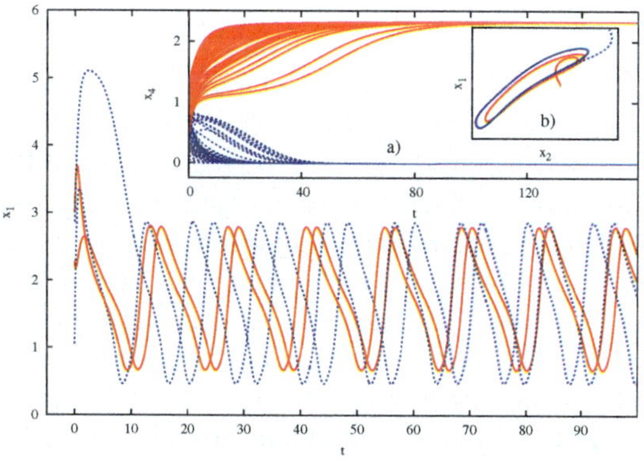

Fig. 3.29 Motif coupling dynamics causing the characteristics of the oscillation that depend on the state of the switch. Only two cases of oscillations for each stable state are shown for clarity, and an inset showing the formation of limit cycles from these oscillations, one case for each for clarity

3.4.5 Discussion

We observe clear connectivity preferences for the combination of two network motifs in two coupling arrangements. For the case of the oscillator controlling the switch, a coupling to the regulatory gene in the switch and a coupling preference from the negative feedback loop in the oscillator are observed. Similar connection outcomes are seen when two different objective functions are used to optimize the coupling parameters. Only the bi-rhythmicity network dynamics are observed in the above experiments, while in Gonze [138] and Thomas and Jin [417] oscillations around a single stable state only were also observed using this coupling arrangement. In [138], a periodic switch between stable states in the toggle switch is also observed. By contrast, these dynamics were not observed in [417], which uses a similar motif combination technique.

In case the switch controls the oscillator, we observe a clear preference over coupling with the positive feedback loop in the oscillator network, favoring a connection from the regulatory gene in the toggle switch network over the target gene. In contrast to the other coupling arrangement, no connection with the negative feedback loop in the oscillator to the regulatory gene of the switch is evolved and only one case of a connection between the negative feedback loop and the target gene of the switch has occurred. These results indicate that the coupling between network motifs exhibits a preference for positive or negative feedback loops that may depend on coupling arrangement rather than the objective setting.

3.5 Evolutionary Reconstruction of Transcriptional Regulatory Networks in *s. coelicolor*

3.5.1 Introduction

In the previous sections, several examples are given to show that evolutionary algorithms are able to synthesize important regulatory network motifs in silico. Another important purpose of modeling gene regulatory networks is to reverse-engineer important biological gene regulatory networks based upon gene expression time series data, which remains a grand challenge in computational and systems biology. Critically, an inference algorithm must be able to accurately and efficiently infer connections between a large number of genes.

It is possible to infer regulatory networks from the gene expression profiles [2, 478]. Network reconstruction from time series data is particularly important for understanding biological systems and can provide an interface between the fields of systems biology and bioinformatics [145]. The reverse-engineering of networks from experimentally determined data using computational techniques can provide insights into network structures, regulatory mode, robustness, evolvability, as well as the prediction of regulatory targets [433]. However, experimental constraints have led to the under-determinism of such problems, where there are insufficient time series data available to statistically reconstruct large networks [216]. Despite the impressive progress made in the past decades, reconstruction gene regulatory networks from expression data remain challenging.

To date, mostly only very small gene regulatory networks have been considered for reverse-engineering due to the high complexity of network reconstruction [122]. For practical use [287], much progress is still required as true gene regulatory networks often contain thousands of genes, which will pose much more serious challenges to the capability and computational efficiency of current algorithms. Thus, there is a high demand for computationally efficient algorithms that can reconstruct large-scale biological networks comprising hundreds of genes from experimental time series data generated under different conditions.

In the following, we introduce an evolutionary framework for network inference using a multi-objective optimization algorithm to reconstruct the gene regulatory pathway responsible for antibiotics production in *s. coelicolor* (PhoP network) from experimental time series data [419, 421]. We will show that this method is also able to reproduce experimental expression profiles to a high accuracy and the identified regulatory relationships are highly consistent with the experimental observations of regulation mode from the literature where known (but not used in the modeling processing).

3.5.2 PhoP Pathway and Gene Expression Data

GENIAL is applied to the PhoP network in *s. coelicolor*, which is linked to antibiotic production, containing 911 genes [7]. Experiments of the STREAM consortium SysMO experiments[1] provide extensive time series data for gene expression of *s. coelicolor*. With up to 32 time points, reverse-engineering the PhoP network with 911 genes [7] remains an extremely under-determined large-scale problem. Additional difficulties arise from the use of experimental data which are inherently noisy compounding the overall problem.

3.5.3 GENIAL

The general network inference algorithm (GENIAL) is a generic algorithm that can reverse-engineer a network's connectivity and connection parameterization based on time series gene expression data, including single measurements. GENIAL employs a multi-objective evolutionary algorithm to identify previously unknown regulatory connections, and the model is based on ordinary differential equations (ODEs).

3.5.3.1 Modeling Gene Regulation

The dynamic expression of all genes is modeled using ODEs [420, 478]:

$$\begin{aligned} \frac{dx_i^a}{dt} &= \frac{\omega_{ij}(x_j)^{n_{ij}}}{(\phi_{ij})^{n_{ij}}+(x_j)^{n_{ij}}} - \gamma_i x_i^a \quad (a) \\ \frac{dx_i^r}{dt} &= \frac{\omega_{ij}}{1+(x_j/\phi_{ij})^{n_{ij}}} - \gamma_i x_i^r \quad . \quad (b), \end{aligned} \tag{3.46}$$

where x^a (x^r) corresponds to an activating (or inhibitory), and ω, n, ϕ, and γ are connection parameters to be evolved. The connection parameters correspond to the regulation weight (w), the nonlinearity of the interaction (n), the threshold of the interaction (ϕ), and the degradation of the protein produced by x_i over time (γ).

[1] http://www.sysmo.eu/index.php?index=62.

3.5.3.2 Multi-objective Evolutionary Algorithm

A multi-objective evolutionary algorithm is applied to reconstruct the gene regulatory network (GRN) based on time series expression data. In particular, the elitist non-dominated sorting genetic algorithm [83], NSGA-II for short, is employed, as it has been successful in a wide range of real-world applications [84]. A common objective for reverse-engineering a network from time series data is to minimize the sum squared error (SSE) between the expression profile of the genes from the model and the data [439, 460]. In addition, a process called multi-objectivization [151, 298] that converts a single-objective problem into a multiple-objective one is adopted for facilitating the solution of complex optimization problems [162, 175, 260]. Multi-objectivization has been used in GRN parameter inference [165], where it is found that better exploration of the parameter space can be achieved. Here, the reconstruction problem is formulated as the following bi-objective optimization problem:

$$\begin{aligned} F_1 &= \textstyle\sum_{t=0}^{N} \left(\tilde{g}_i(t) - x_i^a(t)\right)^2 \quad (a) \\ F_2 &= \textstyle\sum_{t=0}^{N} \left(\tilde{g}_i(t) - x_i^r(t)\right)^2 \quad (b) \end{aligned} \tag{3.47}$$

where N is the number of data points, $\tilde{g}_i(t)$ is the normalized experimentally determined expression level of gene i at time t, and $x_i^a(t)$ and $x_i^r(t)$ are determined by solving Eq. (3.46). Note that the optimization problem remains the same, although the objectives are represented differently. All candidate solutions (parameter sets) now consist of a reduced number of decision variables, i.e., the four real-valued parameters in Eq. (3.46), ω, n, ϕ, and γ. This representation enables the optimization of the regulation mode indirectly, thus reducing the number of decision variables in the problem from five (including mode of regulation), in Eq. (3.46), to four, in Eq. (3.47). It is shown that the accumulative effect of this small difference results in significant performance improvements over a large number of connections. Additionally, a decoupling strategy where each connection in the network is optimized independently, is adopted to obtain a Pareto front of solutions for each connection in our network. This enables the parallel execution of GENIAL across multiple CPUs reducing computational run-times significantly.

3.5.3.3 Network Inference

The 387 direct PhoP regulatory targets identified from [7] are known connections in the network, however, the details of these interactions are unknown. In this case, GENIAL can predict the regulation mode, weighting, protein degradation, and additional connection information providing insight into specific pathways. For the 524 indirect targets of PhoP, GENIAL again not only determines regulatory mode and parameters but also infers connections based solely on the experimental data set. This identifies the unknown pathways from direct PhoP-mediated regulation to the differentially expressed genes. For simplicity, only model regulatory interactions on the gene level are modeled, and only a single connection to each indirect target is inferred.

The reconstructed network consists of PhoP with connections to its direct targets, which in turn have connections to PhoP's indirect targets. The direct targets are known through the experimentation of [7], and the parameter values, including regulation mode, are determined by GENIAL by solving Eq. (3.47). The connections to the indirect targets are inferred by GENIAL and are modulated by specific direct PhoP target genes. Thus, there is a connection from PhoP to a direct target, which in turn may have a connection to an indirect target, yielding an indirect pathway between PhoP and the differentially expressed genes. Due to the complexity of this under-determined problem, additional biological data of [7] is used, which identified the PhoP target genes in *s. coelicolor* that encode DNA-binding proteins and therefore regulate others at the gene level. This both reduces the dimensionality of the problem and removes the inference of biologically implausible connections. This step reduces the number of direct PhoP targets that themselves control genetic pathways from 387 to 55 genes. In other words, each of the 524 genes is regulated by one, or more, of the 55 direct targets of PhoP, with the other 332 of the 387 genes encoding non-regulator genes and thus not being able to regulate other genes in the model.

The connections to the indirect targets are selected from an ensemble of optimizations, where the connection that yields in the lowest SSE is selected. For each of the 524 indirect targets, a connection to all of the 55 direct targets is optimized according to Eq. (3.47). Consequently, 55×524 optimizations are performed in parallel and the best objective value and parameter set (including regulation mode) are obtained in each case. By ranking the 55 s_{best} objective values from these potential connections, the regulatory pathway which yields the lowest SSE can be obtained. Therefore, the one connection out of the 55 potential pathways that best fits the experimental data set is selected. The regulation of each indirect gene is based on the state of its regulator in the model rather than the original data set, ensuring that a plausible model is constructed and the gene expression profiles throughout the network can be reproduced based on the state of PhoP alone.

The expression profile of PhoP for each replicate data set is interpolated to give smooth curves of expression. The interpolated PhoP profiles are used in Eq. (3.46) as x_j for the regulation of x_i for its direct targets. The SSE of the model is only determined when there is a corresponding experimental measurement at that time point and all results quoted are in terms of the number of experimental data points.

3.5.4 *Results and Discussion*

3.5.4.1 Experimental Data

The bacterium *Streptomyces* is an unusual organism that is able to produce a range of compounds used in human and veterinary medicines such as antibiotics and

Table 3.1 Expression error between GENIAL and experimental data

Gene subset	SysMO 1 (%)	SysMO 2 (%)	SysMO 3 (%)
55	0.5592	1.1308	0.7891
332	0.6191	1.1639	0.8456
524	0.6562	1.1644	0.8827
Mean	0.6115	1.1530	0.8391
Std error	0.0489	0.0193	0.0471

anti-tumor drugs [161]. The production of antibiotics can be triggered in the event of phosphate starvation [437]. It is known, however, that the phosphate control system is not a simple regulatory circuit, and that it is part of a complex network with strong links with other stress responses and bacterial virulence [238].

The experimental gene expression data sets collected by the STREAM consortium consist of three replicate experiments comprising 32, 8, and 16 time points, respectively. This experimental investigation captures the genome-wide expression of the wild-type *s. coelicolor* strain during phosphate starvation. This genome-wide data set is reduced to the PhoP sub-network using the ChIP-chip experiments and differential gene expression analysis from [7], identifying 387 direct and 524 indirect regulatory targets of PhoP, respectively. It should be noted that the interatomic (ChIP-chip) data and the transcriptomic time series data used here were obtained from independent experiments from different laboratories.

The expression level of gene i, $g_i(t)$ is normalized so that the area of the expression profile, from initial measurement at t_0 to final measurement at t_f, is unity. This ensures all gene expression profiles lie on the same range of expression levels and direct comparisons between parameters for different connections can be made. Errors are calculated by summing the absolute difference between the experimental gene expression level and the levels produced by GENIAL. The error is then averaged over 50 independent simulations and normalized to the number of genes and time points to compare the replicate data sets and different gene subsets. The mean experimental expression level of a subset of genes can then be used to calculate the percentage error of the model. Note the 387 direct PhoP targets are divided in Table 3.1 into the 55 genes that encode DNA-binding proteins, and the remaining 332.

GENIAL optimizes connection parameters and regulation mode for all 387 direct PhoP targets based on the SysMO time series data only. As indicated in Table 3.1, the optimized parameter values lead to the extremely accurate reproduction of the experimental measurements of gene expressions across all data sets from the SysMO experiments.

Regulatory pathways between PhoP and its indirect targets are reverse-engineered by GENIAL based on the experimentally measured expression profiles. This is the most computationally intensive part of the algorithm due to the extensive search space and level of noise in the data. Despite these difficulties, the resulting network

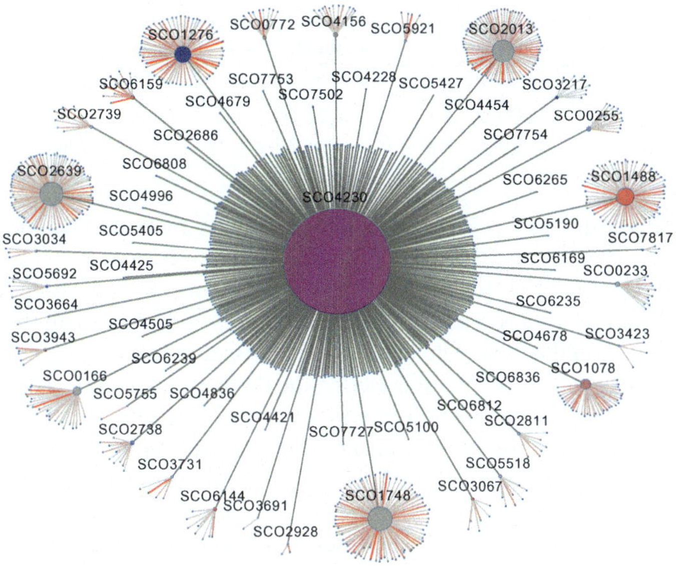

Fig. 3.30 The PhoP regulatory network inferred in this study using the data set of 32 time points from the SysMO experiments

reproduces the original gene expression data of the 524 genes to a high degree of accuracy as for the direct targets (Table 3.1).

The final output of GENIAL is a hierarchical parameterized network for all 911 genes. The reconstructed network shows good agreement with independent experimental observations for regulation mode in addition to accurately reproducing the input data, which supports the biological plausibility of the network determined by GENIAL. The full network produced from the SysMO 1 data set averaged over 100 independent simulations, including mode of regulation and connection confidence, is given in Fig. 3.30. PhoP (SCO4230) is denoted by the big node in the center. Experimentally determined connections for 387 genes from [7] are indicated by dark gray lines and consist of the inner circle of genes and those with labels. Labels here are only for genes that regulate other genes in this model, i.e., PhoP and a subset of the 55 direct targets that encode DNA binding proteins. Inferred connections are those from labeled genes (except PhoP), and those that occur in more than 30 % of simulations are colored red or are otherwise light gray. The size of the nodes is dependent on the number of regulatory targets with a minimum size of 1. The size of all nodes (except PhoP) is determined by the model. Node colors indicate the mode of regulation, blue for activation and red for repression by the regulator gene in at least 70 % of simulations. Genes colored gray indicate no observable preference for regulation type.

It is clear [238, 437] that PhoP exerts diverse control over a wide range of pathways. The breadth of control over many genes in *s. coelicolor* provides insight into

Table 3.2 Percentage of repressing connections from PhoP over 50 independent simulations of network reconstruction. Values quoted with the standard error in the mean

Gene subset	SysMO 1	SysMO 2	SysMO 3
55	54.00 ± 1.29	53.85 ± 1.25	65.71 ± 1.15
332	55.40 ± 1.08	51.42 ± 0.48	57.89 ± 0.55

the role of PhoP in many cellular processes, including morphological development and antibiotic production. Table 3.2 summarizes the overall mode of regulation of the direct targets of PhoP for the 55 and 332 gene subsets, indicating that overall PhoP mainly serves as a repressor. This prediction is consistent with the ChIP-chip study of [7]. It should be noted that although information from [7] is used to identify the repertoire of genes in the PhoP network, the mode of regulation is determined by GENIAL during the reconstruction process from the time series transcriptome data. Therefore, the above observation that PhoP is a global repressor of its regulatory targets is predicted independently of [7]. The overall repression of genes by PhoP is likely to be linked with the physiological transition from primary to secondary metabolism. As growth is not possible during phosphate starvation, expression of genes for primary growth-related processes, such as nitrogen metabolism, are unnecessary [339] and are repressed by PhoP [7, 437].

3.6 Summary

Synthesis of genetic networks in computational environments is of great interest since understanding the key mechanisms underlying successfully synthesizing the most frequently seen genetic motifs may help understand how nature has shaped these functionally important genetic circuits from single genes to genetic motifs, and from simple motifs to more complex motifs. This chapter starts with synthesizing simple motifs such as genetic switches and oscillators by evolving the parameters of a predefined network architecture using an evolutionary algorithm. Then a step further is made to evolve both the regulatory architecture and the parameters, again using an evolutionary algorithm. After that, it is empirically shown that it is possible to evolve in *in silico* complex genetic networks from single motifs. These results, albeit still highly simplistic, are exciting and inspiring, because they imply that in principle, evolutionary mechanisms are able to generate genetic motifs that are of fundamental importance for biological gene networks. Finally, it is shown that a multi-objective evolutionary algorithm is successfully employed to reverse-engineer a biological regulatory network underlying producing antibiotics consisting of 911 genes.

Many biological findings are confirmed in computational environments, such as the properties of the Hill functions and the robustness of the evolved genetic motifs. In the next chapters, we will show that evolutionary algorithms can evolve even more complex gene regulatory networks that are of paramount biological significance.

Chapter 4
Evolution of Morphological Development

4.1 Motivations

Biological morphogenesis is one extremely important process that is indispensable in shaping biological intelligence. It has been widely recognized that biological development is governed by gene regulatory networks, cell-cell interactions, and interactions with the environment. In the previous chapter, we show that evolutionary algorithms are able to generate genetic motifs that are widely seen in biological networks, including genetic oscillators and switches. A more intriguing question is is it possible to evolve genetic network that can result in a particular morphology? What genetic mechanisms are responsible for the stable growth of the biological process, and what are the biological (genetic and cellular) mechanisms that underlie the self-healing capability of biological systems? The main purpose of this chapter is to study the evolutionary, developmental, and cellular mechanisms that underlie the biological development using the multi-cellular models presented in Chaps. 1 and 2.

Understanding cellular growth processes could be of practical importance too, e.g., lightweight porous structures can be designed using a multi-cellular developmental system [403]. Note that the representation of irregular inner structures is non-trivial for traditional representation methods such as B-Spline or freeform deformation widely used in geometry design [204]. By contrast, it has been demonstrated that only a small number of parameters are needed to generate highly complex inner structures that are required for porous structure design. Other properties of biological morphological systems are also desirable in engineering systems, such as self-repairing or self-healing [38]. Engineering systems or artificial intelligence systems that are able to autonomously react to environmental changes or system defects are very beneficial and cost-effective. They would also foster an overall more robust engineering process. Such capabilities cannot be easily achieved by traditional robustness design methods.

This chapter will present a multi-cellular growth model governed by an evolved gene regulatory network [362]. We will describe the cellular model, the gene regulatory network model controlling the cellular development, and the evolution of the

Y. Jin, *Computational Evolution of Neural and Morphological Development*, Natural Computing Series, https://doi.org/10.1007/978-981-99-1854-6_4

cellular model that exhibits self-stabilized growth processes (reaching an equilibrium of cell death and cell division) and self-healing capabilities when a small part of the cellular structure is destroyed. An analysis of the underlying evolved functionality will be provided, which in turn helps us to identify the optimal abstraction level and the best strategy to transfer principles from computational systems into technical solutions.

4.2 Preliminaries

4.2.1 Evolution of Gene Regulatory Networks

As we discussed in Chap. 2, computational modeling of gene regulatory networks is meant for either analysis and synthesis of typical gene regulatory networks or reverse-engineering of biological gene regulatory pathways by approximating gene expression data. For example, the role of morphogen gradients in early development is reviewed in [15], discussing different signaling pathways and strategies for regulating the genes and the number of responses a morphogen gradient can generate. In [69], the types of different feedforward loops in *E. coli* are predicted using gene regulatory network models.

As we already shown in Chap. 3, evolutionary algorithms are successful in synthesizing typical gene regulatory dynamics. Bi-stable switches and oscillators are also evolved in [115] and compared with real biological networks. Gene regulatory networks are evolved and analyzed in [329], and it is found that different cyclic patterns may emerge in the evolution. Further analysis shows that the structure of the evolved networks is between random networks and scale-free networks. It is also shown that the evolved cyclic patterns have different lengths, suggesting that *in silico* evolution can generate complex regulatory dynamics from simple building blocks. Robustness, one critical property of biological gene regulatory networks, has also been analyzed with the help of computational models. A Boolean threshold network is adopted to evolve evolvability by varying the fitness function of the evolutionary algorithm [72]. It is also shown in [205] that evolvability is selectable in evolving genetic oscillators using a computational model of development. By applying duplication and mutation, scale-free and small-world genetic network structures are successfully evolved in [234], which are similar to those observed in *E. coli* and *S. cerevisiae*, significantly differing from random networks.

4.2.2 Multi-cellular Growth Models

To study biological morphogenesis *in silico*, it is essential to develop biologically plausible and realistic computational models for simulating multi-cellular cell growth. A brief review of computational modeling of biological development can be found in [156], where existing approaches to simulation of biological cellular development driven by gene regulatory networks are discussed [72, 98, 206, 207, 404]. Evolutionary developmental systems have also been applied for design of electronic circuits [468], fault-tolerant developmental systems [107], and development of cell lineages [124].

The computational model of development to be presented in the following aims mainly to understand stable growth and regeneration in technical systems. It is hoped, however, that such models can also enhance our understanding of biological development. Actually, computational models have been widely accepted in systems biology [224]. The most challenging question is at which abstraction level the computational models should target so that they are computationally tractable yet biologically meaningful and plausible.

4.2.3 *Stable Growth and Regeneration*

It is fascinating to understand which biological mechanisms have contributed to achieving several most important phenomena in biological development, such as stable growth and automatic self-healing. Biological findings suggest that some position-dependent signals are available for guiding tissue regeneration in the developmental process, which will also be employed in computational models. First, the patterning signals are generated locally as a response to lesions. Then, these signals are used by the regulatory network to trigger cell proliferation and differentiation. However, how these signals are locally generated during lesion remains elusive. Therefore, in the computational model presented here, we focus on how the regulatory system responds to cell loss, assuming that these local signals are already available. To this end, pre-diffused maternal gradients are artificially maintained to guide the cellular development and enable the regeneration capacity. Note that maternal morphogen gradients in the early development of organisms have been observed consistently [385].

In most developmental models in the literature, simulations for cellular development are stopped after a fixed number of iterations, regardless of whether the growth process has stabilized. Thus, by a self-stabilized development, we mean that the development either stops by itself, or a balanced cell growth and apoptosis are achieved. Clearly, a self-stabilized growth is more challenging to achieve, requiring a properly evolved gene regulatory network together with the maternal gradients and signals generated during the growth process.

First discussions of stable growth in artificial development are presented in [293]. Instead of manually stopping the developmental process after a predefined number of steps, the authors aim to generate a developmental model that can achieve a stable state, either static or dynamic, where the artificial system maintains its morphology and/or function. In the static case, cells do not divide anymore, either because of an evolved mechanism in the regulatory system or due to additional cellular properties such as contact inhibition. In the case of a dynamic steady state, cell proliferation and differentiation keep happening; however, they are compensated by cell apoptosis, leading to a state of constant cell turnover like in the cnidarian *Hydra*. Of course, a combination of both mechanisms is conceivable, although supporting biological evidence remains to be revealed.

A gene regulatory model is presented in [12], where development converges to a statically stable state due to contact inhibition. In this framework, the cells are fixed on a grid, and cells that are surrounded by other cells cannot divide anymore. Therefore, the results presented here represent a typical example where stability is reached due to some additional (external) constraints, which is biologically less plausible. Static and dynamic stability are evolved in [23] using a model based on cellular automata. A set of 100 rules are defined, which directly depend on the development time (or steps), the number of neighboring cells, their positions, and the number of divisions the cell already performed. Therefore, the basic ingredients like cell position or the number of cellular divisions are predefined to enable convergence to a stable state. In [360], behaving cellular animats are evolved, whose cell functions are represented using the same gene regulatory network model to be presented in this chapter. On the basis of this model, we can achieve dynamic stability and regeneration without additional constraints using physical cells that interact in a continuous space, which is biologically more plausible.

To briefly recap, the framework reported in [358] will be presented in the next, which is able to evolve individuals that reach a dynamically stable state during development without using additional functions like contact inhibition or additional information like the number of surrounding cells. That is, cell turnover achieved by this model is a dynamic equilibrium when cell proliferation is of the same order as cell apoptosis.

4.3 A Computational Model for Morphological Development

In the following, we provide details of a computational environment in which a model of a gene regulatory network is evolved together with a multi-cellular model. The cellular model also has physical interactions with the environment. Using this evolutionary developmental model, we investigate the genetic and cellular mechanisms that can lead to dynamically stable growth and self-healing.

The cellular model is extended from the one in [98]. Two changes are made to the original model. One most important change is that physical cellular interactions are included, which is important for avoiding fixing the cells in predefined rids. In addition, the fitness for evolving the morphology is redefined to be calculated over the whole developmental process, making it more likely to evolve dynamical self-stabilized growth. Compared to the model in [292], cells in this model are not fixed to a grid and the genome structure controlling the cellular growth is less abstract.

The model presented here allows us to take a deeper look into the evolved genetic networks that realize dynamically stabilized growth processes. Meanwhile, this will also enable us to analyze the structural properties of the evolved network motifs [10].

One basic hypothesis is that the morphological development is governed by a gene regulatory network, in addition to physical cellular interactions and morphogen gradients [360]. The morphological development starts with a single cell put in the center of a two-dimensional computational area specified by the user. A cell is circular shaped and defined by its center and a radius ($r = 1$). Cells are not fixed to a grid and can divide or die, depending on their related gene expression. Nonlinear adhesion and repulsion forces control the physical interactions between cells. For distances that are smaller than the sum of the two radii, i.e., if two cells overlap, they will repel each other. On the opposite, cells that have a small overlap adhere to each other. If two cells do not overlap, they attract each other with a decreasing force as the distance increases. No force exists when the distance between two cells is larger than five times the radius. To simulate the randomness in biological systems, the position of all cells is perturbed with a small random number.

As mentioned before, the developmental process of the cellular model is governed by a gene regulatory network. The genetic network consists of a set of genes, each composed by regulatory units (RUs) and structural units (SUs), which is similar to the virtual gene presented in Chap. 2. An example is given in Fig. 4.1. In the representation, SUs define cellular behaviors, such as cell division, cell death, or the production of transcription factors (TFs) for intra- and inter-cellular interactions. The activation of the SUs is determined by its associated RUs. Note that single or multiple RUs may regulate the expression of a single or multiple SUs and that RUs can be activating (RU^{+}) or inhibitory (RU^{-}).

Whether a TF can influence a RU is determined by the affinity value of each RU and TF. If the difference between the affinity values of a TF and a RU is smaller

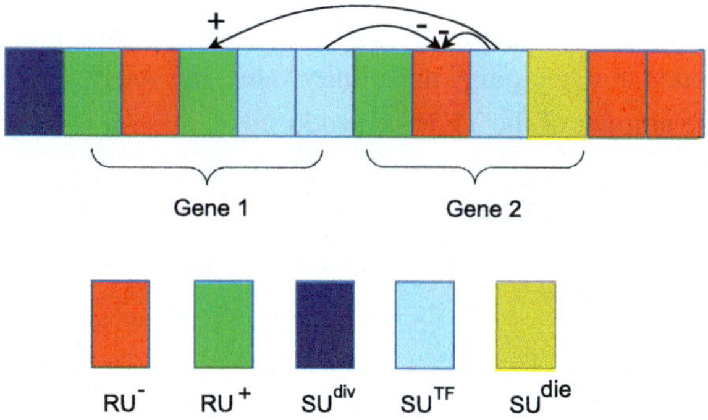

Fig. 4.1 An example genetic representation for the cellular development model. The first gene (gene 1) starts with a regulatory unit (RU). Each transition from an structural unit (SU) to an RU indicates a boundary between two genes. The connections between SUs^{TF} and RUs are determined by their affinity overlap $\gamma_{i,j}$ as defined in Eq. 4.1. Whether the interaction is activating or inhibiting is defined by the sign of the RU

than a predefined threshold ε, the TF can bind to the RU. The affinity overlap ($\gamma_{i,j}$) between the ith TF and jth RU is defined by

$$\gamma_{i,j} = \max\left(\varepsilon - \left|\mathrm{aff}_i^{\mathrm{TF}} - \mathrm{aff}_j^{\mathrm{RU}}\right|, 0\right). \tag{4.1}$$

The concentration c_i of the ith TF is checked to see if it is above a threshold ϑ_j defined in the jth RU if $\gamma_{i,j} > 0$:

$$b_{i,j} = \begin{cases} \max(c_i - \vartheta_j, 0) & \text{if } \gamma_{i,j} > 0 \\ 0 & \text{else} \end{cases}. \tag{4.2}$$

Thus, we can calculate the contribution of the jth RU (denoted by a_j, $j = 1, ..., N$) to the activation:

$$a_j = \sum_{i=1}^{M} b_{i,j}, \tag{4.3}$$

where M is the number of TFs that regulate the jth RU. Assuming the kth structural gene is regulated by N RUs, the expression level of the gene can be defined by

$$\vec{\alpha} = \vec{g}(\vec{c}), \tag{4.4}$$

$$g_k(\vec{c}) = 100 \sum_{j=1}^{N} l_j a_j (2s_j - 1), \; s_j \in (0, 1), \tag{4.5}$$

where s_j is a real number between 0 and 1, the result of $(2s_j - 1)$ determines the jth RU will be activating or inhibitory, and l_j represents the strength of the jth RU. If $\alpha_k > 0$, then the kth gene is activated, i.e., $\delta_k = 1$, resulting in the expression of all its the SUs behind these RUs, and therefore, the implementation of the functions of these SUs.

An $\mathrm{SU}^{\mathrm{TF}}$ denotes an SU that produces a TF, and all regulatory properties related to this TF are encoded in it, including the affinity value, the decay rate D_i^c, the diffusion rate D_i^f, and the amount of the TF to be produced:

$$\vec{A} = \vec{h}(\vec{\alpha}),$$

$$h_i(\alpha_k) = \begin{cases} \beta\left(\frac{2}{1+e^{-20 \cdot f \cdot \alpha_k}} - 1\right) & \text{if } \alpha_k > 0 \\ 0 & \text{otherwise} \end{cases}, \tag{4.6}$$

where f and β are both encoded in the $\mathrm{SU}^{\mathrm{TF}}$.

An $\mathrm{SU}^{\mathrm{TF}}$ may produce both internal and external TFs. Internal TFs mainly influence the expression of the genes in the cell, while the external TFs may diffuse out of the cell and influence the expression of the genes in other cells. To determine the amount of the external TF to be produced, a percentage ($p^{\mathrm{ex}} \in [0, 1]$) is also encoded in the corresponding gene. Thus, $p^{\mathrm{ex}} \cdot A$ and $(1 - p^{\mathrm{ex}}) \cdot A$ denote the amount of external TF and internal TF to be produced, respectively.

In the implementation, external TFs are put on four grid points around the center of the cell, which will undergo a diffusion and decay process governed by the following equation:

$$\text{Diffusion: } \vec{u}_i^*(t) = \vec{u}_i(t-1) + 0.1D_i^f(\vec{G}\vec{u}_i(t-1)), \tag{4.7}$$

$$\text{Decay: } \vec{u}_i(t) = (1 - 0.1D_i^c)\,\vec{u}_i^*(t), \tag{4.8}$$

where $\vec{u}_i$ is a vector of the concentrations of the ith TF at all grid points and matrix $\vec{G}$ defines which grid points are adjoining. Internal TFs will have a decay process only:

$$c_i^{\mathrm{int}}(t) = (1 - 0.1 \cdot D_i^c)\, c_i^{\mathrm{int}}(t-1). \tag{4.9}$$

The concentration of all internal and external TFs is defined to be between [0, 1].

To simulate what happens in biological systems, maternal morphogen gradients are also implemented in the model, which are represented by predefined TFs. These maternal morphogen gradients play an important role in the early stage of biological development. In the model implemented in [362], the first TF has a constant gradient in the x-direction and the second in the y-direction, simulating the head-tail and dorsal-ventral axes in biological development.

The SUs encode cellular behaviors and the related parameters. For example, the SU for cell division encodes the angle of division, indicating at which position the daughter cell will be placed. When an SU for cell apoptosis is activated, the cell will die at this developmental time-step. If both cell apoptosis and cell division are activated at the same developmental step, only cell apoptosis is performed. An example of the genome structures can be seen in Fig. 2.10 in Chap. 2.

In each developmental step, the following actions will be performed. First, the concentrations of all internal and external TFs simulating diffusion and decay are computed. Second, the activation level of all cells is determined according to their genes and the concentrations of the TFs. Finally, all cellular and genetic behaviors of the activated SUs are executed and the physical interactions between the cells will be implemented.

As already discussed in Chap. 2, the cell dynamics can become coupled through external transcription factors, which undertake a diffusion and decay process depending on their position. The number of TFs involved in gene regulation of the cellular behaviors is defined by the genome and the parameters in the resulting GRN as well. The model starts with one single cell and two maternal morphogen gradients, and the number of cells will change (usually increase in the beginning) during development, depending on the activation of cell division or cell apoptosis. However, a maximum number of cells may need to be specified in the simulation due to the limitation of the computational resources. In other words, if the developmental process does not reach a dynamic stable status, the simulation will stop when the maximum number of cells is reached.

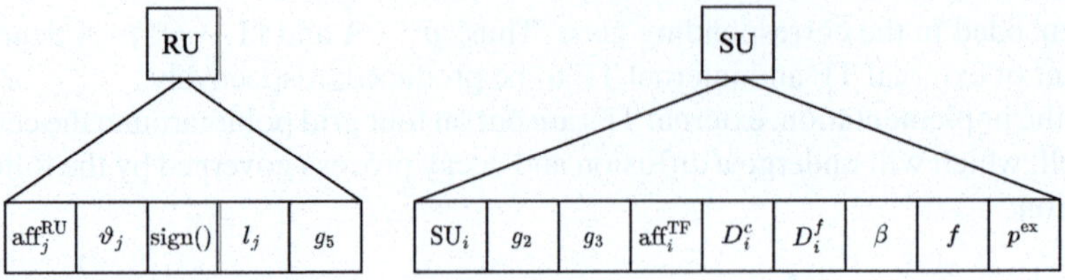

Fig. 4.2 Encoding of the parameters in RUs and SUs

The most relevant parameters encoded in the RUs and SUs are shown in Fig. 4.2, where in the RU, aff_j^{RU} is the affinity value of RU, ϑ_j is the threshold for the concentration, sign() $\in [0, 0.5]$ is inhibitory, sign() $\in (0.5, 1.0]$ is activating, and l_j is the strength of the RU_j. In the SU, $\text{SU}_i \in [0, 0.2]$ produces a TF, $\text{SU}_i \in (0.2, 0.4]$ encodes a cell division, $\text{SU}_i \in (0.4, 0.6]$ encodes a cell death, aff_i^{TF} is the TF affinity, D_i^c is the decay rate, D_i^f is the diffusion rate, β is the expression rate, f the division angle defined in Eq. 4.6, and p^{ex} indicates an external TF. Other genes are unused in this model, as provided in [358].

4.4 Evolutionary System

4.4.1 Genetic Operators

We use a variant of evolution strategy with a parent population size of $\mu = 30$ and an offspring population size of $\lambda = 200$. The non-elitist selection scheme (μ, λ) is adopted, referring to Chap. 1. However, a small number of the best individuals in the parent population can be kept to speed up the convergence. Another change is that the step-size σ may need to be fixed to a very small positive value. Let x_i be a real-valued parameter encoded in the chromosome, a mutation operation changes its value as follows:

$$x_i' = x_i + z_i, \tag{4.10}$$

where

$$z_i \sim \mathcal{N}(0, \sigma_i), \tag{4.11}$$

where $\mathcal{N}(0, \sigma_i)$ denotes a normal distribution. In addition to mutation, three other biologically more plausible operators, namely gene duplication, gene transposition, and gene deletion are introduced in the evolutionary system to be able to change the structure of the resulting gene regulatory networks. Gene duplication copies a randomly chosen piece of RUs and SUs sequences in the chromosome and then inserts it into the chromosome at a different randomly chosen position. For gene transposition, a gene sequence is randomly picked out from the RUs and SUs and

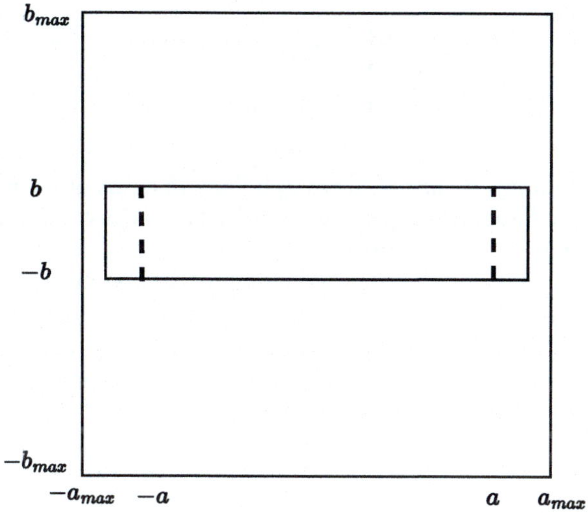

Fig. 4.3 Definition of a desired elongated shape. All cells should be connected and sit between $[-a, a]$ and $[-b, b]$. Cells that lie outside the black rectangle will be penalized. $[-a_{max}, a_{max}]$ and $[-b_{max}, b_{max}]$ define the simulation field

moved to another randomly chosen position in the chromosome. Finally, gene deletion removes a piece of randomly chosen genes in the genome. These three genetic operators will be performed at a predefined, usually small probability, p_{dup}, p_{trans}, and p_{del}. An illustrative example of gene duplication and gene transposition can be found in Fig. 2.11 in Chap. 2.

4.4.2 Fitness Function

It is non-trivial to evolve a desired body shape using the multi-cellular model presented above. Without defining a particular shape, it is very likely that a sphere (circular in 2D case) will be developed. If one aims to develop some elongated shape, for instance, to simulate swimming behaviors, the fitness function of the evolution strategy must be properly defined.

For a desired elongated shape defined in Fig. 4.3, the fitness can be defined as follows, assuming $a < b_{max}$:

$$f = p_1 - p_2 - \min\left\{\min_i\left\{y^i\right\}, a_{min}\right\} + \max\left\{\max_i\left\{y^i\right\}, a_{max}\right\}, \tag{4.12}$$

where (x^i, y^i) represents the position of the ith cell, and

$$p_1 = \begin{cases} (a + b_{max}) - \min_i\left\{x^i\right\} & \text{if } \min_i\left\{x^i\right\} < -b_{max} \\ -a & \text{if } -b_{max} < \min_i\left\{x^i\right\} < -a \\ \min_i\left\{x^i\right\} & \text{otherwise} \end{cases} \tag{4.13}$$

and

$$p_2 = \begin{cases} -(b_{max}) - \max_i \{x^i\} & \text{if } \max_i \{x^i\} > b_{max} \\ a & \text{if } b_{max} > \max_i \{x^i\} > a \; . \\ \max_i \{x^i\} & \text{otherwise} \end{cases} \quad (4.14)$$

To limit the size of the morphology in terms of the number of cells, a penalty term can be applied if the number of cells n_c is larger than a predefined number, say 500. However, the number of cells should not be smaller than a particular number, say 10, to generate a sensible morphology. A large fitness value (here the fitness function is to be minimized) will be assigned if n_c is outside the range. In addition, if the individual is composed of unconnected parts, i.e., one or several cells have a large distance to all other cells, the fitness is set to predefined value too, which should not be too large or too small.

4.5 Experimental Results and Discussions

4.5.1 Experimental Settings

To empirically assess the evolutionary developmental model, three experiments are set up in order to examine the system's properties including stable growth and regeneration. Equation 4.12 denotes the fitness at one developmental step. To properly assess the dynamic stability, three different evaluation procedures are defined that calculate the fitness at different developmental stages. The first setup evaluates individuals at one particular, predefined maximum number of developmental steps. In the second setup, more developmental steps are simulated and the mean of several evaluations after a certain number of developmental steps will be calculated as the fitness value. In the third setup, the regeneration capacity of the multi-cellular model is examined by removing a certain number of cells at a certain developmental step.

- **Setup 1 (S1)**: Simulation of a maximum of 16 developmental steps and evaluation of the fitness.
- **Setup 2 (S2)**: Simulation of a maximum of 30 developmental steps and evaluations of the fitness at steps 16, 17, 18, 19, 20 and 26, 27, 28, 29, 30. Then, the mean fitness over the ten evaluation steps will be the fitness value.
- **Setup 3 (S3)**: Same as S2; however, cells between $-5 < x < 5$ of the individuals are deliberately removed at step 20.

According to the fitness function defined in Eqs. 4.12–4.14, the best achievable fitness is -50 if a, b, a_{max}, and b_{max} are set to 30, 5, 50, and 40, respectively. For S2 and S3, the same fitness is computed ten times and averaged considering the limited computational resources. The simulation will be stopped if the maximum fitness is achieved, or if a maximum number of 1000 generations are reached. To accelerate the simulation, the code can be parallelized.

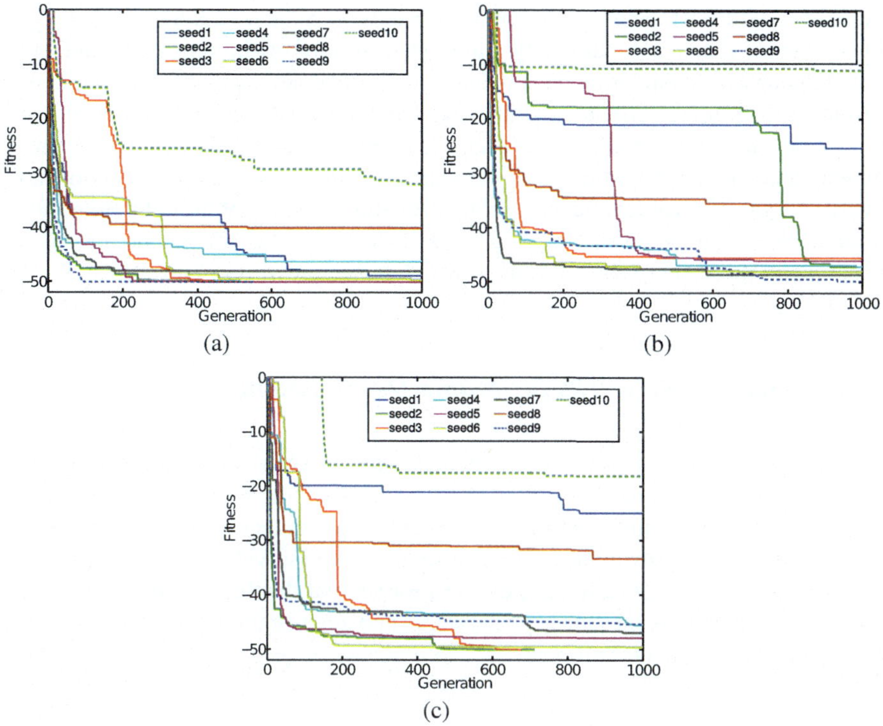

Fig. 4.4 Fitness profiles of the best individual in each of the ten independent runs for three different settings of the fitness function [358]. **a** S1, **b** S2, and **c** S3

4.5.2 Evolvability of Stable Growth

The resulting fitness curves of the three settings in the ten independent runs are plotted in Fig. 4.4. From these results, we find that the mean fitness of setting S1 is better than that of S2 and S3, indicating that it is easier to evolve the desired shape compared to the other two. In S1, four out of ten experiments reach the desired shape, and only one experiment fails completely with a fitness of −31.94. During the experiments without removing cells (S2), only one experiment reaches the global optimum, but only in two runs, it fails completely with a fitness worse than −40. The experiments with removing cells in setting S3 show that only two of the ten runs have successfully found the global minimum.

While the artificial organisms are evolved using different fitness functions, it is of interest to know whether a stabilized growth evolved in one setting can also achieve stabilized growth in another. It has been found in [362] that an individual can most likely develop into a desired elongated shape in the same setting as it has been evolved, which is expected. It is also apparently shown that individuals evolved for S2 and S3 can still achieve good fitness (although slightly worse) in setting S1, while those evolved for S1 cannot develop well for settings S2 and S2. Individuals evolved in S1 cannot reach a dynamically stable state in any of the ten runs, confirming that dynamically stable growth is evolvable only under an explicit selection pressure toward self-stabilized growth. On the other hand, the results in [362] indicate that in three cases individuals evolved for S2 also perform well for S3, meaning that once the genetic mechanisms for reaching a dynamically stable state have been evolved, the evolved organism is able to regenerate (self-healing) if some of the cells are removed after the development process is completed. In their experiments, regeneration is also observed in three out of the ten runs in which stable growth is evolved.

4.5.3 *Genetic Networks for Stable Development*

It is of great interest to know which gene regulatory networks can successfully result in stable growth. To this end, we analyze the dynamics of the gene regulatory networks evolved in the above three settings, in particular the structure behind the process of dynamically stable growth and regeneration for three individuals. Therefore, one of the best individuals that have been evolved for each of the three setups is chosen for analysis. We then look into the gene networks and analyze the activation of the genes during the developmental process.

4.5.3.1 Evolved GRN in S1

Eight snapshots from the cellular development of the morphology of the best individual evolved from S1 are shown in Fig. 4.5. At the beginning of the development, the cells in the middle (y-direction) divide and cells in the upper part die. The cells in the lower part keep stationary. After developmental step 12, cell apoptosis is activated in the lower part. At developmental step 15, the cellular function changes and every two cell dies. Then each cell first divides and dies in the next developmental step.

The full gene regulatory network (GRN) shown in Fig. 4.6a is highly connected. However, many of its connections have never been activated. The regulatory connections that have been activated in the developmental process are shown in Fig. 4.6b. Note that all regulatory connections used in the development are activating, among which one gene is for cell division (gene 8) and three genes are for cell death (genes 3, 24, and 35).

Comparing the individual gene activation status with the growing phenotype reveals that the gene for cell division is active in cells above a fixed x-position; see also Fig. 4.5. Since apoptosis is implemented before cell division if the GRN activates both functions within the same cell, gene 3 (coding cell death) is also activated by the first maternal TF and becomes active above a horizontal line. Gene 3 and Gene 35 are activated simultaneously, which therefore has no further influence on the development. The activation of gene 24 is more complicated, which depends on the activation of gene 4 that can be activated by gene 23. The activation of gene

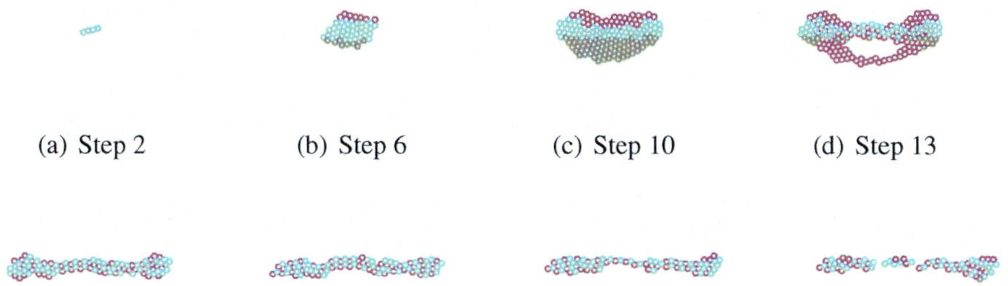

(a) Step 2 (b) Step 6 (c) Step 10 (d) Step 13

(e) Step 15 (f) Step 19 (g) Step 25 (h) Step 29

Fig. 4.5 The development of the best individual of S1 from run 9 [358]. Cells that will divide in the next developmental step are marked in blue, dying cells in magenta, and stationary cells are marked in gray

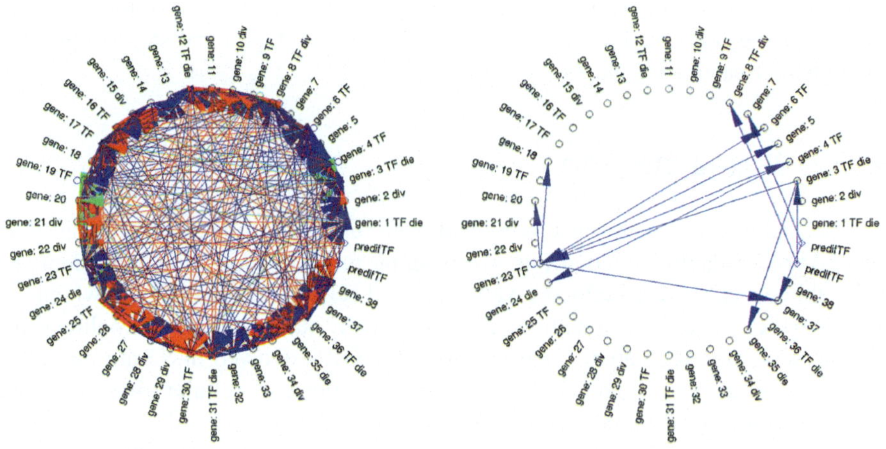

(a) Full connections (b) Activated connections

Fig. 4.6 The GRN of the best individual of S1 from run 9 [358]. Dots represent different genes and diamonds denote the two pre-diffused maternal TFs. Arrows denote connections between different TFs and genes, denoting regulatory relationships. In **a**, we show all possible interactions coded in the genome, and in **b**, we illustrate only the interactions that are used during development. Red arrows represent an inhibiting connection, blue an activating, and green arrows represent both, an activating and inhibiting influence

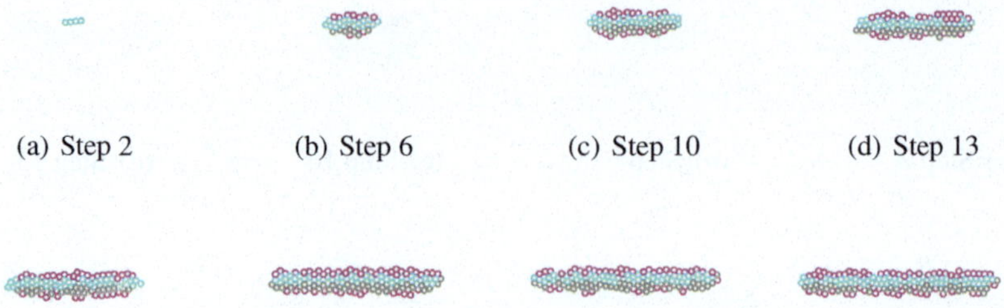

(a) Step 2 (b) Step 6 (c) Step 10 (d) Step 13

(e) Step 15 (f) Step 19 (g) Step 25 (h) Step 29

Fig. 4.7 The development of the best individual of S2 from run 2 [358]. The color coding of the cells is the same as in Fig. 4.5

4 starts at developmental step 10, producing a high concentration of the TF in the middle of the cell it resides. Every cell needs two developmental steps to produce enough TF to activate gene 24.

4.5.3.2 Evolved GRN in S2

The second example is the best individual from run 2 of setting S2, and its developmental process is illustrated in Fig. 4.7. Cellular function is position dependent. All cells above and below a defined y value die, and cells in between divide, which is a simple but effective way of achieving a dynamically stable state in the y-direction.

The activating regulatory connections of the GRN evolved in S2 are plotted in Fig. 4.8. Of the genes in the GRN, only gene 15 represents cellular division, which is activated by the first predefined TF and deactivated by the second. Similarly, gene 15 is active above a horizontal line, but deactivated at the right side of the individual. Genes 6 and 7 (cell apoptosis) are activated below and above the horizontal line, respectively.

4.5.3.3 Evolved GRN in S3

The development of the third example individual from S3 is shown in Fig. 4.9. The development of this individual is similar to that of the second individual, even if the activated GRN is considerably different, as shown in Fig. 4.10. Since the activated GRN connections are still rather complex, we further simplified it by removing all connections that do not influence the fitness of the individual even though they are activated during development. The activated GRN and its pruned version are given in Fig. 4.10. Among the activated genes, gene 13 encodes cellular division, while genes 1, 9, and 20 encode cell apoptosis. Gene 13 is activated by the second predefined TF and is active in all cells except those on the left end. Gene 1 (apoptosis) is

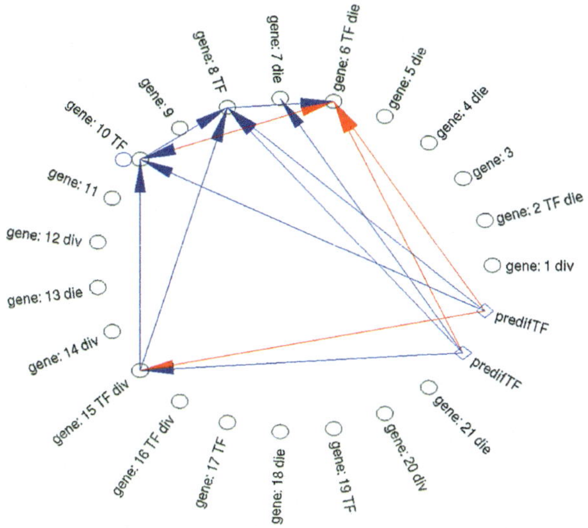

Fig. 4.8 The GRN of the best individual of S2 from run 2 [358]

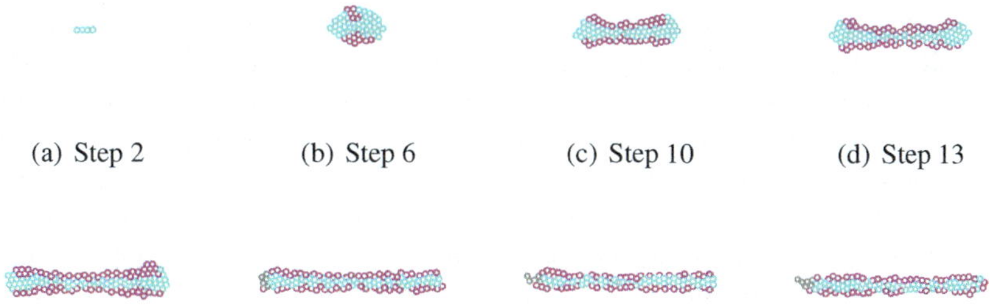

(a) Step 2 (b) Step 6 (c) Step 10 (d) Step 13

(e) Step 15 (f) Step 19 (g) Step 25 (h) Step 29

Fig. 4.9 The development of the best individual of S3 [358]. The color coding of the cells is the same as in Fig. 4.5

also activated by the second predefined TF and is only active in cells located at the right end of the individual. Gene 9 encoding apoptosis is active at the bottom of the individual. It is activated by the TF produced by gene 15 and deactivated by the first predefined TF. Gene 15 is activated in all cells, which, however, will be activated by gene 20 if they are above a horizontal line. Gene 20 (apoptosis) is active only in cells at the top, which is activated by the first predefined TF.

Note, however, that the pruned GRN is less stable against mutations than the pruned GRN, indicating that the seeming redundant genetic connections are functionally also important, since robustness to mutations is a key feature of biological GRNs, as discussed in Chaps. 1 and 2. Therefore, the redundancy improves the

to discover dynamically stable growth in the first place. The detailed analysis of the genetic regulatory networks revealed that most individuals achieve dynamically stable growth and regeneration by means of position-dependent control of cellular apoptosis using pre-diffused transcription factors simulating maternal morphogen gradients. Stable growth and controlled cell removal during the evolution can lead to high morphological plasticity and to the desired regeneration capability. It should be noted, however, that self-healing is evolved at the expense of higher energy costs. Two motifs in the evolved gene regulatory networks can be identified that occur in the most successful evolved organisms that are responsible for the self-healing capability.

Note that the individuals evolved in the evolution scenario S1 do not reach a dynamically stable state in any of the ten runs. This implies that it is more difficult to evolve dynamical stability than a predefined specific shape. It should be noted, however, other evaluation schemes (selection pressure) may also be conceivable, e.g., if the individual is evaluated after different steps and then the best values are recorded. In this way, the fitness will no longer depend on a specific development step, although only one step is evaluated.

One important assumption of the model is that pre-diffused morphogen gradients (maternal morphogen gradients) are available to guide both the developmental and the regeneration processes. By introducing and artificially maintaining the maternal gradients, the scaffolding for regeneration is provided, working as a global coordinate system for the development of the animat. It has been suggested that it is very difficult, if not impossible, to evolve gene regulatory networks that can lead to dynamical stability and regeneration without the presence of the maternal morphogen gradients. However, since maternal morphogen gradients have been observed in many developmental systems, such a hypothesis does reduce the biological plausibility of the presented model.

Overall, the following observation can be made from the results presented above.

- It is possible to evolve cellular growth of a simple morphology without introducing biologically less plausible hard constraints such as fixing cells to a grid and contact inhibition. Moreover, such findings may also have implications to engineering optimization such as topology optimization. It remains to be validated, though, whether it can scale up to more complex geometries.
- The evolved GRNs are highly complex, while only a small number of regulatory connections have been activated for governing the process. This is, however, consistent with findings in biology and also in other work on evolving gene regulatory network [362] that a certain level of redundancy is required for the evolutionary process, simply because the gene regulatory network must be robust to random genetic mutations in evolution.
- Regeneration is a fascinating topic and the finding based on this model just touches upon the understanding of the involved self-healing capability. The mechanisms behind self-stabilized growth and self-healing are important not only for biological systems, but also for engineering design, too. The high morphological plasticity

observed in the experiments may lead to highly dynamic simulation meshes in engineering simulations.

4.6 Summary

Self-stabilized growth and self-healing are the two most important features in many biological systems. However, the genetic and cellular mechanisms underlying self-healing and regeneration remain elusive [11, 119]. Understanding regeneration in computational systems based on biologically plausible models is of interest not only to biologists but also to engineers for designing self-healing materials [70].

This chapter presents a multi-cellular developmental model controlled by a gene regulatory network. Several efforts are made to make the model biologically more plausible, e.g., by removing the requirements on fixing the cells to particular grids, contact inhibition, or information about the number of neighboring cells. The virtual genome used contains regulatory and structural units, which can be changed using not only mutations, but also gene duplication, gene transposition, and gene deletion. One important information used in the model is the maternal morphogen gradients, which are critical in biological morphogenesis and have been shown to be essential for achieving self-stabilized growth and regeneration. Another finding is that self-stabilized growth and self-healing are embedded properties in evolution, meaning that such properties will emerge only if they are subject to a selection pressure already during the evolution. It has also been shown that the evolved gene regulatory networks are highly redundant, confirming that robustness is a required property also in computational environments like in natural evolution.

Chapter 5
Evolution of Neural Development

5.1 Introduction

Neural and morphological developments are closed coupled, and both are important for cognitive development. In the previous chapter, we present a model for multi-cellular development, which, albeit at a very high abstraction level, is able to reveal interesting findings that are valuable for understanding biological morphological development, and probably also for engineering design. In this chapter, we aim to introduce some research efforts that aim to understanding organizational principles in evolving neural development, including early neural development and activity-dependent neural development. In the evolution of early neural development, we focus on the principles underlying different neural architectures we observe in nature. In evolving neural plasticity, we concentrate on the ability of neural plasticity to improve the learning performance, in particular how to reduce interference in learning the sensor input structures. It should be noted that both early neural development (mainly gene-driven) and evolution of neural plasticity (activity-dependent) depend more or less on the environment, as already mentioned in Chap. 1.

5.1.1 Major Transitions in Neural Evolution and Development

Three aspects of the evolution of the nervous system are of interest to both biologists and computer scientists. The first is the evolutionary transition in the architecture of the nervous systems. Second is the change in neural functions, and the third is the relationship between the neural architecture and its neural function, as well as the coupling between neural and morphological development.

Generally speaking, nervous systems can be divided into two categories according to their architecture, namely diffuse nervous systems and centralized nervous

Y. Jin, *Computational Evolution of Neural and Morphological Development*, Natural Computing Series, https://doi.org/10.1007/978-981-99-1854-6_5

systems. The diffuse nervous system architecture can usually be found in the most primitive nervous systems, in which neurons (nerve cells) are distributed throughout the organism [281]. Large concentrations of nerve cells—as in the brain—are not found in these systems, though there may be ganglia or small local concentrations of neurons. Diffuse systems are found in cnidarians, such as hydras, the existing animals that have the oldest nervous system. The nervous systems of cnidarians distribute radially symmetrically in their bodies, in which similar parts are arranged symmetrically around a hollow body wall.

However, the functional complexity of diffused nervous systems is limited. For example, they cannot perform undulatory swimming. By contrast, undulatory movements can be achieved by bilaterally symmetric nervous systems, which is one type of centralized nervous systems. Centralized nervous systems include simple bilaterally symmetric nervous systems (e.g., the nervous systems of flatworms), moderately cephalized systems, and vertebrate systems.

It is believed that various nervous systems are all evolved from primitive, diffused nervous systems. This means there have been major transitions in the evolution of the structure of the nervous systems. It is thus particularly of interest to understand what genetic and cellular mechanisms and what selection pressure (environmental influences) have shaped the major transitions in the evolution and development of nervous systems.

In this chapter, we are going to present computational models that aim to gain some insights into the organization principles in the evolution and development of primitive nervous systems using the hydra as the model animal. We then focus on the evolution of neural plasticity rules that are used to self-organize the structure of a particular type of neural network model, namely reservoir computing for investigating the gene regulation and locality in enhancing their functional performance. The coupling between neural and morphological developments will be elaborated in Chap. 6.

5.1.2 Principles of Neural Self-Organization

Understanding the principles behind the self-organization of biological nervous systems is key to understanding human cognition. Generally speaking, neural self-organization can be studied from the evolution and development point of view. In the history of evolution, there were a number of major transitions or divergences, e.g., from the diffused nervous structure in cnidaria to the bilaterally symmetric one in flatworm [129]. Computational models have been built up for co-evolving the development of the neural system and body plan of an animate based on primitive organisms such as hydra and flatworms and the results suggest that energy efficiency might be the most important constraint in neural self-organization [196, 208]. In addition, a strong coupling between the evolution of neural systems and body plan is revealed [213, 362].

Meanwhile, increasing evidence suggests that adult human brains undergo intensive rewiring [168], which involves neural plasticity including the strengthening or weakening of existing connections, or even the formation of new synapses and elimination of existing ones. Seminal studies by Merzenich and Kaas [288, 289] demonstrated that once sensory nerves are severed, the cortical maps to which they are projected are subsequently reorganized to accept nerves from surrounding nerves. This topographic adaptation can only be realized through neural plasticity and indicates the experience-dependent nature of neural plasticity and its central role in forming the basis of continual learning.

5.2 Evolution of Primitive Neural Development

5.2.1 Motivations

Understanding the evolution of biological neural organization using computational models has attracted increasing attention. Two issues are considered essential for this body of research to be biologically plausible. First, the evolution of neural organization should be coupled with that of the body plan [90, 210]. Second, the influence of neural development on the evolution of the nervous system should also be taken into account [328].

Several computational models for neural development have been published. Cangelosi et al. [51] suggested a developmental model for cell division and cell migration using a rule-rewriting system. A model for neurogenesis that includes metabolic reactions and diffusion was proposed by Kitano [223], although no functionality of the developed neural network has been considered. A recurrent artificial neural network was used for modeling the development of a spiking neural network for the control of a Khepera robot [106]. A good overview of computational models for neural development can be found in [229].

The computational model for primitive neural development [203] to be introduced in the following aims to investigate if the fundamental requirements for neural development, in particular the correct developmental order, i.e., first cell division, then cell migration, and finally axon growth [352] can evolve using a randomly initialized gene regulatory network (GRN). Similar to the GRN for morphological development [405] used in Chap. 4, the gene regulatory network in the neural development model presented here is composed of a number of structural genetic units whose expression is controlled by regulatory genetic units. To evolve the neural development in a meaningful environment, we use the hydra as a model animal, since hydra has the oldest biological nervous systems having a diffused structure. In the development, the neurons are distributed over the surface of a cylinder that simulates the body of hydra. Neurons are then connected according to a probability inversely proportional to the distance between them. The function to be accomplished by the developed nervous system is to perform a food-catching behavior by swaying its body (and also the tentacles), while its foot is attached to the ground, as happened

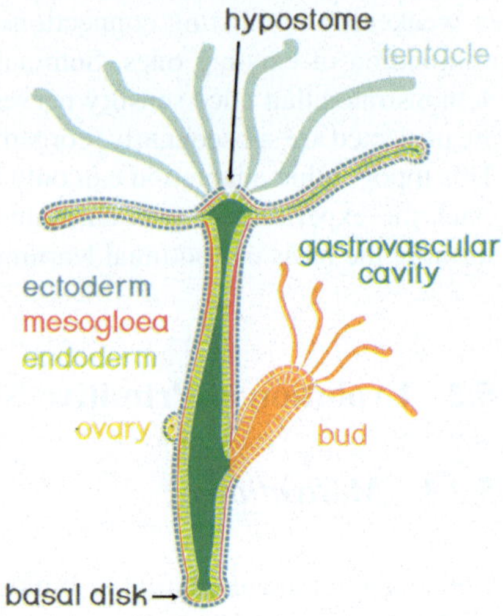

Fig. 5.1 Schematic diagram of the hydra and its epithelial layers. Illustration 162611419 © Luayana—https://www.Dreamstime.com

to biological hydra. the connectivity and weights of the developed neural system are further adapted.

5.2.2 Hydra-Like Model

The freshwater hydrozoa (hydra) is a small cnidarian organism, about 0.5–2 cm long, which lives in freshwater. Biologically, it belongs to the phylum Cnidaria so it is related to many jellyfish species. For biologists, the hydra is an interesting animal because of its enormous ability to regenerate. If it is cut into several pieces, a small 'head' and a small 'foot' (basal disk) are formed at each piece. This happens with little cell division; the small hydras grow through a reorganization of the existing pieces, a process known as morphallaxis [130]. When they get enough food, the small hydras grow, eventually reaching normal size. If the pieces are put into a centrifuge, they can actually re-aggregate into a full hydra [228]. The body of the hydra is essentially a tube with tentacle extensions at the top as shown in Fig. 5.1. The body wall consists of an inner endoderm, the mesoglea, and an outer ectoderm. The endoderm and the ectoderm form an epithelial layer. The neurons of the hydra reside within the mesoglea. Its tentacles are used to catch food, relying on an olfactory sensor to detect food (prey). The movements of the tentacles and of the body are controlled by the developed neural network, which is a spiking neural network.

The physical body of the hydra-like animat simulates hydra at a very high abstraction level, which consists of a cylinder with four tentacles equally distributed on its top. Each tentacle is driven by a muscle cell controlled by a neural network composed of motor neurons, inter-neurons, and sensory neurons, all of which are modeled with integrate-and-fire (IAF) neurons, as shown in Fig. 5.2. To simplify the experimental

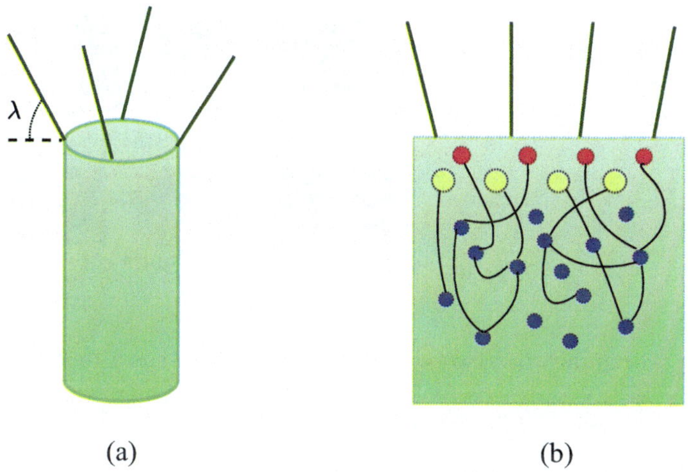

Fig. 5.2 The hydra-like animat. **a** The body plan with four tentacles; **b** The nervous system shown on the 2D area representing the unfolded body wall, where the red, yellow, and blue dots mean motor, sensory, and inter-neurons

setup, four sensory neurons and four motor neurons are put on the top of the body, while the number and position of the inter-neurons are determined by the developmental process. During the evolution of the GRN controlling the neural development, four pieces of food drop sequentially around the body and the closest sensory neuron will generate a number of spikes as the input to the neural network.

5.2.3 A Genetic Model for Early Neural Development

The genome in our model consists of structural units (SUs) representing genes that can result in certain cellular behaviors, such as cell division, cell migration, axon growth, and cell death, or can produce transcription factors (TFs) for intra- and intercellular interactions. The SUs are preceded and controlled by a number of regulatory units (RUs). An example is given in Fig. 5.3. Note that multiple RUs may regulate the expression of a single or multiple SUs and that RUs can be activating (RU^+) or repressive (inhibitory) (RU^-). Three different cell types are simulated in the model, namely stem cells, interstitial cells (after division), and neurons (after axon growth), depending on the activation status of the structural units of the cell. The general structure of the genes here is very similar to that of the genes used in the multi-cellular model in Chap. 4.

Similar to the multi-cellular model presented in Chap. 4, each RU and TF has a certain affinity value that determines whether a TF can regulate a RU. If the difference between the affinity values of a TF and a RU is smaller than a predefined threshold ε (ε is usually a positive number smaller than 0.5), the TF can be bound to the RU

Fig. 5.3 An example of the genes for regulating early neural development

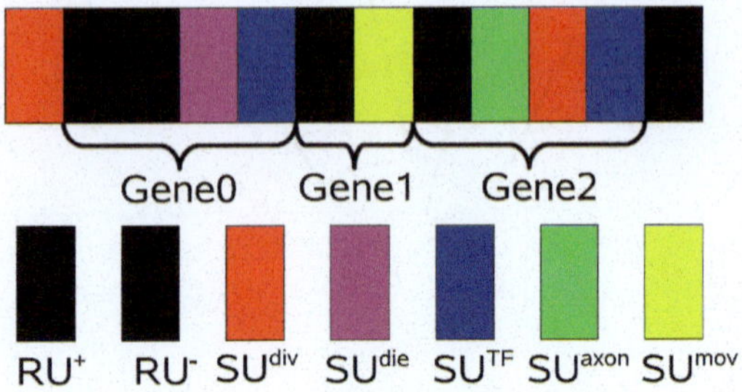

to regulate. The affinity overlap ($\gamma_{i,j}$) between the i-th TF and j-th RU is defined by

$$\gamma_{i,j} = \max\left(\varepsilon - \left|\text{aff}_i^{TF} - \text{aff}_j^{RU}\right|,\ 0\right). \tag{5.1}$$

If $\gamma_{i,j}$ is greater than zero, then the concentration c_i of the i-th TF is checked to see if it is above a threshold ϑ_j defined for the j-th RU:

$$b_{i,j} = \begin{cases} \max(c_i - \vartheta_j,\ 0) & \text{if } \gamma_{i,j} > 0 \\ 0 & \text{else} \end{cases}. \tag{5.2}$$

Thus, the activation level contributed by the j-th RU (denoted by a_j, $j = 1, \ldots, N$) can be calculated as follows:

$$a_j = \sum_{i=1}^{M} b_{i,j}, \tag{5.3}$$

where M is the number of TFs that regulate the j-th RU. Assuming the k-th structural gene is regulated by N RUs, the expression level of the gene can be calculated by

$$\alpha_k = 100 \sum_{j=1}^{N} h_j a_j (2s_j - 1), \tag{5.4}$$

where $s_j \in (0, 1)$ denotes the strength and sign (positive for activating and negative for inhibitory) of the j-th RU, which is subject to evolution, and h_j is an additional parameter representing the regulatory strength of the j-th TF. If $\alpha_k > 0$, then the k-th gene is activated and its corresponding behaviors will be implemented.

An SU that produces a TF encodes all parameters related to the TF, such as the affinity value, the decay rate, the diffusion rate, as well as the amount of TF to be produced:

$$A = \beta \frac{2}{1 + e^{-20 \cdot f \cdot \alpha}} - 1, \tag{5.5}$$

where f is a parameter also encoded in the SU, and β is a parameter for expression rate of the SU.

A TF produced by an SU can be partly internal (for inter-cellular interactions) and partly external (intra-cellular interactions). Internal TFs stay inside a cell and do not diffuse. External TFs can diffuse into the whole simulation area, a process depending on the diffusion parameter of the TF. To determine how much of a produced TF is external, a percentage ($p^{ex} \in (0, 1)$) is also encoded in the corresponding genome. Thus, $p^{ex} \cdot A$ is the amount of external TF and $(1 - p^{ex}) \cdot A$ is that of the internal TF.

External TFs are put on four grid points around the center of the cell, which undergoes first a diffusion and then a decay process governed by the following equation:

$$\text{Diffusion: } \mathbf{u}_i(t) = \mathbf{u}_i(t-1) + 0.1 \cdot D_i^f \cdot (\mathbf{G} \cdot \mathbf{u}_i(t-1)), \tag{5.6}$$

$$\text{Decay: } \mathbf{u}_i(t) = \min((1 - 0.1 \cdot D_i^c)\,\mathbf{u}_i(t),\ 1), \tag{5.7}$$

where $\mathbf{u}_i$ is a vector of the concentrations of the i-th TF at all grid points, matrix $\mathbf{G}$ defines which grid points are adjoining, D_i^f is the diffusion parameter, and D_i^c is the decay rate.

There are also a few parameters defined by the SUs that generate cellular behaviors. The SU for cell division specifies the angle of division, indicating whether the daughter cell is placed above or below the mother cell. For cell migration, two parameters are encoded, one for direction (moving up or down) and the other for moving velocity. The SU for axon growth encodes the expected lifetime (t_{ttd}) of a neuron and three parameters (c_1, c_2, c_3) determining the probability threshold for axon growth. The limited lifetime of the neuron, t_{life}, is defined by the following Gaussian distribution:

$$t_{\text{life}} \sim \mathcal{N}(20t_{ttd}, 4). \tag{5.8}$$

The threshold for whether the i-th neuron is to be connected to the j-th neuron is calculated as follows:

$$\varphi_{ij} = \frac{c_1}{1 + e^{c_2 \cdot (d_{ij} - 10c_3)}}, \tag{5.9}$$

where d_{ij} is the distance between the i-th and j-th neurons. The distance is computed in the 2D region, referring to Fig. 5.2b, which is the distance between the two neurons along the surface of the cylindrical body. Then, a random number p ($p \sim \mathcal{N}(0, 1)$) is generated, and if $p < \varphi_{ij}$, a connection between the two neurons will be generated. The connectivity between the inter-neurons and the sensory neurons, as well as between the inter-neurons and motor neurons, is determined in the same way. However, no direct sensory-sensory, motor-motor, and sensory-motor connections are modeled.

5.2.4 Spiking Neural Network Model

The dynamics of the neural network are modeled by IAF spiking neurons [78]:

$$\tau_m \frac{dV(t)}{dt} = -(V(t) - V_{\text{resting}}) + R_m I(t), \tag{5.10}$$

where τ_m is the membrane time constant, R_m is the membrane resistance, $I(t)$ is the external current, $V(t)$ is the membrane potential, and V_{resting} is the resting potential. When the membrane potential $V(t)$ is larger than a given threshold V_{th}, a spike is generated. The spiking event is characterized by its firing time t^f. After a spike is generated, the membrane potential is reset to V_{reset} for a refractory period τ_{ref}. In an IAF neural network, the membrane potential of a neuron (V_k) can be calculated by

$$V_k(t) = \sum_{i=1}^{H} w_{ik} y_i(t), \tag{5.11}$$

where w_{ik} is the weight between neurons k and i, N is the total number of its presynaptic neurons, and $y_i(t)$ is the unweighted contribution of the i-th presynaptic neuron:

$$y_i(t) = \varepsilon(t - t_i^f - d_{ik}), \tag{5.12}$$

where ε is a spike response function modeling the post-synaptic potential, t_i^f is the firing time of neuron i, and d_{ik} is the synaptic delay.

The spiking neural network is simulated using the Neural Simulation Tool [127].

5.2.5 Evolving the Developmental Order

5.2.5.1 The Virtual DNA

Three different cell types are simulated in the model, namely stem cells, interstitial cells, and neurons. All cells have the potential to divide, move, and build axons. The different cell behaviors define the cell type. All cells begin as stem cells and are intended to divide, change position, and generate axons when the final position is achieved. While moving, the cell is termed an interstitial cell; a cell with axons is a neuron. In this model, a developmental order is required, first cell division, then cell movement, and at the end, axon growth is needed, as shown in Fig. 5.4. In biology, axon growth is not apparent until the cell reaches its final position.

At the beginning of a neural developmental process, some stem cells are put in the area at $y = 0$. In the x-direction, they are evenly distributed.

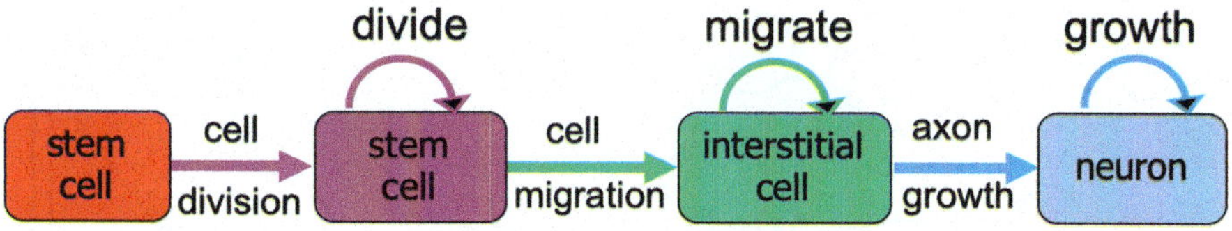

Fig. 5.4 The order of three events in early neural development, cell division, cell migration, and axon growth

5.2.5.2 Manually Designed GRNs

A DNA sequence (GRN) is hand-coded to simulate the growth process as a simple means to test the simulation model. The first manually designed GRN, as shown in Fig. 5.5a, represents an open loop control and can be seen as a SIM network motif. The GRN produces an increasing TF, activated by the pre-diffused TF, and activates the genes in the correct developmental order. This implies that the thresholds of all activating and deactivating RUs have to be set in this order. Therefore, the lowest threshold is responsible for activating the cell division, the next lowest is responsible for deactivating it, then the movement has to be activated before it is stopped, and finally, the axon growth has to be initialized. There is a RU to deactivate the axon growth at the end. The structures and values for the different gene types keep constant, and only other parameters are allowed to be changed by the EA. The overall aim was to evolve a high variance in the z-positions of the neurons to assess the flexibility of the genetic representation. With only one SU for cell movement, it is of course impossible to let the neurons move in different directions; nevertheless, it is still possible, in principle, for the neurons may stop at different places in the z-position. However, the results indicate that the EA fails to evolve the right regulatory order with this setting. This might be attributed to the fact that there is only one increasing TF, making it hard for the EA to find the optimum, since the phase space is only one-dimensional. That is, the trajectories can only lead to fixed points [305], and for complex behaviors, a minimum of three dimensions is necessary.

To get a more flexible genome, a second hand-coded GRN with negative autoregulation is designed, as shown in Fig. 5.5b. In this GRN, the genes with the structural units for division, movement, and axon growth each have their 'own' SU to produce TFs. This means, when one of the actions is performed, a particular TF is produced. So every action can deactivate itself and activate the following action. With this predefined GRN, the correct developmental order can be found, although no variation in the y-position of the neurons has occurred. Therefore, a third hand-coded GRN is examined, as shown in Fig. 5.5c. In this GRN, there is another TF for every genome, making the interactions more flexible. In the second GRN, TF2 cannot further increase when for example the division is deactivated. As a result, the cell movement must be activated with the same threshold that deactivates the division. So another TF is added at this point, which will increase provided that the concentration

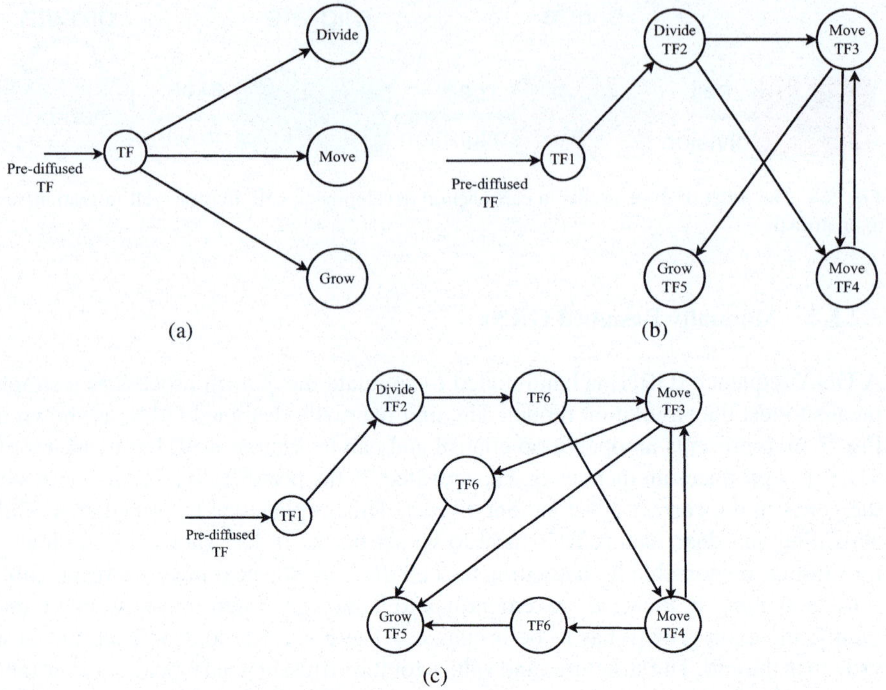

Fig. 5.5 Three manually designed gene regulatory networks for neural development

of TF2 is sufficiently high there. This holds true only when there are internal TFs only, which is biologically sensible.

In the next, we will evolve the GRN using an evolutionary algorithm to achieve the correct regulatory dynamics for primitive neural development.

5.2.5.3 Evolutionary Algorithm

An evolutionary algorithm is applied to evolve the gene regulatory network to achieve the desired developmental order in terms of the corresponding gene expression levels, i.e., the genes for cell division, cell migration, and axon growth should be activated sequentially during the development. In addition, the gene for cell division should be deactivated before the gene for migration is activated, as the gene for migration should be deactivated before the gene for axon growth is activated. Gene transposition, gene duplication, and mutations are employed as the genetic variations. If gene transposition is performed, two randomly chosen units (both SUs and RUs are possible) are marked, and all units between these two marked units are cut out and pasted at another randomly chosen position. A gene duplication differs from gene transposition in that all units between the markers are copied and pasted at another randomly chosen position. All real-valued parameters (scaled between [0,1]) in the genome are adapted using an evolution strategy (ES) [369].

In all experiments, $\mu = 100$ and $\lambda = 300$. Three types of genetic variations are employed:

- Mutation. It changes the double values in the RUs and SUs with a Gaussian probability function with $\sigma = [1e^{-4}, 1e^{-6}]$. The initial value of σ is set to $1e^{-6}$. All values in the genome are restricted to the interval [0, 1].
- Gene duplication, which copies randomly selected RUs and SUs at a probability of 0.05. Therefore, three probability values to define the beginning and the end of the duplicated genes and the position where they are inserted are needed.
- Gene transposition, which moves a section of RUa and SUs in the genome at a probability of 0.05. Three probability values to define the beginning and the end of the transposed genes and the position where they are inserted are also needed.

A repairing operator is introduced so that a single gene contains only one type of SU related to cellular behaviors. For example, if one gene contains two or more different SUs to divide, move, or bring about axon growth, one of them is deleted. SUs for cell death may also be removed.

5.2.5.4 Fitness Definition

A fitness function that counts the cells having an incorrect developmental order is first adopted. This fitness function is tested with a randomly initialized genome and with some that are hand-coded with restricted genetic variations. The simulation is stopped if the number of cells is more than 300 to limit the computational costs. On the other hand, a penalty of 500 is assigned to the fitness when there are less than 14 cells. When there are enough cells, but no neurons, the fitness is set to 450 if there are only stem cells, 400 if there are cells that have divided, or 350 if there are interstitial cells. When there are neurons, but less than 10, 100 minus the number of

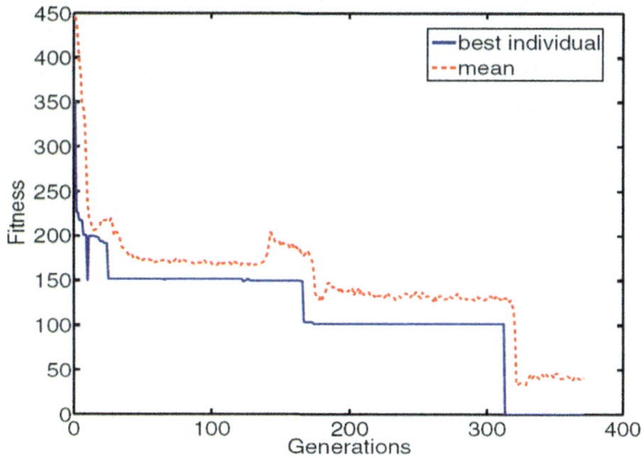

Fig. 5.6 Fitness changes in evolving the growth order. The mean as well as the best of the individuals are plotted [203]

5.2.6 Evolution of Food-Catching

Twelve stem cells with the genome that performs the cellular behaviors in the correct order are distributed on the simulation area so that a neural network is developed. For the neural network to perform the food-catching behavior, we employ an ES to adapt its connectivity and weights. Here, we assume that only one tentacle is needed to catch a piece of food. Thus, the target of the behavior adaptation is that the motor neuron closest to the dropping food should fire as strong as possible to maximize the possibility to catch the food, while the activity of other motor neurons should be minimal to reduce energy consumption.

To achieve the above-mentioned target, the following fitness function is defined for the i-th motor neuron (output neuron):

$$F_i = -f_i + \sum_{j!=i} ns_j, \text{ for } i, j = 1, ..., 4, \tag{5.14}$$

where ns_j is the number of spikes generated by other motor neurons in the simulated time period, penalizing unnecessary energy loss, and f_i reflects the performance of the i-th neuron measured by the time (t_i^c) for the corresponding tentacle to stand vertically, i.e., when the angle $\lambda = 90$, refer to Fig. 5.2, or by the maximum λ during the simulation period if λ never reaches 90 degree during that time. The angle of the i-th tentacle is calculated as follows:

$$\lambda_i(t) = \lambda_i(t - 0.1) - g * 0.1 + s_i(t), \ \ \lambda_i \in [0, 90]. \tag{5.15}$$

The second term on the right side of Eq. 5.15 simulates the passive dropping-down of tentacle due to gravity (g), $t \in [0, 30]$ is the simulation time in millisecond in behavior adaptation, and $s_i(t)$ equals 0 or 1 depending on whether there is a spike at time t. The final fitness is summed over four runs with different food dropping conditions.

We use a (30, 200)-ES and the initial mutation step-sizes are set to 0.1 with a lower bound of $1e^{-6}$. The maximal weight is set to 3000, which can be activating or inhibitory, and the synaptic delays are fixed to 1. All parameters are scaled between zero and one. The fitness curve for behavior optimization is shown in Fig. 5.9 [203]. The spikes of the input and output neurons of the best-adapted individual are shown in Fig. 5.10. From the figure, we can see that in three of the four cases, the neural network performs optimally. In one case (last panel in Fig. 5.10), outputs 2 ($o2$) and 4 ($o4$) also fire, though only output 3 ($o3$) is assumed to spike. Nevertheless, $o3$ does fire the strongest.

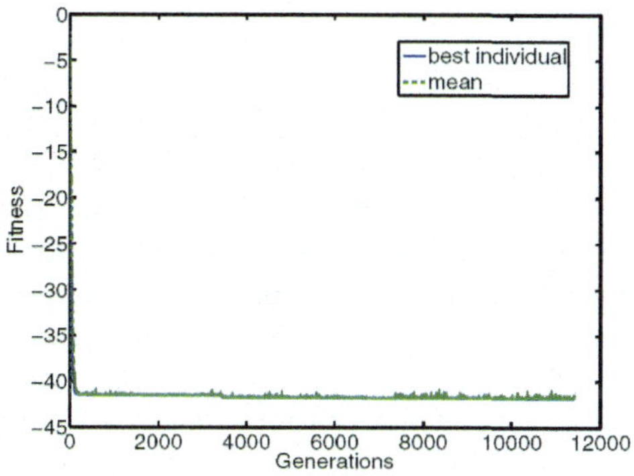

(a)

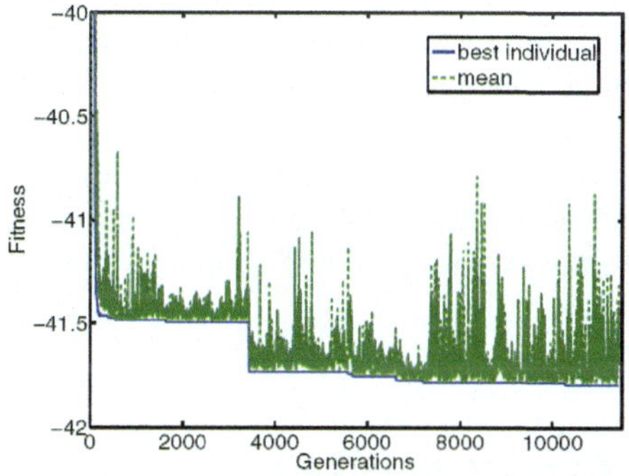

(b)

Fig. 5.9 **a** Fitness profile for evolving food-catching behavior, and **b** fitness after zoomed in

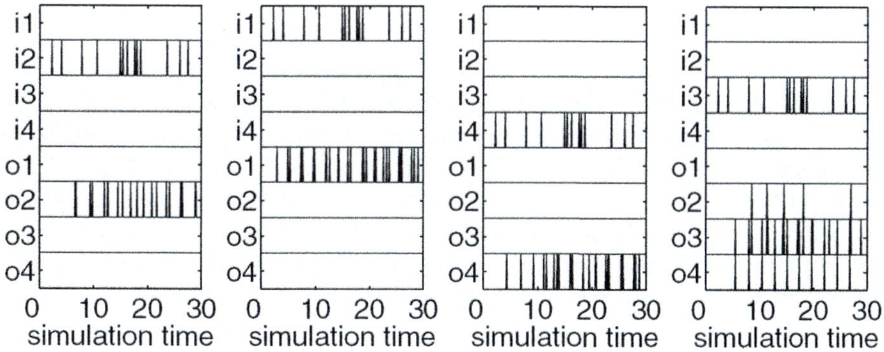

Fig. 5.10 Spikes at the input and output neurons for four different inputs [203]

5.2.7 Discussion

The neural developmental model presented above is based on a gene regulatory network that ensures the temporal order needed for primitive neural development. The gene regulatory is subject to an evolutionary process that changes its structure and parameters using biologically plausible genetic operators such as gene duplication, gene transposition, and gene mutation. The empirical results verify that the gene regulatory network has evolved successfully to achieve the correct developmental order, namely cell division, cell migration, and axon growth in sequence in a simulated evolution. After the gene-regulated neural development is complete, the connectivity and weights of the neural network are further adapted using an evolution strategy for performing food-catching behaviors of a hydra-like animat.

The current developmental model suffers from several limitations. Since no gene deletion is implemented in the model, the complexity of the genome tends to become more complex due to gene duplication. The cell migration behaviors generated by the model are deterministic since there are not much local interactions between the cells. In addition, the current gene regulatory model can produce neural development behaviors without considering the morphological development. It will be more interesting if neural and morphological developments are co-evolved to investigate their coupling effects in the evolution and development, which will be discussed in Chap. 6.

5.3 Evolution of Gene-Regulated Neural Plasticity

5.3.1 Introduction

Neural plasticity plays a paramount role during and after neural development of biological systems. Consequently, many efforts have been made to computationally model various types of neural plasticity, as discussed in Chap. 1. There has been a number of trend changes in the investigation of plasticity models. In the early days, the focus was to provide a computationally stable, self-regulated formulation of Hebbian learning [35, 157, 312]. Then, the focus was shifted toward biologically more plausible spiking neural networks and related neural plasticity rules that depend on the timing of spikes between connected neurons [60, 396]. These models, however, are considered to be phenomenological approaches [384], and therefore, more biological and molecular bases were targeted [49, 139]. In addition, other neural adaptation mechanisms, such as neuro-modulators, were included in spike-timing models to add reinforcement capabilities on top of purely associative learning [183, 245].

While the high-level functions of neuroplasticity such as learning and memory are taken for granted, the suggested roles of plasticity in formally defined neural network models are varied and sometimes contradictory. In some cases, simply applying models of plasticity to existing paradigms, such as reservoir computing, has yielded improved results [214, 365, 461]. Meanwhile, other studies [158, 257, 425] suggest neural plasticity is able to increase the mutual information in the signals between connected neurons. Hebbian plasticity has also been found to increase the correlation between neurons in a reservoir [436], while others suggest that plasticity can decorrelate neural activities, which is actually a desired property [16, 186].

In the following, we start with an empirical analysis of interference in learning sensory structures in the data based on neural plasticity embedded in a reservoir computing model. Then, we hypothesize that such interference may be attributed to the fact that existing plasticity models are all global, meaning that all neurons in the neural network use the same plasticity model and the same parameters, which is biologically implausible and computationally less powerful. To address this issue, two main revised models are described, one based on gene-regulated neural plasticity and the other on locally evolved plasticity rules.

5.3.2 Interference in Plasticity-Based Learning

Recurrent neural networks consisting of biologically based spiking membrane models have been applied to real-world learning tasks under a framework called reservoir computing [47, 269], which is often known as the third generation of artificial neural network models. The models of this kind use a recurrently connected set of neurons driven by an input signal to create a nonlinear, high-dimensional temporal transformation of the input that is used by single-layer perceptrons [340] to produce desired outputs. This restricts the training algorithms to linear regression tasks, while still allowing the potential to work on temporal data in a nonlinear fashion.

Given an initially fixed, unchanging connectivity, reservoir computing is based on the principle of random projections of the input signal in which the network structure is completely independent of the input patterns. In these models, the only features learned by the trainable parameters of the perceptron readout are correlations between the randomly projected features and the desired output signal.

Although supervised learning has been extremely successful in machine learning, in particular in deep learning, such learning paradigms are biologically implausible. By contrast, all neural synapses should be able to learn to encode the structure of the input signal. Ideally, synaptic learning should not rely on the presence of a desired output signal, from which an error is calculated by comparing the output of the neural network with the desired output. The neural activities generated by the input signal should provide enough information for the synapses to adapt and encode structural properties in the input signal. Such learning capabilities can be seen as unsupervised learning, which is highly desirable for tuning artificial neural models where labeled data is limited. For example, the structure of the reservoir in echo state networks or liquid state machines (both reservoir computing models) is usually randomly initialized, which may not be optimal for efficiently learning different problems. By applying unsupervised adaptation to the synapses in the form of biologically derived plasticity rules [34, 35, 458], it is hoped that the recurrently connected neurons in the reservoir can learn a structure that generates more effective features than a completely random projection that is not specific to the input data.

To investigate the effectiveness of various plasticity rules, we will explore the impact of applying these rules to a liquid state machine (LSM), which consists of spiking neurons, on three real-world speech recognition and human motion recognition problems. Two forms of spike-timing dependent plasticity (STDP) [34, 458] are examined, along with the Beinenstock-Cooper-Munro (BCM) rule [35]. Synaptic plasticity will be applied in an unsupervised pre-training phase before the supervised learning of the readout weights is carried out. We compare the performance of the LSMs whose reservoir structures are tuned by various plasticity rules on these tasks. Specifically, we focus on the structural adaptation of the weight matrices when the LSM is trained on samples (inputs only) from different classes. To make it possible for quantitative analysis, we will introduce a measure to evaluate to which extent the synaptic changes encode class-specific features in the network structure. Interference between different samples is a well-established phenomenon in sequentially trained learning models [117, 147, 336], which is also known as catastrophic forgetting [218]. As this directly concerns neural networks trained on sensory recognition tasks, a quantification of the interference between the synaptic parameters will be made for each plasticity model while applied to each task.

Existing studies report that adapting neural networks with plasticity improves their performance on pattern recognition tasks [147, 149, 461]. However, there lacks a detailed analysis of how the adaptation of synaptic parameters leads to the performance improvement. For a review of work applying plasticity models to improve the function of neural networks, the reader is referred to [60, 218].

5.3.2.1 Liquid State Machines

The neural network model is illustrated in Fig. 5.11. Recurrently connected neurons, indicated by L (the reservoir), are stimulated by the inputs directly as injected current, I, that perturbs the membrane potential modeled with a simple model [181]. This method for modeling the spiking activity of a neuron is shown to reproduce the most naturally occurring patterns of activity [182]. The real-valued inputs are normalized between 0 and 1, which are multiplied by a scaling factor of 20 before being injected as current into L. The number of input connections is usually set to one-fifth of the reservoir size (the number of neurons in the reservoir) projected randomly to the network nodes. The network activity dynamics are then simulated for 30 ms. Then, the resulting spike trains produced by each of the neurons are passed through a low-pass filter, f, to produce a real-valued vector used to train a linear readout with the iterative, stochastic gradient descent method.

In the following experiments, the reservoir consists of from $N = 35$ to $N = 135$ spiking neurons with the ratio of excitatory to inhibitory as 4:1, where N is the reservoir size. Neurons are connected with static synapses, i.e., the delta impulse (step) function. Connectivity is formed by having $N^2 \cdot C$ synapses, each having

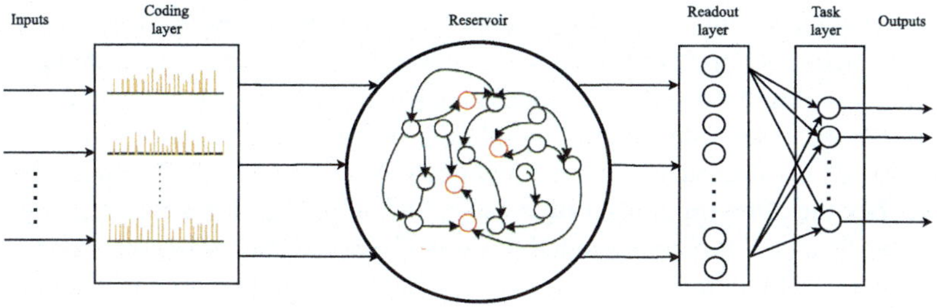

Fig. 5.11 A diagram of a liquid state machine, which consists of an input layer, a spiking coding layer, a reservoir, a readout layer, and an output task layer. The reservoir is composed of spiking neurons, and the output layer has an activation function such as the softmax function for classification or regression. The red circles in the reservoir denote inhibitory neurons and the black circles excitatory neurons

source and target neurons drawn according to a uniform random distribution, where C is the probability of a connection between any two neurons. Weights are drawn from two Gaussian distributions; $\mathcal{N}(6, 0.5)$ for excitatory and $\mathcal{N}(-5, 0.5)$ for inhibitory. When the plasticity rule adapts the reservoir weights, they are set to $[w_{\min}, w_{\max}] = [-10, 10]$. All parameters for excitatory and inhibitory neuron membranes are taken from [181]. The equations for the membrane model are as follows:

$$v' = 0.04v^2 + 5v + 140 - u + I, \tag{5.16}$$

where v denotes the membrane potential, v' denotes its derivative with respect to time, I is the input current, and u represents a membrane recovery variable, which describes the activation of the K^+ ionic currents and inactivation of the Na^+ ionic currents, providing negative feedback to v. Note that the constants in the above equation were obtained by fitting the spike initiation dynamics of a cortical neuron to scale the membrane potential v to the mV scale and the time t to the millisecond scale. The resting potential ranges between −70 and −60 mV depending on the value of b. The threshold potential can be set between −55 and −40 mV according to the history of the membrane potential before the spike is triggered.

$$u' = a(bv - u). \tag{5.17}$$

With the spike firing condition,

$$\text{if} \quad v > 30 \text{ mV} \quad \text{then} \quad \begin{cases} v \leftarrow c \\ u \leftarrow u + d \end{cases}, \tag{5.18}$$

where a, b, c, d are parameters to be defined. According to [181], parameter a represents the time scale of the recovery variable u. The smaller u is, the slower the recovery will be. Typically, $a = 0.02$. Parameter b denotes the sensitivity of the recovery variable u to the subthreshold fluctuations of the membrane potential v. The larger b is, the stronger coupling between u and v, possibly resulting in subthreshold oscillations and low-threshold spiking dynamics. Typically, $b = 0.2$. Parameter c specified the after-spike reset value of the membrane potential v caused by the fast high-threshold K^+ conductance. c is recommended to set to -65 mV. Finally, parameter d describes the after-spike reset of u caused by slow high-threshold Na^+ and K^+ conductance. $d = 2$ is recommended. It should be noted that tuning these parameters is non-trivial, although the spiking network model should be able to exhibit firing patterns of all known types of cortical neurons if the parameters are set according to the above specifications [181].

To generate a real-valued output from the discrete spiking activity, the spike train from each neuron is low-pass filtered according to Eq. 5.19. The vector of values produced is then weighted with the readout weight matrix and summed to produce a single output value, as given in Eq. 5.20.

$$x_i = f(S(t)) = \sum_{t=1}^{T} \exp\left(\frac{-t}{\tau}\right) w_{\max}, \tag{5.19}$$

$$y = \sum_{i=1}^{n} x_i w_i. \tag{5.20}$$

The state vector for a neuron is denoted by x_i, the filter function is $f()$, and the spike train is $S(t)$. The maximum number of time-steps in $S(t)$ is T, i.e., 50. The decay constant τ is 6.

The output weights are updated according to the least mean square method, as given in Eq. 5.21.

$$w_i \longleftarrow w_i + \mu(d - y)x_i, \tag{5.21}$$

where d is the desired output, y is the actual output, x_i is the input taken from a neuron's filtered state, and μ is a small learning rate. The weight from x_i to the output is w_i.

5.3.2.2 Description of Sensory Inputs

Complex sensory signals are projected through a common set of nerve fibers to cortical regions that must learn to distinguish between them based on differences in their spatial-temporal features.

Three sensory recognition tasks are selected, among which two of them consist of real audio signals of human speech and videos of human motion. For all tasks, the neural network output is trained to respond uniquely to each of the different types of input samples and therefore be able to perform effective recognition between them.

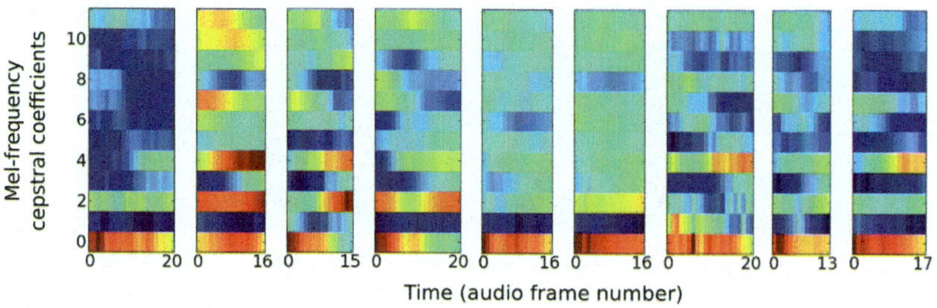

Fig. 5.12 Vowel samples from the nine speakers in the speaker recognition task. The audio signals in the data set are pre-processed into 12 Mel-frequency cepstrum coefficients (MFCC) features. Samples from each speaker have variable time durations in the number of audio frames they contain

Also, sample-specific synaptic adaptations are analyzed to determine if a unique structure is learned within the network due to synaptic plasticity.

The auditory task is to distinguish between nine different speakers based on short utterances of the vowel **/ae/**. Each of the 640 samples consists of a frequency 'spectrogram' that plots frequency intensity over a sequence of audio time frames. An example sample from each of the nine speakers is plotted in Fig. 5.12.

A synthetic benchmark task is taken from a study performed with Echo State Networks [187], a similar type of network model to the one we employ here, but using artificial neurons instead. The task is to predict which of the three signal-generating functions is currently active in producing a time-varying input signal. To generate a sample of the signal at a given time-step, one of the three following function types is used. First, a sine function of a randomly selected period; second, a chaotic iterated tent map; and third, a randomly chosen constant. The generator switches at a low probability of 0.05, from the current function to another one at each time-step [187].

The KTH data set [366] is a visual task consisting of 2391 video files of people performing one of the following six actions: boxing, clapping, waving, walking, and jogging. All samples are video sequences of many different subjects performing those six behaviors. The videos are converted into a sparse representation before being inputted into the reservoir model. Extracted still frames and processed features are plotted in Fig. 5.13, where one subject is performing each of the six behaviors.

There are 25 different subjects and samples, covering a range of conditions as detailed in [366]. Each video sample is taken at 25 frames per second and down-sampled to a resolution of 160 $\times$ 120 pixels. We process the raw video sequences according to the following equations:

$$M(t) = \|[\Delta(I_1, I_2), ..., \Delta(I_{N-1}, I_N)]\|, \tag{5.22}$$

$$M(t,i) = \begin{cases} 1 & \text{if } M(t,i) \geq 0.2 \cdot max(M(\cdot)), \\ 0 & \text{else.} \end{cases} \tag{5.23}$$

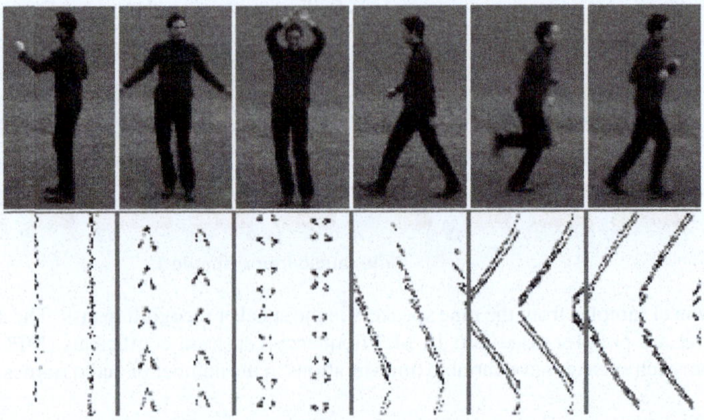

Fig. 5.13 Human motion samples for the six types of behavior in the KTH visual discrimination task. This illustration consists of different behaviors from a single person, while the whole data set contains 25 persons. Top row: Still frames from example video samples; boxing, clapping, waving, walking, running, and jogging. Bottom row: Features extracted corresponding to the samples above. Features are the raw time series activity used as input to the neural network

The final input matrix M is indexed by time frames, t, and spatial samples i. Column vectors I_n are individual frames, re-shaped into one dimension. Each sample contains up to a total of N frames. In other words, this process essentially further down-samples by a factor of 0.2 and calculates the difference between pixels in consecutive frames, which are then used as the new input features. Each frame is then re-shaped into a single-dimensional column vector, which is then appended together to form an input matrix in which each column is used as the neural network input at consecutive time-steps. Figure 5.13 shows frames extracted from an example of each type of motion along with the corresponding processed features.

A speaker recognition task is a classification problem dealing with mapping time series audio input data to target speaker labels. We use a data set taken from [232], which contains utterances of nine male Japanese speakers pronouncing the vowel **/ae/**. The task is to discriminate each speaker based on the speech samples. Each sample is comprised of a sequence of 12 feature audio frames. The features of each frame are the LPC cepstrum coefficients. The sample sequence ranges between 7–29 frames. The data set is divided into training and test sets containing, respectively, 270 and 370 samples. Unlike the benchmark data used here, the samples are not in a consecutive time series. However, each sample is composed of a time series sequence of audio frames.

This synthetic data set [187] is generated to model a low spatial dimension but very high-frequency temporal structure, in contrast to the previous two sensory tasks. Three functions generate time-varying single-dimensional signals that the network learns to distinguish between.

The auditory and visual tasks are described in [232] and [366], respectively, with data availability also provided.

5.3.2.3 Input-Specific Structure Learning

To validate the hypothesis that synaptic plasticity helps learn the encoding of a distinct structure of different labels in the input signals. For the speech task, these labels consist of different speakers, and for the video recognition task, the labels consist of different human behaviors.

The data sets are divided evenly into two subsets, each being used to train a liquid state machine model for 10000 iterations. In the training, a sample is randomly selected in each iteration. The changes to the weight matrix resulted from neural plasticity are recorded for each sample presentation, which is then used to create a class-specific average weight change for each of the class labels in both of the sample subsets. Finally, we calculate the Euclidean distance between each class in one set and each class in the other as follows:

$$\text{Dist}(C^X, C^Y) = \sum_{i=1}^{N} |\varDelta W_i(C^X) - \varDelta W_i(C^Y)|, \tag{5.24}$$

where C denotes class labels, X and Y distinguish the separated sets of samples, $\varDelta W$ is the change in weight matrix for a presented sample, N is the number of synapses, and i the synapse index. This will result in a confusion matrix of similarity in the synaptic weight change for different classes of input. Having lower values on the descending diagonal means that there is a structural adaptation that is specific to the class of that column compared with the similarity between structural adaptations of two different classes.

Figure 5.14 shows the 'weight change confusion matrices' described above, for each plasticity model applied to all sensory tasks (nine experiments in total). All of the experiments show at least some stronger similarity in the descending diagonals and most are stark in this manner. It is certainly a strong enough pattern to show that through the many iterations of training, each of the plasticity models has become sensitive to the particular structure of the sensory input signals so that each different class of sample will give rise to changes in synaptic strength that are distinct from other classes compared with the similarity to themselves. We re-iterate that the class labels were not used in any way in the plasticity models themselves and so the differences in the weight change arise from the input signals alone.

A few exceptions to the strong diagonal patterns can be observed in Fig. 5.14, which implies that the model is not able to distinguish some classes from each other after the reservoir is adapted using the plasticity rule. For example, speakers 8/9 with bi-phasic STDP, behaviors 1/2 with BCM, and behaviors 1/2/3 and 4/5/6 with tri-phasic STDP. The latter confusion corresponds to the behaviors of boxing/clapping/waving and walking/running/jogging. The reason becomes apparent if we look at the similarity of those input features of these actions, as shown in the lower panels of Fig. 5.13.

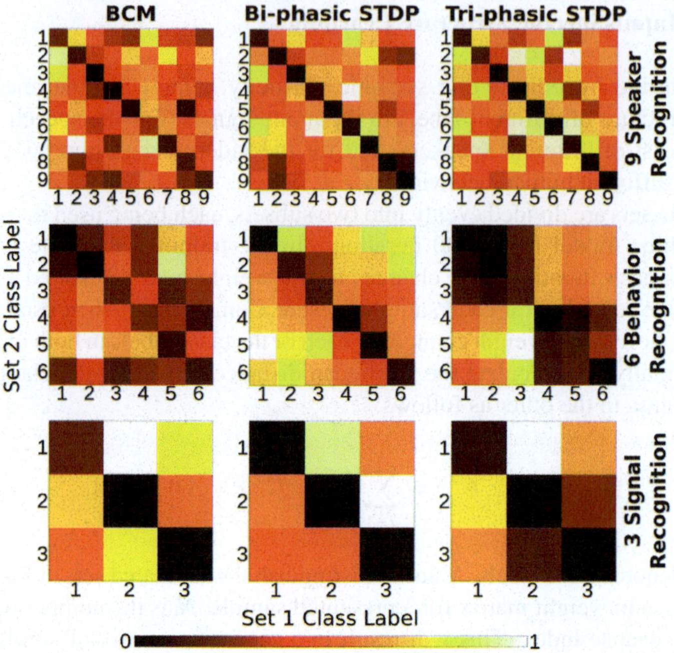

Fig. 5.14 Class correlation of structural synaptic adaptation. Heat map plots indicate the structure learned on each class for the three tasks under each of the plasticity rules. Essentially, it is a confusion matrix of the geometric distance between the weight matrix adaptation of each class of sample. The training data for each task is divided into two sets. Class-average adaptation is found for each set. There is then a distance calculated between each class of the two sets. Lower values on the descending diagonal indicate higher correlation within a class adaptation and therefore strong class-specific structure learned

5.3.2.4 Classification Performance with Plasticity

Recall that the ultimate goal of the reservoir model is to accurately distinguish different types of input samples according to their patterns. Therefore, we compare the error rates achieved by the liquid state machine on the three sensory tasks, with and without the different forms of plasticity under investigation here. Table 5.1 lists the error rates obtained on each of the learning tasks with the different plasticity rules active in a pre-training phase, together with a static network with fixed internal synapses.

From the error rates in Table 5.1, we note that pre-training the network with synaptic plasticity cannot always reduce the error rate, and surprisingly, sometimes even increase the error, e.g., on the KTH human behavior data set. The three plasticity models even increase the error rate by a percentage of between 1.7 and 10%. The best improvement was observed on the tri-function signal recognition task, with an improvement of 1.5% only when using the tri-phasic STDP.

Table 5.1 Classification error rates

	Static	BCM	STDP	TP-STDP
Tri-func	0.153	0.157	0.204	**0.138**
KTH	**0.283**	0.3	0.333	0.383
Vowels	0.089	**0.086**	0.092	**0.086**

It is clear from the previous results that pre-training with synaptic plasticity does not necessarily improve the learning performance of the liquid state machine, which suggests that the plasticity rules are not able to learn the structure as we expect. To shed some light on this seemingly contradictory observation, we investigate interference between synaptic changes to determine if the above problem can be attributed to interference in sequentially learning the sensory structure.

5.3.2.5 Synaptic Interference

When a model adapts incrementally to sequentially presented input, existing patterns that have been learned by the model parameters are prone to be overwritten by learning new patterns. This is known as interference. The work that has studied this effect [117, 279, 336] tests the ability to recognize previously presented input after the model has been trained on new ones in order to estimate how much learning has been undone. When new training leaves the model unable to recognize old patterns, it is said there has been catastrophic interference and forgetting.

We introduce a method of measuring interference directly in synaptic parameters instead of the model output. Our measure is described in detail in the Methods section. I^{total} directly quantifies all synaptic changes that are overwritten. Our formulation of synaptic interference is based on the synaptic changes from sequentially presented samples. Changes for a given type of sample are called ΔW_t and changes for all others are ΔW_o. Interference must be calculated individually for each type of sample, I_t^{class}, and averaged together to get the overall interference, I^{total}. The equations are as follows:

$$I_t^{\text{class}} = \sum_{i=1}^{N} \frac{1}{N}[\Delta W_{ti} \cdot \Delta W_{oi} < 0][|\Delta W_{ti}| < |\Delta W_{oi}| \cdot C], \tag{5.25}$$

$$I^{\text{total}} = \sum_{t=1}^{C} \frac{I_t^{\text{class}}}{C}, \tag{5.26}$$

where I is interference, N is the number of synapses, C is the number of competing sample types, and ΔW is a vector of synaptic changes. Subscript i denotes the

Table 5.2 Synaptic interference

	BCM	STDP	TP-STDP
Tri-func	0.82	**0.8**	0.88
KTH	**0.92**	0.93	0.96
Vowels	0.96	**0.58**	0.9

parameter index, subscript t denotes samples of a given type, and subscript o denotes samples of all other types.

Table 5.2 lists the interference obtained in each of our experiments. In all but one of the experiments, the interference level is between 82 and 96%, which is surprisingly high. These results reveal that most of the learned structure for each class of input is forgotten in that the consecutive samples overwrite each other's previous changes. It should be noted that Bi-phasic STDP applied to speaker recognition has the lowest interference level of 58%.

To further explore interference and visualize the impact of plasticity, synaptic changes will be analyzed directly. Figure 5.15 gives an example for the speaker recognition task with the BCM plasticity, showing the adaptation of the synaptic weight matrix resulting from each speaker in the voice recognition task. This is plotted against the activity level for each neuron, **S**, and its readout weights, **R**, that are trained to generate an output that is sensitive to the given speaker. Each sub-plot presents the average response taken over all sample presentations from that speaker. This visualizes a whole process from the synaptic change of an internal network connection, to the average neuron state for a given speaker, and to the selective weights of the readout for that speaker. We expect that a positive weight change should correspond to a neuron activation that is unique to the class which should then in turn improve the recognition ability of the readout to identify that class if the plasticity is supposed to help improve the learning performance.

In the class weight matrix, the green parts in Fig. 5.15 highlight an example where synaptic interference is occurring between different types of patterns. The directly opposing features in the weight matrix adaptations show the samples overwrite each other's changes. Nevertheless, the same features are also most distinctively class specific.

Recall that any synapse can only increase or decrease, implying that a single synapse can only adapt to distinguish between two mutually exclusive kinds of input patterns. If n synapses are considered in combination, then the number of input patterns that can be discriminated becomes 2^n theoretically. Figure 5.15 illustrates this principle in practice with respect to the nine-speaker recognition tasks. The adapted synapses labeled (a) can clearly distinguish speaker {#1} from speakers {#2, #3}, but fail to distinguish {#2} from {#3}. Similarly, the adapted synapses labeled (b) can distinguish speakers {#1, #6, #8} from speakers {#3, #4, #9} but cannot distinguish speakers within either of those sets. However, if the synapses (a), (b), (c), and (d) are considered in combination, all speakers can be recognized by synaptic

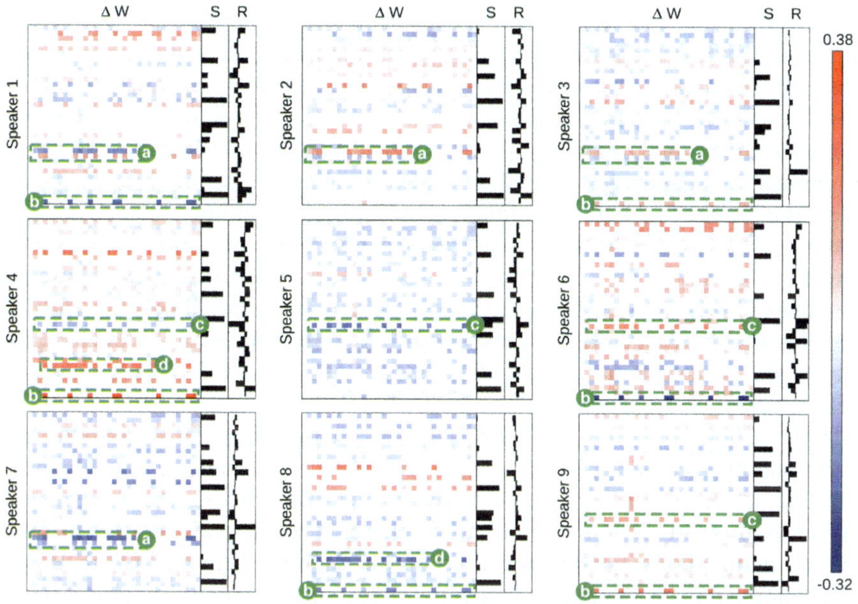

Fig. 5.15 The class-specific synaptic adaptation for the nine class speaker recognition tasks under BCM plasticity. The main heat maps in each sub-plot show the adaptation of the weight matrix (synapses) after the presentation of voice input data from each speaker. Blue values show a reduction in synaptic strength and red values show an increase. The bar-chart, **S**, shows the average neuron activation for each class. The bar-chart, **R**, shows the learned readout weights. Labeled synapses **a, b, c, d** indicate key structural changes that are selective between different speakers. Each label alone can distinguish between two sets of speakers. Taken all together, the labeled synapses adapt specifically to each speaker in a unique pattern, learning a distinct network structure for each one

plasticity changes alone. Figure 5.15 indicates that the changes in the synapses are not correlated with the neural activity or readout weights. For the plasticity to enhance the accuracy of sensory discrimination, the synapses are expected to strengthen for class-specific neural activities and weaken for common neural activities, which is not the case in these results.

Both STDP and BCM adapt the synapses of a network in distinctive patterns according to which type of sample is being presented to the network. Thus, the label of the samples is not required for plasticity rules to learn the specific structure information of sensory inputs. Thus, we hypothesize that this is achieved by learning to drive the synaptic parameters to a structure representing the average of all input samples. Once converged, any further input stimulus will drive the synaptic parameters in a direction away from this average structure. We show synaptic plasticity spends most of its action counter-acting previous changes and overwriting learned patterns. The same patterns of synaptic adaptation that distinctly characterizes each class of input are the same ones that reverse adaptations made by other inputs.

5.3.3.2 Gene Regulatory Network

The GRN for regulating the parameters in the BCM rule is a simplified version of the ordinary differential equation [94]:

$$\eta(t+1) = (1-\gamma_\eta)\eta(t) + \alpha_\eta f(c_s(t), c_c(t)), \tag{5.30}$$
$$\varepsilon(t+1) = (1-\gamma_\varepsilon)\varepsilon(t) + \alpha_\varepsilon g(c_s(t), c_c(t)), \tag{5.31}$$
$$\lambda(t+1) = (1-\gamma_\lambda)\lambda(t) + \alpha_\lambda h(c_s(t), c_c(t)). \tag{5.32}$$

where η, ε, and λ are three parameters in the BCM rule; γ_η, γ_ε, and γ_λ are the decay factors of the gene concentration controlling the three plasticity parameters; and $c_s(t)$ and $c_c(t)$ are the concentrations of sodium and calcium ions, whose expression levels are influenced by the i-th neuron's weights and weight plasticity:

$$c_s(t+1) = (1-\gamma_s)c_s(t) + \alpha_{c_s} \sum x_i(t) w_i(t), \tag{5.33}$$
$$c_c(t+1) = (1-\gamma_c)c_c(t) + \alpha_{c_c}\theta(t), \tag{5.34}$$

where γ_s and γ_c are decay factors of the sodium and calcium ion concentrations. The activation functions are defined as follows:

$$f(c_s, c_c) = \frac{1}{1+e^{-(k_1 c_s + k_2 c_c)}}, \tag{5.35}$$
$$g(c_s, c_c) = \frac{1}{1+e^{-(k_3 c_s + k_4 c_c)}}, \tag{5.36}$$
$$h(c_s, c_c) = \frac{1}{1+e^{-(k_5 c_s + k_6 c_c)}}, \tag{5.37}$$

where $k_1, k_2, k_3, k_4, k_5, k_6$ are parameters of the activation functions.

The gene regulatory network has altogether 16 parameters, namely γ_η, γ_ε, γ_λ, γ_c, γ_s, α_η, α_ε, α_λ, α_c, α_s, and $k_i, i = 1, 2, ..., 6$, will be tuned by an evolutionary algorithm. The evolution strategy adopted in [285] is the covariance matrix adaptation evolution strategy (CMA-ES), which evolves the above parameters to optimize the classification or regression performance of the BCM spiking neural network model.

The overall structure of the GRN-regulated BCM spiking neural network (GRN-BCM) is given in Fig. 5.16. The parameters of the BCM plasticity of each neuron are regulated by a gene regulatory network, with its states being influenced by the neural activities and its parameters being evolved by a CMA-ES.

Empirical results in [284] indicate that GRN-BCM outperforms its variant without gene regulation, the standard liquid state machine, and several traditional object recognition algorithms on three human behavior recognition tasks. Note that for human behavior detection tasks, spatiotemporal features must be extracted to achieve good recognition performance, while the spiking neural network with gene-regulated plasticity relies only on spatial features.

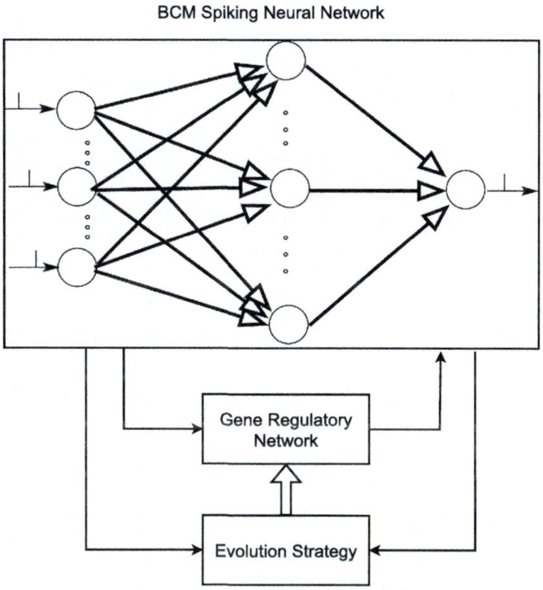

Fig. 5.16 The overall framework of the GRN-BCM, in which the plasticity parameters of each neuron depend on its activity

There are two developmental components in the GRN-regulated LSM model. First, the dynamics of the reservoir are tuned by a local BCM plasticity rule and further by the weight pruning process. Second, the plasticity parameters are regulated by the gene regulatory network, whose states are in turn influenced by the neuron dynamics.

5.3.4 Self-Organization of Liquid State Machines

5.3.4.1 Overall Framework

The previous GRN-regulated plasticity rule is applied to feedforward spiking neural networks. An extension of that model was reported in [463], where GRN-tuned BCM plasticity rules are applied on the reservoir of the liquid state machine (LSM). The main purpose of the BCM plasticity is to tune the structure and parameters of the reservoir in the LSM. Contrary to the existing methods with a fixed and randomly wired structure in the reservoir, the reservoir is initialized with full connectivity, but all synaptic weights are allowed to grow or shrink subsequently according to the GRN regulation model. A predefined threshold of 0.005 is adopted to prune the synaptic weights in that work, i.e., those synaptic connections whose weights are lower than the predefined threshold will be deleted from the reservoir, thereby tuning the structure of the reservoir. As a result, an optimal structure of the reservoir can be generated for different tasks in a self-organizing way.

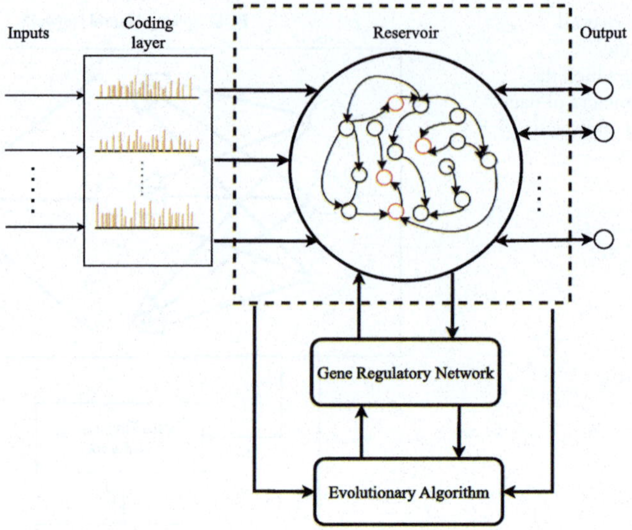

Fig. 5.17 The overall framework of the gene-regulated liquid state machine, whose parameters are evolved by an evolution strategy

Pruning very weak connections is biologically plausible, as findings observed in biological neural development reveal that weak synaptic connections that are very rarely used will die out. The pruning of the weak connections is an important part of activity-dependent neural self-organization. In biological systems, maintaining the weak connections consumes additional energy without any benefits, while in computational systems, more computational resources are needed to keep these weak connections. Some empirical studies on the relationship between the performance and complexity of neural network models imply that overly complex neural network models may even degrade the learning performance.

The BCM plasticity rule and the GRN for regulating the parameters in the plasticity rule are the same as those introduced in Sect. 5.3.3. The overall structure of the model is given in Fig. 5.17. Note that there is an additional change to the standard LSM in that the output neurons also have recurrent connections to the readout neurons, making it more powerful in handling spatiotemporal patterns. Detailed discussions will be given in the next section.

5.3.4.2 LSM with Recurrency

As we previously noted, there are recurrent connections between the output neurons and readout neurons. The output of the LSM with recurrent connections can be written as follows:

$$\mathbf{x}(t+1) = W_{\text{in}} f(\mathbf{u}(t+1)) + W f(\mathbf{x}(t)) + W_{\text{back}} f(\mathbf{y}(t)), \tag{5.38}$$

where W_{in}, W and W_{back} are the weight matrix between the inputs and the reservoir, the weights in the reservoir and the recurrent connections, respectively; $\mathbf{u}$, $\mathbf{x}$, and $\mathbf{y}$ are the input, state, and output vectors; and $f(\cdot)$ is the activation function of the neurons in the reservoir,

$$f(t) = \frac{t}{\tau}\exp(1 - \frac{t}{\tau})H(t), \tag{5.39}$$

which implies that the spiking neurons in the reservoir are leaky-and-fire (LIF) neurons, τ is a time constant, and $H(t)$ is the Heavy-side step function:

$$H(t) = \begin{cases} 1 & \text{if } t \leq 0, \\ 0 & \text{if } t > 0. \end{cases} \tag{5.40}$$

The output of the LSM can be calculated as follows:

$$\mathbf{y}(t+1) = f_{\text{out}}(W_{\text{out}}\mathbf{x}(t+1)), \tag{5.41}$$

where f_{out} is the activation function of the output neurons.

Since the back-propagation method cannot be directly applied to the recurrent connections, the back-propagation through time (BPTT) method [454] is the training approach for the recurrent weights. BPTT transforms a recurrent neural network (RNN) into a feedforward neural network (FFNN) by 'unfolding' the RNN through time. When the network starts processing a data sequence, identical copies of the RNN are created and the connections within the RNN are redirected to generate new feedforward connections over time from one network to the subsequent network. Eventually the RNN is converted to a large FFNN. Then the training proceeds in a similar manner to training a FFNN with back-propagation. The loss function is the summer square error:

$$E = \sum_{t=1}^{T} \parallel \mathbf{d}(t) - \mathbf{y}(t) \parallel^2, \tag{5.42}$$

where $\mathbf{d}(t)$ is the desired output vector at time instant t. Assuming δ^j_{out} and δ^j_{back} are the back-propagated error of the j-th feedforward and recurrent weights between the readout neurons and the output neurons, respectively, we have

$$\delta^j_{\text{out}}(t) = \left[(d^j(t) - y^j(t)) + \sum_{i=0}^{L} \delta^i_{\text{back}}(t+1) w^{ij}_{\text{back}} \right] \frac{\partial f(\mathbf{u})}{\partial \mathbf{u}}, \tag{5.43}$$

$$\delta^i_{\text{back}} = \left[\sum_{j=1}^{N} \delta^j_{\mathbf{out}} w^{ji} + \sum_{j=1}^{L} \delta^j_{\text{out}}(t) w^{ji}_{\text{out}} \right] \frac{\partial f(\mathbf{u})}{\partial \mathbf{u}}, \tag{5.44}$$

Table 5.3 Comparison of different methods on KTH data set

GRN-LSM	Spatial features	LSM+BCM+GRN	88.15
GRN-BCM	Spatial features	BCM+GRN	84.81
Dollar et al. [88]	Spatiotemporal (Gabor filters)	1-nearest neighbor	81.17
Schuldt et al. [50]	Spatiotemporal features	Support vector machine	71.83
Antonios et al. [14]	Spatiotemporal features	Boost+Relevance vector machine	80.80
Le et al. [331]	Spatiotemporal features	Independent subspace analysis	93.90
Jhuang et al. [190]	Spatiotemporal features	Feedforward hierarchical NN	91.7

where i, j indicates the i-th inter-neuron (neuron in the reservoir) and the j-th output neuron. Thus, the weights can be updated as follows:

$$w_{\text{in}}^{ij}(k+1) = w_{\text{in}}^{ij}(k) + \gamma \sum_{t=1}^{T} \delta_{\text{back}}^{i}(t) u^{j}(t), \tag{5.45}$$

$$w_{\text{out}}^{ij}(k+1) = w_{\text{out}}^{ij}(k) + \gamma \sum_{t=1}^{T} \delta_{\text{back}}^{i}(t) x^{j}(t-1), \tag{5.46}$$

$$w_{\text{back}}^{ij}(k+1) = w_{\text{back}}^{ij}(k) + \gamma \sum_{t=1}^{T} \delta_{\text{back}}^{i}(t) y^{j}(t-1), \tag{5.47}$$

where γ is the learning rate. Note that the above learning rules cannot be directly applied to weights in the reservoir since neurons in the reservoir are spiking neurons.

Comparative studies are performed to examine the performance of the LSM with local BCM plasticity rules. The algorithms under comparison include the gene-regulated BCM spiking model described in the previous section and five other algorithms based on spatiotemporal features extracted in advance. Note that the models described here and in the previous section use spatial features only. From Table 5.3, we can see that the recognition accuracy of GRN-LSM outperforms all other compared models except for the models reported in Le et al. [331] and Jhuang et al. [190]. These two models, however, have more advanced spatiotemporal feature extraction methods.

5.3.5 Co-Evolving Sub-Reservoirs of Liquid State Machines

The two plasticity rules described in the previous sections achieve the local regulation by means of a gene regulatory network whose states are influenced by the neuronal activities. An alternative to achieve local regulation is to divide the reservoir into sub-structures, each of which is regulated by a different plasticity rule of the same plasticity rule with different parameters. In this section, we introduce a co-evolutionary approach to evolve the structure of the reservoir [477], whose connectivity is further tuned by plasticity rules. The benefit of using a co-evolutionary approach is that separate sub-reservoirs can be evolved, which finally form a complete reservoir.

5.3.5.1 Overall Framework

Different from the standard LSM, the generative liquid state machine (G-LSM) divides the reservoir into several blocks to achieve local plasticity, which are then organized into sub-reservoirs [477]. A diagram of the overall architecture of the G-LSM is given in Fig. 5.18. The encoding layer converts the input data into the spike trains, which are subsequently injected into the reservoir layer. The reservoir states are extracted by the readout layer and then utilized by the softmax to calculate classification accuracy. The top of the diagram shows the block, the sub-reservoir, and the pathway of the reservoir. The reservoir contains four sub-reservoirs and each sub-reservoir is composed of four blocks. The pathway connects the output neurons and the input neurons of two blocks with a group of the synapses. The spiking neurons in each block are connected based on four complex network models, namely random network model, scale-free network model, small-world network model, and a three-layer network model. Each sub-reservoir contains two types of spiking neurons, i.e., excitatory neurons and inhibitory neurons. The excitatory neurons create positive post-synaptic potentials (PSPs), whereas the inhibitory neurons produce negative PSPs. The number of excitatory and inhibitory neurons follows the ratio of 4 : 1 [269]. Every four blocks are organized as a sub-reservoir, and the reservoir consists of four sub-reservoirs.

There are four block types and four sub-reservoir types, as shown in Fig. 5.19. The blocks are connected by a set of synapses, which is referred to as a pathway. The block chooses several neurons as the output and input neurons based on topological sorting. The input and output blocks of the reservoir layer are determined by the topological sorting of the four sub-reservoirs. The pathway creates the synapses between the output neurons and the input neurons of two blocks at a probability of p_{res}.

5.3.5.2 Block-Based Reservoir Representation

Block-based architecture representation is an effective indirect encoding scheme for NAS [101, 261]. Each block holding an unique network structure is responsible for fundamental information processing in the reservoir. The network structure of the block can be represented by a directed graph $G(V, E)$, where vertices $V = \{v_i, i \in 1, ..., N\}$ denote N spiking neurons and edges $E = \{e_{i,j}, i \in 1, ..., N, j \in 1, ..., N\}$ denote synapses from the spiking neuron v_i to v_j. The network structure can be mathematically represented by an adjacent matrix $A = \{a_{i,j}, i \in 1, ..., N, j \in 1, ..., N\}$, where $a_{i,j} = 1$ denotes existing synapse $e_{i,j}$, and $a_{i,j} = 0$ denotes no synapse $e_{i,j}$.

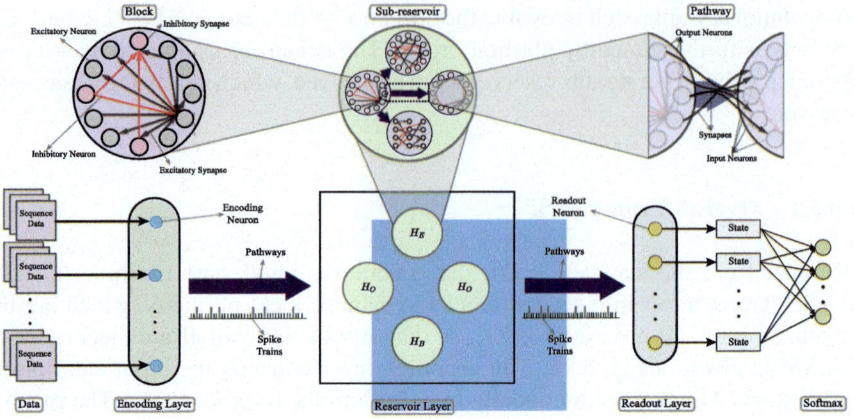

Fig. 5.18 A diagram of the G-LSM architecture

The network structure of the block is highly constrained if only the random blocks are used. G-LSM employs various complex network models to explore more diverse and effective modular structures. Due to the limited computational resources, we employ four typical complex network models to generate the network structure of the block. The network structures of the four complex network models have different properties, which we believe is sufficient for this study to validate the impact of network structure on the classification performance of the G-LSM. The four block types are random blocks, scale-free blocks, small-world blocks, and three-layer blocks. The examples of the four block types are shown on the left of Fig. 5.19a, with twelve neurons in each block. The network structures of different blocks show different characteristics. The synapses of the random block are uniformly distributed. The scale-free block contains a major hub with a larger degree. The neurons of the small-world block connect only with neighbors, and the three-layer block has no synapses between the neurons in the same layer. In the following, we introduce the details of the four block types. The input and output neurons of the block are determined by topological sorting and strongly connected components.

Refer to Sect. 2.2.2 for details of generating random, small-world, and scale-free network structures. The three-layer network is the most common network structure for conventional neural networks, which is also adopted in the block as a candidate network model. The neurons of three-layer blocks are divided evenly into three layers. Each neuron can only connect to the neurons in other layers with the input connection probability p_{in} and output connection probability p_{out}. The connection probabilities p_{in} and p_{out} decrease with the increase of the in-degree d_{in} and out-degree d_{out} of the neurons. For each two neurons v_i and v_j in the different layers, the synapse $e_{i,j}$ is connected following the connection probability $p_{out} \times p_{in} \times \sigma^{(d_{out}(v_i)+d_{in}(v_j))}$, where σ is the probability decay coefficient.

Once the network structure is confirmed, the input and output neurons that communicate with other blocks must be specified. To ensure that the input spikes flow

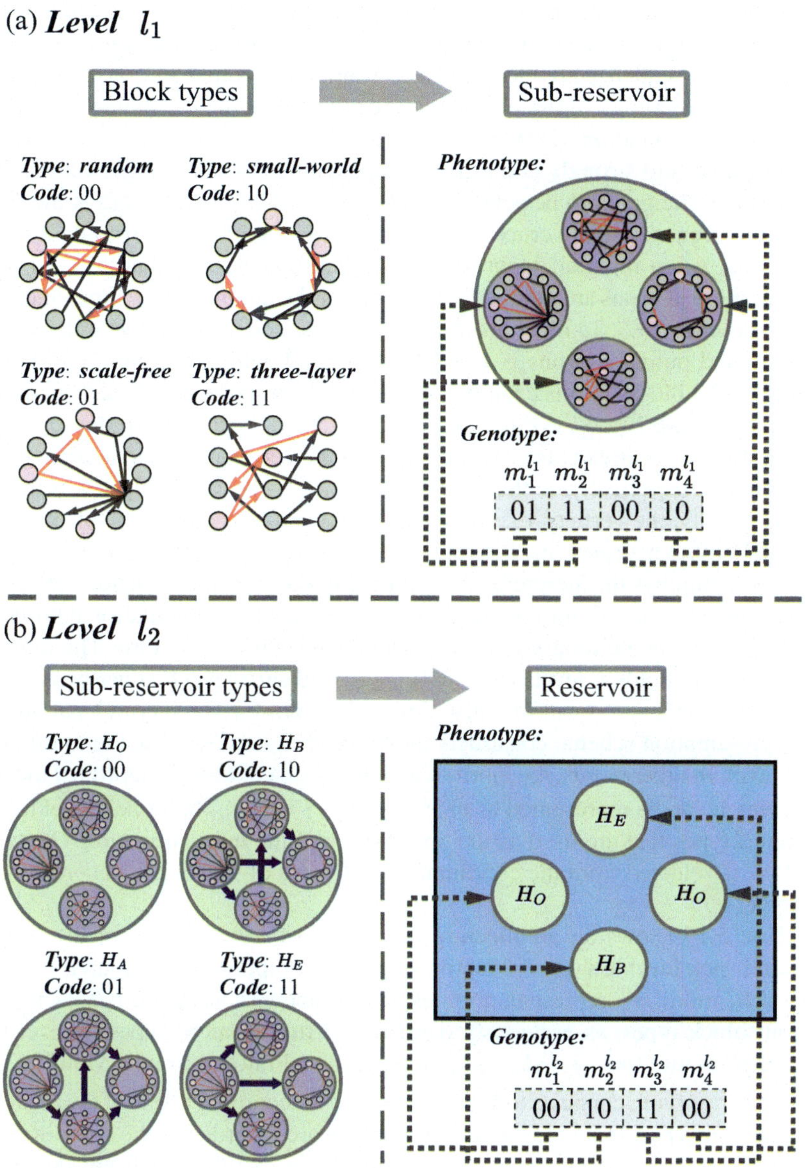

Fig. 5.19 An illustration of the level of the hierarchical architecture representation. There are two levels, where the l_1-th level contains four block types and the l_2-th level includes four sub-reservoir types. Each level shows an example of mapping the genotype to the phenotype

through as many neurons as possible, topological sorting is performed to determine the sequence of the neurons. The vertices at the start and the end of the sorted topological sorting sequence are identified as the input and output neurons, respectively. However, the topology of the network is a cyclic graph, which should guarantee to be directed acyclic graph (DAG) for topological sorting. The cycle in the graph can be aggregated into strongly connected components [414] as a single vertex cluster V_c to transfer the graph to be acyclic graph $G(V, V_c, E)$. If the V_c are at the start or end of the sequence, one vertex of the V_c is chosen as the input or output neuron. In addition, when the total input and output neurons are less than 30% of all the neurons, more neurons are randomly chosen until the number of the input and output neurons reaches 30%. To illustrate the proposed method, an example of determining the input and output neurons is shown in Fig. 5.20, which is composed of twelve neurons in the block. In this example, the graph is composed of 12 vertices and two cycles, which are $\{v_0, v_1, v_2\}$ and $\{v_6, v_7, v_8, v_9, v_{10}\}$. According to the topological sorting sequence, vertex cluster $\{v_0, v_1, v_2\}$ and the vertex v_4, v_{12} are regraded as the head and tail, respectively. In the vertex cluster $\{v_0, v_1, v_2\}$, v_0 is selected as the input neuron. Finally, the vertices v_2, v_5, v_8 and v_7, v_{11} are also randomly selected as the input and output neurons, respectively.

G-LSM employs the hierarchical architecture representation scheme [259] to further organize the blocks into the reservoir and reduce the computational burden for the block-based evolutionary optimization of reservoir architecture. The hierarchical architecture holds L levels $\{l_i, i = 1, ..., L\}$ and each level l_i consists of M basic components $\{m_j^{l_i}, j = 1, ..., M\}$, also denoted as motifs. The hierarchical architecture representation scheme constructs the higher level motif using the motifs in the lower level. In other words, the motif of a higher level contains M motifs of the lower level, which can be represented as $m_j^{l_{i+1}} \leftarrow \{m_j^{l_i}, j = 1, ..., M\}$. Each level has several motif types that are used as the coordinates of the motif structure. The type of motif $m_j^{l_i}$ is selected from the coordinate motif types of the level l_i according to the genotype.

Figure 5.19 shows an architecture consisting of $L = 2$ levels in the reservoir layer and each level has $M = 4$ motifs since our computational resource is limited. In level l_1, motif $m_j^{l_1}$ represents a block whose network structure is selected from the four block types, as previously described. After all motif types in level l_1 are confirmed, motifs $\{m_j^{l_1}, j = 1, ..., 4\}$ are aggregated into motif $m_j^{l_2}$. Motif $m_j^{l_2}$, also called sub-reservoir, connects to $\{m_j^{l_1}, j = 1, ..., 4\}$ according to one of the predefined sub-reservoir types. In level l_2, four DAG graphs are regarded as the coordinate sub-reservoir types, as shown on the right side of Fig. 5.19b, which are denoted by H_O, H_A, H_B, and H_E, respectively. Notably, the four sub-reservoir types are chosen from the typical DAG graphs when $M = 4$. If level l_2 contains more than $M = 4$ motifs, the coordinate sub-reservoir types also increase. However, considering the computational burden, $M = 4$ is selected to illustrate the effectiveness of the proposed hierarchical architecture representation scheme. Finally, all sub-reservoirs are aggregated into a whole reservoir without inner pathways. The input of the sub-reservoir is the head block of the DAG graph according to the topological sorting.

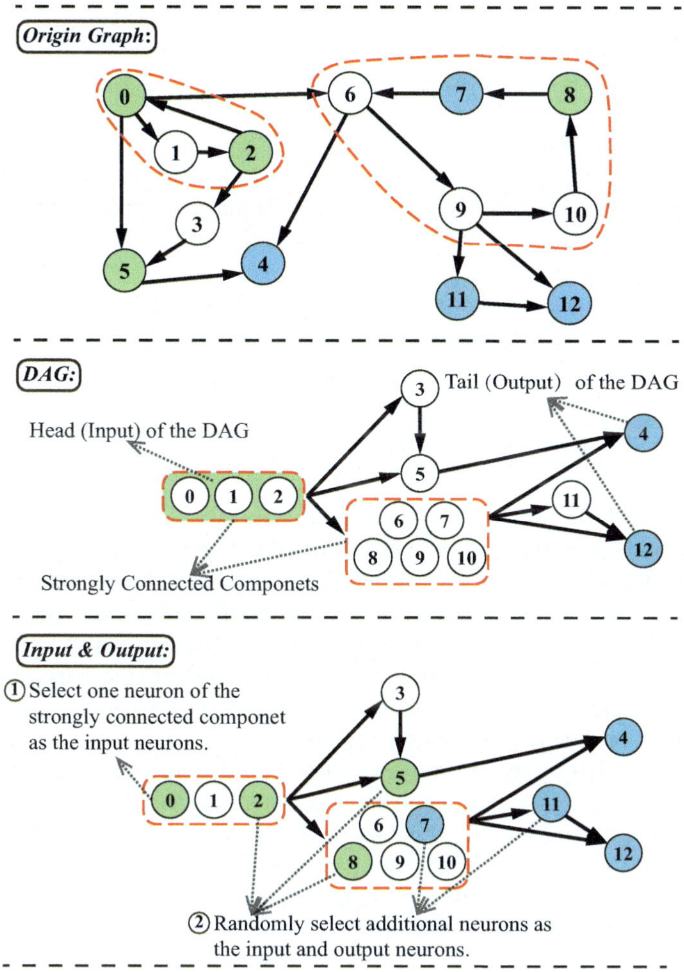

Fig. 5.20 An illustrative example of determining the input and output neurons of the block based on strongly connected components and topological sorting

The pathways between the encoding layer and the reservoir layer actually connect the encoding block to the input blocks of the sub-reservoirs.

Figure 5.19 also shows an example of mapping the genotype to the phenotype based on the hierarchical architecture representation. Motifs $\{m_j^{l_1}, j = 1, ..., 4\}$ are generated according to the genotype |01|11|00|10|, which means the phenotype of the sub-reservoir consists of the scale-free block, the three-layer block, the random block, and the small-world block. In level l_2, each motif $m_j^{l_2}$ contains the same motifs $\{m_j^{l_1}, j = 1, ..., 4\}$. Meanwhile, according to the genotype |00|10|11|00|, the motifs of the sub-reservoirs $\{m_j^{l_2}, j = 1, ..., 4\}$ are connected based on the type of H_O, H_B, H_O, and H_E, respectively.

5.3.5.3 Plasticity Rules

In G-LSM, the intrinsic plasticity rule [27, 447] and the synaptic plasticity rules [396, 435] are adopted to adjust the neural threshold $T_i(t)$ and the synaptic strength w^{ij}, respectively.

Intrinsic plasticity is a homoeostatic learning rule for spiking neurons to adjust the firing state. According to the input spike trains, the intrinsic plasticity rule adjusts the neural threshold $T_i(t)$ following (5.48),

$$\tau_{th}\frac{dT_i(t)}{dt} = -T_i(t) + T_0 + \Delta T \cdot \delta(t - t_i^{(f)}), \tag{5.48}$$

where the τ_{th} is the time constant of the dynamic threshold, T_0 is the basic threshold, and ΔT denotes the threshold increment when the neuron v_i fires.

The crucial contribution of the intrinsic plasticity rule is to prevent the saturation of the firing state. The more neurons fire, the more difficult it is for a post-synaptic neuron to fire the next spike in the decay time τ_{th}. Thus, it maps the input spikes and the output spikes with a nonlinear relationship. Similar to the activation function of conventional neural networks, the intrinsic plasticity rule can be regarded as the spiking activation function of spiking neurons.

Synaptic plasticity plays an important role in learning and cognition in neural networks. Synaptic plasticity rules change the synaptic strength to enhance or inhibit the stimuli injected into the post-synaptic neuron. The most widely used synaptic plasticity rule for SNNs is spike-timing-dependent plasticity (STDP) [396]. The STDP rule modifies the synaptic strength depending on the temporal correlation between presynaptic and post-synaptic spikes. The synaptic strength increases when the presynaptic neuron fires earlier than the post-synaptic neuron; otherwise, the synaptic strength decreases. Similar to STDP, a more stable plasticity rule, denoted by the multiplicative STDP rule, considers not only the temporal correlation but also the upper and lower bounds of synaptic strength [435]. Both the STDP rule and the multiplicative STDP rule represented in (5.49) are adopted in the G-LSM.

$$\Delta w_{ij} = \begin{cases} A_P \cdot (w_{i,j}^{max} - w_{i,j})^{\gamma} \cdot exp\frac{-\Delta t}{\tau_p} \\ -A_D \cdot (w_{i,j} - w_{i,j}^{min})^{\gamma} \cdot exp\frac{-\Delta t}{\tau_p} \end{cases}, \tag{5.49}$$

where $\Delta w_{i,j}$ is the strength increment of $w_{i,j}$, and $w_{i,j}^{max}$ and $w_{i,j}^{min}$ are the upper and lower bound of $w_{i,j}$, respectively. $A_P = 0.01$ and $A_D = 0.0105$ are the incremental coefficient, τ_p is the time constant of the plasticity rule, and γ is the plasticity type parameter. In particular, when $\gamma = 1$, (5.49) represents the STDP rule, while when $\gamma = 0$, (5.49) represents the multiplicative STDP rule. To clearly illustrate the effect

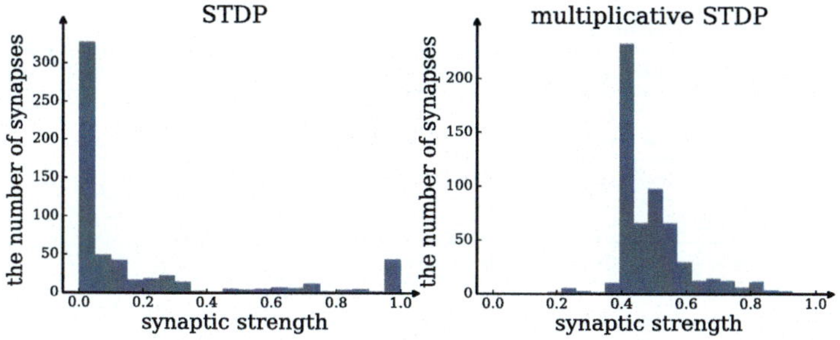

Fig. 5.21 An illustrative example of the synaptic strength distribution after being trained by the STDP rule and the multiplicative STDP rule

of the synaptic plasticity, examples of the synaptic strength distribution of the G-LSM after being trained are shown in Fig. 5.21. These examples show that the synaptic strength is distributed around $w_{i,j}^{max}$ and $w_{i,j}^{min}$ after being trained by the STDP. By contrast, the synaptic strength is distributed around $(w_{i,j}^{max} + w_{i,j}^{min})/2$ after being trained by the multiplicative STDP.

5.3.5.4 Cooperative Co-Evolutionary Algorithm

The Cooperative co-evolutionary algorithm (CCEA) decomposes the original problem into several low-dimensional sub-problems [324]. Each sub-problem is independently optimized with a subpopulation to obtain partial solutions. The individual of the subpopulation is evaluated in cooperation with the other subpopulations. The best partial solutions are combined to be a complete solution of the original problem. The fitness of the complete solution is then utilized to determine the fitness of the corresponding partial solutions [268, 462].

Accordingly, the essential aspects of the CCEA include problem decomposition, collaborator selection, and individual fitness assignment. CCEA decomposes the optimization of the reservoir architecture into three or six subpopulations according to the genotype settings of the G-LSM. The original population and the subpopulations are represented as (5.50)

$$\begin{aligned}\mathbf{X} &= \{\mathbf{x}_i, i \in 1, 2, ..., N_g\} \\ &= \{x_{i,j}, i \in 1, 2, ..., N_g, j \in 1, 2, ..., N_p\}, \\ &= \{x_j, j \in 1, 2, ..., N_p, N_g = 1\}\end{aligned} \tag{5.50}$$

where $\mathbf{x}_i$ denotes the subpopulation and N_g is the number of the subpopulations. x_j and $x_{i,j}$ denote the individual of the original population and subpopulation, respectively, and N_p is the number of the individuals. The subpopulation is evolved by a genetic algorithm. The best individual b_i of each subpopulation $\mathbf{x}_i$ is selected to form the complete solution $\mathbf{b} = \{b_i, i \in 1, 2, ..., N_g\}$ and the fitness

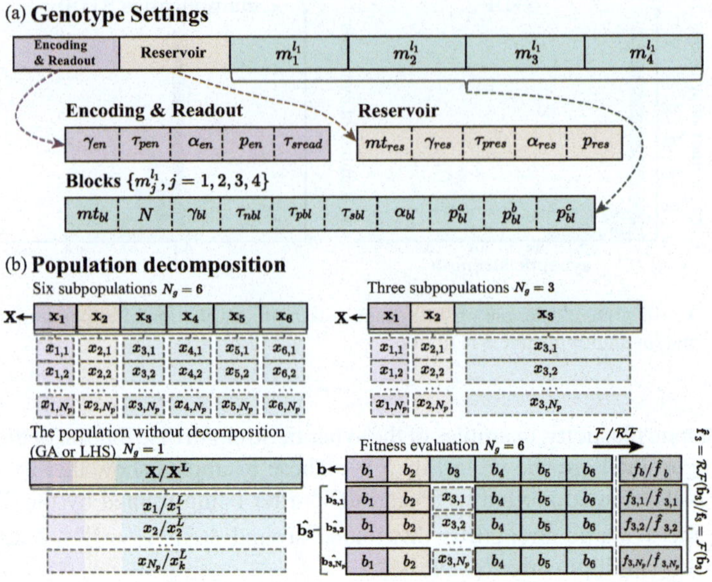

Fig. 5.22 An illustration of the genotype settings and the population decomposition for evolving G-LSM

of $\mathbf{b}$ is $f_b = \mathscr{F}(\mathbf{b})$, which is short for $f_b = \mathscr{F}(\mathbf{b}, \mathscr{D})$. The fitness of the individual $x_{i,j}$ is estimated by replacing the corresponding segment of complete solution $f_{i,j} = \mathscr{F}(x_{i,j}) = \mathscr{F}(\hat{\mathbf{b}}_{i,j})$, where $\hat{\mathbf{b}}_{i,j} = \{b_1, b_2, ..., x_{i,j}, ..., b_{N_g}\}$. The fitness of all individuals of subpopulation $\mathbf{x}_i$ is expressed as $\mathbf{f}_i = \mathscr{F}(\mathbf{x}_i) = \mathscr{F}(\hat{\mathbf{b}}_i)$.

The problem decomposition is based on the genotype settings, which is distinguished by different colors in Fig. 5.22b with $N_g = 6$ and $N_g = 3$. The motifs $\{m_j^{l_1}, j = 1, 2, 3, 4\}$ are regarded as one subpopulation when $N_g = 3$. $N_g = 1$ means the population without decomposition, and the CCEA is therefore equivalent to the genetic algorithm. An example of the fitness evaluation for the subpopulation $\mathbf{x}_3$ ($N_g = 6$) is also shown in Fig. 5.22. In $\hat{\mathbf{b}}_3 = \{\hat{\mathbf{b}}_{3,j}, j = 1, ..., N_p\}$, b_3 is replaced by $\{x_{3,j}, j = 1, ..., N_p\}$. Therefore, the fitness of $\mathbf{x}_3$ is $\mathbf{f}_3 = \mathscr{F}(\hat{\mathbf{b}}_3)$.

5.3.5.5 Random Forest as a Surrogate

To reduce the computation consumption, the individual predictor estimates the fitness of the individuals $\mathbf{x}_i$ by a random forest surrogate, which is represented as $\hat{\mathbf{f}}_i = \mathscr{RF}(\hat{\mathbf{b}}_i)$ [195, 409]. The random forest is an ensemble machine learning method by training numerous decision trees [163]. Each decision tree is trained with the data randomly sampled from the whole data set. The mean and variance of the decision tree outputs are used as the predicted value and the uncertainty of the random forest output. The random forest performs better than other machine learning methods on discrete data. Since the genotype of the G-LSM is hybrid coded, the random forest is well suited for establishing the surrogate of the G-LSM.

The surrogate-assisted CCEA uses $D = 100$ classification and regression trees (CARTs) [264] in the random forest regression method to create the surrogate. The surrogate employs the lower confidence bound (LCB) [376] described in (5.51) as the prediction value of $\hat{\mathbf{b}}_{i,j}$.

$$\mathscr{RF}(\hat{\mathbf{b}}_{i,j}) = \mu(\hat{\mathbf{b}}_{i,j}) - \rho \cdot \sigma(\hat{\mathbf{b}}_{i,j}), \tag{5.51}$$

where $\rho = 2.576$, and $\mu(\hat{\mathbf{b}}_{i,j})$ and $\sigma(\hat{\mathbf{b}}_{i,j})$ are the mean and variance of the random forest output for $\hat{\mathbf{b}}_{i,j}$, respectively.

The details of the surrogate-assisted CCEA are given in Algorithm 3. The algorithm firstly samples k coordinate solutions $\mathbf{X}^L = \{x_j^L, j \in 1, 2, ..., k\}$ by Latin hypercube sampling to create the random forest surrogate. During the evaluation process, $\mathbf{x}_i^O$ denotes the offspring individuals of the subpopulation. The surrogate-assisted CCEA evaluates the most promising N_s offspring individuals $\mathbf{x}_i^S$ by the original function $\mathscr{F}$ in every η generation. Then the surrogate $\mathscr{RF}$ is updated according to the newly evaluated results, which can increase the prediction accuracy of $\mathscr{RF}$. The best fitness f_b and individual $\mathbf{b}$ are then chosen from $\{f_b, \mathbf{f}_i^S\}$ and $\{\mathscr{F}(\mathbf{b}), \mathscr{F}(\hat{\mathbf{b}}_i^S)\}$, respectively. The corresponding recombination and mutation are employed depending on the coding type. The subpopulation of the next generation is selected from combining the parent and the offspring subpopulation $\{\mathbf{x}_i, \mathbf{x}_i^O\}$ using tournament selection.

Empirical results in [477] demonstrate that the surrogate-assisted CCEA exhibits the best performance on three widely used data sets, the KTH action recognition benchmark [367], the HAR action recognition benchmark [338], and the dynamic vision sensor (DVS) [120]-based neuromorphic benchmark (N-MINIST) [315].

5.3.6 Evolving Local Synaptic Plasticity in Echo State Networks

In this section, we will introduce a method for evolving local plasticity rules that allow different neurons to use different types of plasticity rules as well as different parameters, which is achieved by optimizing the parameters of the local plasticity rules using the evolution strategy with covariance matrix adaptation. It will be shown that evolving neural plasticity will result in a synergistic learning of different plasticity rules, which plays an important role in improving the learning performance. Meanwhile, evidence will be given showing that the evolved local plasticity rules can effectively alleviate synaptic interference in learning the structure in sensory inputs.

Algorithm 3: Surrogate-Assisted CCEA

Input: The population size N_p, the number of evaluated individuals N_s, the evaluation interval η, and the number of coordinate solutions for Latin hypercube sampling k.
Output: The best fitness f_b, the best solution $\mathbf{b}$.

Latin hypercube sampling $\mathbf{X}^L = \{x_j^L, j \in 1, 2, ..., k\}$;
Evaluate $\mathbf{f}^L = \mathscr{F}(\mathbf{X}^L)$;
Create the surrogate $\mathscr{RF}$ based on $(\mathbf{f}^L, \mathbf{X}^L)$;
Update $f_b \leftarrow \min \mathbf{f}^L$, $\mathbf{b} \leftarrow \arg\min \mathscr{F}(\mathbf{X}^L)$;
for $i \leftarrow 1$ **to** N_g **do**
    Generate an initial subpopulation $\mathbf{x}_i$;
    Estimate $\hat{\mathbf{f}}_i = \mathscr{RF}(\hat{\mathbf{b}}_i)$;
end
$loop = 1$;
while *The termination is not satisfied* **do**
    $loop = loop + 1$;
    for $i \leftarrow 1$ **to** N_g **do**
        Generate an offspring sub-population $\mathbf{x}_i^O$ by the recombination and the mutation of $\mathbf{x}_i$;
        Estimate $\hat{\mathbf{f}}_i^O = \mathscr{RF}(\hat{\mathbf{b}}_i^O)$;
        if $loop == \eta$ **then**
            Select the most promising N_s offspring $\mathbf{x}_i^S$ from $\mathbf{x}_i^O$, where $\hat{\mathbf{b}}_i^S = \arg\min \mathscr{RF}(\hat{\mathbf{b}}_i^O)$;
            Evaluate $\mathbf{f}_i^S = \mathscr{F}(\hat{\mathbf{b}}_i^S)$;
            Update the surrogate $\mathscr{RF}$ based on $(\mathbf{f}_i^S, \hat{\mathbf{b}}_i^S)$;
            Update $f_b \leftarrow \arg\min \{f_b, \mathbf{f}_i^S\}$, $\mathbf{b} \leftarrow \arg\min \{\mathscr{F}(\mathbf{b}), \mathscr{F}(\hat{\mathbf{b}}_i^S)\}$;
            Replace the corresponding the evaluation values of $\mathbf{x}_i^S$ with $\hat{\mathbf{f}}_i^S$ in $\hat{\mathbf{f}}_i^O$;
            $loop = 0$;
        end
        Rank and select best N_p offspring based on $\arg\min\{\hat{\mathbf{f}}_i^O, \hat{\mathbf{f}}_i\}$ from $\{\mathbf{x}_i, \mathbf{x}_i^O\}$ as the sub-population of the next generation $\mathbf{x}_i$;
    end
end
return f_b, $\mathbf{b}$;

5.3.6.1 Echo State Networks

Echo State Networks (ESNs) can be seen as a particular type of recurrent neural networks (RNNs). The original ESN belongs to one class of reservoir computing models, similar to liquid state machines (LSMs). The main difference to LSMs lies in the fact that all neurons in the reservoir are analog neurons instead of spiking neurons. An architecture of the canonical ESN model is shown in Fig. 5.23.

The number of neurons for the input layer, reservoir, and output layer is denoted as K, M, and L, respectively. The weights from the input to the reservoir neurons, between the internal neurons in the reservoir, and from the reservoir neurons to the output layer are denoted as W_{in}, W_{res}, and W_{out}, respectively. The activation states of

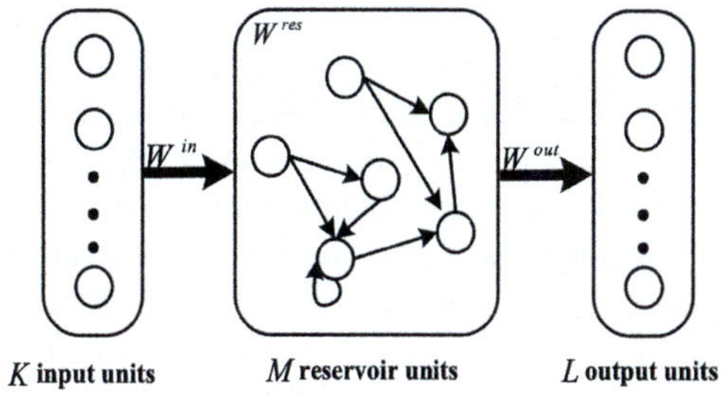

Fig. 5.23 The architecture of a canonical echo state network model

the neurons in the reservoir $x(t+1)$ and the output layer $y(t+1)$ at time $t+1$ are updated using the following equations [186]:

$$\mathbf{x}(t+1) = f(W_{\text{in}}\mathbf{u}(t+1) + W_{\text{res}}\mathbf{x}(t)), \tag{5.52}$$

$$\mathbf{y}(t+1) = W_{\text{out}}\mathbf{x}(t+1), \tag{5.53}$$

where $\mathbf{u}(t+1)$ represents the input signal at time $t+1$, and $f(\cdot)$ denotes the activation function of the neurons in the reservoir, which typically is the sigmoid function.

In the canonical ESN model, W_{in} and W_{res} are fixed after random initialization, while W_{out} is trained by using a supervised learning algorithm. The goal of the training process is to minimize the difference between the desired output ($\mathbf{y}^d$) and the actual output of the ESN ($\mathbf{y}$). Specifically, for prediction or regression problems, we minimize the normalized root mean square error (NRMSE) to train the readout weights (W_{out}), which is defined as follows:

$$L_{\text{NRMSE}} = \sqrt{\frac{\sum_{t+1}^{T} \parallel \mathbf{y}^d - \mathbf{y} \parallel}{T\sigma^2}}, \tag{5.54}$$

where σ^2 represents the variance of the desired outputs. The readout weights W_{out} are computed by using least square estimation method:

$$W_{\text{out}} = (X^T X)^{-1} X Y. \tag{5.55}$$

In training W_{out}, the activation states of the reservoir neurons and the desired outputs at each time instant will be stored in X and Y over time $t = 1, 2, \ldots, T$, respectively [187]. For classification problems, the classification accuracy can be defined as follows if the softmax function is used [265]:

$$p_i = \frac{\exp y^i(t)}{\sum_{i=1}^{L} \exp(y^i(t))}, \tag{5.56}$$

where $y^i(t)$ represents the actual output of the i-th output layer neuron at time t, L is the number of the classes (output neurons), and p^i represents the probability in which $y^i(t)$ belongs to i-th class. The output node with the highest probability and the corresponding class is used as the actual class label. The classification accuracy can be calculated by comparing the desired class label and the actual class label.

There are a few parameters in ESNs that are particularly important for their performance, including the number of neurons in the reservoir, the connection density, and the input scaling. The number of neurons in the reservoir determines the complexity of the ESN, and therefore, the optimal complexity heavily depends on the complexity of the task. In general, the memory capacity of the ESN enhances as the number of neurons increases, which may, however, also increase the risk of overfitting the data. The connection density represents the percentage of the neurons that are interconnected within the reservoir. In general, the connection density is chosen within the range 0.01 and 0.2 [395]. Similar to LSMs, particular structures such as small-world or scale-free networks can be of interest. Input scaling can be of great importance for the performance of ESNs, and different types of data samples need to be properly scaled before being inputted into the network.

Once the parameters of the ESN, together with its neuronal dynamics, are determined, one generic performance indicator, called the spectral radius, can be used to measure its potential performance on a given task. Spectral radius is the largest absolute eigenvalue of the connection weight matrix W_{res} of the neurons in the reservoir. In order to ensure that the ESN demonstrates echo state properties, the spectral radius should be less than one [185].

5.3.6.2 Global Plasticity Rules

Plasticity rules inspired from findings in computational neuroscience have widely been employed to adjust the weights within the reservoir according to the activation states stimulated by the input. Most of the existing synaptic plasticity mechanisms can be seen as the extension and development of the Hebbian plasticity rule. The Hebbian rule can be described by the following expression:

$$\Delta W_{\text{res}}^{ji} = \eta x^i(t) x^j(t), \tag{5.57}$$

where $\Delta W_{\text{res}}^{ji}$ is the change of the synaptic weight between the pre-(x^i) and post-(x^j) synaptic neuron at time t, and η is the learning rate that is a small positive constant in general [16]. Because the Hebbian rule only describes how the synaptic weights are increased, it does not give a mathematical description of how the synaptic weight is decreased. Therefore, the Hebbian learning rule is unstable in nature. Oja [312] proposed a modification of the original Hebbian rule to solve the stability problem by normalizing all weights connected to post-(x^j) synaptic neuron:

$$\sqrt{\sum_{i=1}^{n} (W_{\text{res}}(t)} = 1, \tag{5.58}$$

where W_{res} represents the synaptic weight between the pre-(x^i) and post-(x^j) synaptic neuron at time t. The normalization of the synapses can be normalized in the following form:

$$W_{\text{res}}^{ji}(t+1) = \frac{W_{\text{res}}^{ji} + \eta x^i(t)x^j(t)}{\sqrt{\sum_{i=1}^{n}[W_{\text{res}}^{ji} + \eta x^i(t)x^j(t)]}}, \tag{5.59}$$

where the learning rate η is usually a small positive constant. When η goes to 0, the above equation can be expanded as a power series [312]. The learning rule with a weight depression term can be derived, which is called the Oja learning rule [60, 312]:

$$W_{\text{res}}^{ji}(t+1) = W_{\text{res}}^{ji}(t) + \Delta W_{\text{res}}^{ji}(t), \tag{5.60}$$

$$W_{\text{res}}^{ji}(t) = \eta x^j(t)[x(t) - x^j(t)W_{\text{res}}^{ji}], \tag{5.61}$$

where $W_{\text{res}}^{ji}(t)$ is the change of the synaptic weight between the pre-(x^i) and post-(x^j) synaptic neuron at time t. $x^j(t)W_{\text{res}}^{ji}(t)$ is the depression term.

The anti-Oja learning rule is an inverted form of the Oja learning rule. The only difference between them is the update of the synaptic weights [16]. That is, the learning rate of the anti-Oja rule is a negative constant. According to the anti-Oja rule, the weight is updated in the following form:

$$W_{\text{res}}^{ji}(t+1) = W_{\text{res}}^{ji}(t) - \Delta W_{\text{res}}^{ji}(t). \tag{5.62}$$

It has been suggested that with the weights between the neurons in the reservoir being trained using the synaptic plasticity rules, the network will have stronger homeostasis [436]. In other words, the state of the reservoir neurons can be better stabilized if the weights are tuned with the plasticity rules compared with randomly initialized weights.

It has been recognized that using global plasticity rules is ineffective in learning the structure in the sensory input due to interference, as discussed in Sect. 5.3.2, and they may not be able to enhance the learning performance in tuning the weights in the reservoir. In addition, in most existing work, only one single type of plasticity rule is used. Thus, the global plasticity rules fail to simultaneously capture the needs of correlation and decorrelation in learning complex tasks.

5.3.6.3 Evolving Local Synaptic Plasticity Rules in Echo State Networks

To evolve local plasticity rules from a single neuron aspect for plasticity modeling, we introduce in this section the algorithm presented in [449], which aims to design local synaptic plasticity rules with different learning rates or completely different plasticity rules to tune the connections between different neurons within the reservoir. Here, a plasticity rule is called local if the parameter in the plasticity rule varies from neuron to neuron or if different types of plasticity rules are used.

The overall evolutionary process consists of the following main steps:

- Input and scale the data through the input layer;
- Pre-train the connection weights within the reservoir by using local plasticity rules;
- Train the readout weights by the least squared method. Two types of local plasticity rules are considered, one is the local Oja learning rule and the other is the local anti-Oja learning rule. The local Oja learning rule is in the following form:

$$\Delta W_{\text{res}}^{ji} = \eta^j x^j(t)[x^i(t) - x^j(t)W_{\text{res}}^{ji}], \tag{5.63}$$

where W_{res}^{ji} is the change of the synaptic weight between the pre-(x^i) and post-synaptic (x^j) neuron at time t, and η^j is the learning rate of the post-synaptic neuron (x^j).

The local Oja learning rule used here is the same as the original Oja learning rule. Following the local Oja learning rule, we can easily create the local anti-Oja learning rule. Similarly, the only difference between them is also in synaptic weight update, i.e., the synaptic weight update of the local anti-Oja learning rule is negative.

It can be seen from the above equation that the only difference between the local and the global plasticity rules is that the local plasticity rule applies a different learning rate η^j to different post-synaptic neurons (x^j) in the reservoir. The CMA-ES (refer to Sect. 1.1.2) is used to evolve the parameters of the Oja rules for each neuron so that different neurons are allowed to use different parameters. Note that the main difference between the Oja rule and the anti-Oja rule is the update of the synaptic weights. In other words, the learning rate of the Oja rule is a positive constant, while the anti-Oja rule is a negative constant. Thus, we can investigate whether the Oja rule or anti-Oja rule can more effectively enhance the learning performance.

5.3.7 *Evolving Local Intrinsic Plasticity in Echo State Networks*

5.3.7.1 Intrinsic Plasticity Rule

Most present research of neuroscience focuses primarily on synaptic plasticity, which is able to learn the structural information of input signal based on the activities stimulated by the neurons. In addition, biological experiments show that a single neuron also has the ability to alter its intrinsic excitability to match the received synaptic input by changing voltage gated ion channels [473]. This neural plasticity mechanism is known as intrinsic plasticity (IP), which has also been considered in [477] for spiking neurons in G-LSM, as described in Sect. 5.3.5.3. To be consistent with the descriptions of the ESNs, we provide further discussions and a slightly different formulation of the IP rule below.

IP is a common physiological phenomenon observed in many species and brain regions [426]. IP plays an important role in maintaining the activity level of individual neurons within a certain range of homeostatic control while maximizing information transmission to downstream targets [427]. The homeostatic characteristic can effectively maintain a stable metabolic cost of the system. Maximizing information transmission can maximize the mutual information between neuron's input and output, so that the neuron's response curve can match the synaptic input distribution. In other words, the IP rule can efficiently represent the synapse input in the condition of a stable metabolic cost. However, neither Triesch's rule [427] nor the information maximization rule [407] could efficiently resolve the contradiction between minimizing the firing rate and maximizing the sensitivity in the IP model. In further study, Li pointed out that the Weibull distribution is more appropriate in describing the real activity distribution of a single biological neuron [253].

Findings in neurobiology experiments indicate that responses of neurons to natural scenes are approximately exponential spike-count distributions in both the primary visual cortex and the inferior temporal cortex [20]. Note that exponential distributions have the highest information entropy among all non-negative distributions while keeping the same mean [426]. However, the adjustment process and the underlying mechanism of IP in neural circuits still remain elusive. In addition to experimental research, computational modeling of the IP offers a novel option to better understand these neural plasticity mechanisms. Inspired by neuroscience and information theory, Triesch [426] proposed an IP learning rule built by computational modeling. It is assumed that the output distribution of the single sigmoid neuron is an exponential distribution with a fixed mean, and the IP rule can be derived by minimizing the Kullback-Leiber divergence between the actual output distribution and the desired exponential distribution of neurons.

The goal of the IP rule is to obtain an approximately desired distribution of the firing rate level while obeying constraints on its metabolic costs. To investigate the impacts of IP mechanism on functions and dynamics of neural circuits [253], Li's rule is chosen as an adaptation rule for the following experiment and theoretical analysis. The desired output distribution of Li's rule is also based on the non-negative distribution; therefore, the sigmoid function is employed as the activation function of the neurons:

$$y = \frac{1}{1 + \exp^{(-ax+b)}}, \tag{5.64}$$

where a and b are two parameters of the function. The probability density function of the Weibull distribution is described as follows [452]:

$$f(x) = \begin{cases} \frac{\beta}{\alpha}(\frac{x}{\alpha})^{\beta-1}\exp(-\frac{x}{\alpha})^{\beta}, & \text{if } x \geq 0, \\ 0 & \text{if } x < 0, \end{cases} \tag{5.65}$$

where $\alpha > 0$ and $\beta > 0$, respectively, represent the scale parameter and the shape parameter of the Weibull distribution. Shape parameter β is the most important parameter that determines the basic shape of the probability density function. Scale

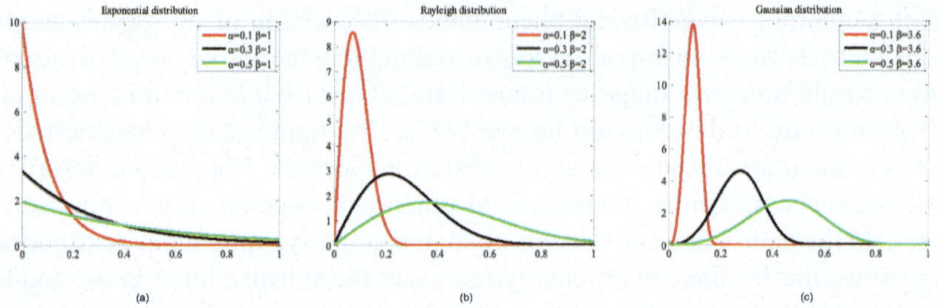

Fig. 5.24 The shape of the Weibull function when α and β change, which is **a** the exponential distribution, **b** Rayleigh distribution, and **c** Gaussian distribution, respectively, when β is set to 1, 2, and 3.6

parameter α scales the curve, but does not affect the shape of the distribution. In addition, other types of distribution can be represented or approximated by changing the shape parameter of the Weibull distribution, for example, the exponential distribution, the Rayleigh distribution, and the Gaussian distribution. The shape of the function is illustrated in Fig. 5.24 when the two parameters change.

The derivation of the IP rule in [253] is similar to the original IP rule, which is also based on the KL divergence. For a sigmoid neuron model, the IP rule is derived by minimizing the KL divergence between the actual output distribution and the Weibull distribution. The gain a and bias b of the sigmoidal neuron are updated by the stochastic gradient descent rule at each time-step which is as follows:

$$\Delta a^i(t+1) = \frac{\eta}{a^i(t)} + \eta \Delta b^i(t) x^i_{\text{in}}(t), \tag{5.66}$$

$$\Delta b^i(t+1) = \eta[\beta(1 - x^i(t)) - x^i(t) - \frac{\beta}{\alpha^\beta}(x^i(t))^\beta(1 - x^i(t))], \tag{5.67}$$

$$a^i(t+1) = a^i(t) + \Delta a^i(t+1), \tag{5.68}$$

$$b^i(t+1) = b^i(t) + \Delta b^i(t+1), \tag{5.69}$$

where $\Delta a^i(t+1)$ and $\Delta b^i(t+1)$ are the change of the gain and the bias of the i-th reservoir neuron's activation function at the time t, respectively. η is the learning rate of the IP rule. $\alpha > 0$ and $\beta > 0$ represent the shape parameter and the scale parameter of the Weibull distribution, respectively.

5.3.7.2 Local Intrinsic Rules

Existing IP rules for altering the intrinsic excitability of the reservoir neurons are mainly applied in the pre-training process of the ESN to achieve global plasticity optimization of the activation function within the reservoir [365]. The parameters of the IP rules such as the learning rate, the scale parameter, and the shape parameter in the whole reservoir neurons are the same in the global plasticity optimization.

Note, however, that the output distributions of real biological neurons may have different forms in different brain regions and species [253]. Therefore, existing IP rules apparently lack a certain degree of flexibility.

To further improve the flexibility of the IP rule and the learning performance of the ESN, we introduce a local IP rule, where the learning rates, scale parameters, and shape parameters for different neurons can be different [448]. The local IP learning rule is expressed as follows:

$$\Delta a^i(t+1) = \frac{\eta^i}{a^i(t)} + \eta \Delta b^i(t) x^i_{\text{in}}(t), \tag{5.70}$$

$$\Delta b^i(t+1) = \eta^i[\beta^i(1 - x^i(t)) - x^i(t) - \frac{\beta^i}{\alpha^{i,\beta^i}}(x^i(t))^{\beta^i}(1 - x^i(t))]. \tag{5.71}$$

From the above equations, we can see that the only difference between the global and local intrinsic plasticity rules is that the parameters in the local rule become individualized for each neuron, which are tuned by the evolutionary algorithm.

5.3.7.3 Empirical Evidence

Empirical studies in [448, 449] on several regression and classification problems show that the ESNs with local plasticity rules outperform their counterparts with global plasticity rules. In the following, we give illustrative examples of the necessity and rationality of having individual plasticity rules for the neurons in the reservoir on the sunspot time series prediction problem. For this problem, the smoothed monthly mean sunspot numbers, which are one of the typical time series, are usually used as the benchmark prediction problems. This data set includes 3174 samples of sunspot activity numbers from 1750 to 2013. The training, validation, and test data sets contain 1600, 500, and 1074 data, respectively, and the first 200 training samples are only used to washout historical data; $y(t-20)$, $y(t-10)$, and $y(t)$ are used to forecast the future data $y(t+1)$. That is, the number of the input layer neurons and output layer neurons is set to 3 and 1.

The convergence profiles of the parameters in the local plasticity rules are given in Figs. 5.25 and 5.26, respectively, for the local synaptic rules described in Sect. 5.3.6 and the local intrinsic plasticity rules presented in Sect. 5.3.7. These results clearly show that the parameters in the local plasticity rules converge to different values, indicating the necessity of having individual plasticity values for each neuron. Naturally, results in [448, 449] also convincingly show that local plasticity rules are able to enhance the learning performance of the ESNs compared to global rules.

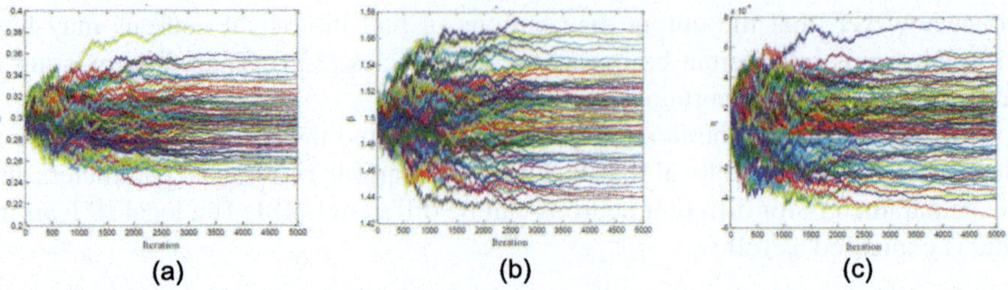

Fig. 5.25 The convergence profiles of the parameters of the local intrinsic plasticity rule on the sunspot problem

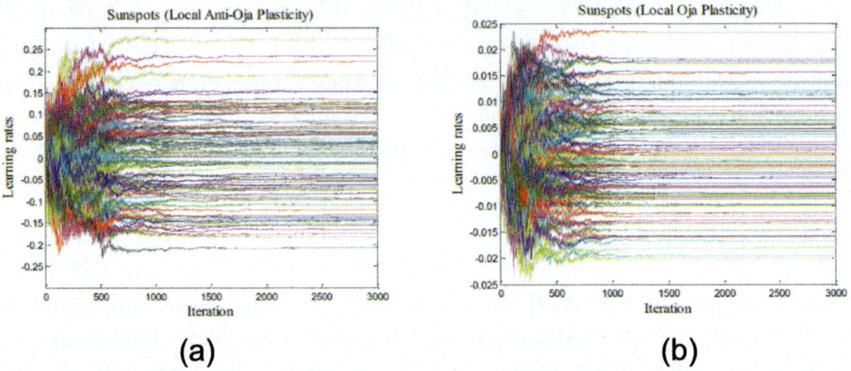

Fig. 5.26 The convergence profiles of the parameters of the local Oja and anti-Oja rules

5.4 Summary

This chapter presents different approaches to evolving neural development governed by gene neural networks. A computational model for gene-driven development of primitive nervous systems using the hydra as the model animal indicates that it is possible to evolve a gene regulatory network that produces the correct gene expression sequence required during the early neural development. Then, we analyze the interference in learning sensory input structure using plasticity rules, revealing that catastrophic forgetting occurs if a global plasticity rule is applied to all neurons in the neural network model, in particular reservoir-based computing models including echo state networks and liquid state machines. Then, we present two algorithms to generate local neural plasticity rules whose parameters are regulated by a gene regulatory network whose states are controlled by the activity of the individual spiking neurons in the reservoir of a liquid state machine. Finally, two plasticity models where the type and/or parameters are evolved using an evolutionary algorithm in an echo state network are introduced. Empirical results confirm that gene-regulated neural plasticity evolving parameters or the type of the plasticity rules will result in local plasticity rules, leading to better learning performance.

Several questions remain open. For example, it is of interest to understand the role neural plasticity has played in the modularization of the brain, and if it is possible to use local plasticity rules to achieve modularity in a computational neural system such as the reservoir of an echo state network or a liquid state machine. Second, it is unclear if neural plasticity has played a role in memory consolidation, and if yes, in which way neural plasticity can resolve catastrophic forgetting effectively and efficiently. Finally, it will be exciting if it is possible to fully automatically develop highly complex deep neural networks, such as in a similar complex to the human brain. This will be the ultimate goal of the evolutionary developmental approach to artificial general intelligence.

Chapter 6
Computational Brain-Body Co-Evolution

6.1 Body Is Key to Understanding Human Intelligence

Understanding how nature has shaped intelligent systems like the human brain is of paramount significance for creating artificial general intelligence. Although deep learning is in principle based on artificial neural networks inspired by information processing in the human brain, deep neural networks still lack several fundamental properties of human intelligence, such as learning many tasks based on small data, reasoning, planning, anticipation, and learning to learn. Considering the fact that natural intelligence has been shaped by three main adaptation mechanisms in nature, namely evolution, development, and learning, understanding the roles of both evolution and development becomes essential. Over the past decade, epigenetic robotics has enjoyed great success and achieved significant progress in understanding mental (cognitive) development. Despite its significant success, developmental robotics without looking into the influence of the body can be challenged both from ontogenetic and phylogenetic points of view.

First, the development of biological organisms consists of both physical and mental development. Thus, developmental robotics that concentrates on mental development is incomplete. Ontogenetically, mental development is based on and closely coupled with the physical development of an organism, including the development of both the body plan and the nervous systems. For example, findings in neuroscience reveal that early development of nervous systems is also considerably driven by neural activity [398], suggesting that activity-dependent and activity-independent development of neural networks cannot be separated. Morphogenetic robotics [198, 200], inspired by biological morphogenesis, was proposed that deals with the physical development of robotic systems, thus complementing epigenetic robotics to provide a full picture of developmental robotics.

Second, biological evidence suggests that autonomous mental development is driven by intrinsic motivational systems [85] among others. Such findings have triggered one key research area in epigenetic robotics that studies the role of intrinsic motivation systems in mental development [318]. Typically, an intrinsic motivation

Y. Jin, *Computational Evolution of Neural and Morphological Development*, Natural Computing Series, https://doi.org/10.1007/978-981-99-1854-6_6

system is predefined in the robotic system. However, autonomous mental development in the living system was not hard-wired from scratch; rather, it has been gradually shaped by a brain-body co-evolution embedded in a changing environment. Thus, it is indispensable to integrate mental and physical development in understanding the evolutionary emergence of natural intelligence. In this chapter, we discuss ideas of co-evolving the neural network controller (brain) together with the evolution of the body, revealing a close coupling between the body and brain in the evolution and development of computational systems.

6.2 Emerged Functional Bilateral Symmetry in Motor Configuration

6.2.1 Motivation

There are major transitions in the evolution of nervous systems, and understanding the underlying mechanisms of these transitions is of great interest for understanding the organization principles in the evolution of nervous systems. One major transition, as discussed in Sect. 5.1.1, is the divergence into a diffused structure (also known as radially symmetric) and a bilaterally symmetric structure. Actually, bilaterally symmetric structures are often seen in most higher organisms, and they can be partitioned into both dorsal and ventral halves. By contrast, most primitive organisms are radially symmetric, and it is believed that the bilateral properties of higher organisms could have evolved from such radiata [281]. In addition, it is assumed that the nervous systems evolved together with their body plans. For example, the jellyfish, which is a radial organism, has a diffused nervous structure, while the flatworm, a bilateral organism, has a bilaterally symmetric nervous system. It should be pointed out that such divergence process and their coupling with the evolution of the body plan have not been fully understood.

Some attempts have been made to understand the emergence of the bilaterally symmetric nervous structure and its coupling with the emergence of a bilaterally symmetric body plan. In [210], a simple computational model simulating the basic swimming behavior of the flatworm has been proposed in a relatively simple yet biologically plausible environment. Although the body plan is not rigorously evolved, the proposed model allows evolving the configuration of the motor neurons, thereby simulating the evolution of the body plan. This is based on the assumption that if all muscles around the model organism are evolved to play a part in movement, the muscle configuration can be regarded as radially symmetric. By contrast, the muscle configuration can be regarded as being bilaterally symmetric if only opposite muscles that are on the dorsal or ventral, or on the left or right parts of an animal are evolved for movement. Note that these symmetric properties are not predefined, but rather emerged from an evolutionary process.

Similar work investigating the coupling between the morphology and neural controller can be divided into two general categories. The first category is concerned with modeling two-dimensional models of undulating organisms for replicating undulating locomotive behaviors (e.g., swimming) and neural controllers that can lead to such behaviors [99, 100, 177, 476]. These studies focus on understanding locomotion from a neuroscientific perspective. The second category of research along this line is concentrated on the evolutionary emergence of such behaviors. For example, one early work was reported in [388, 389], which has been followed up by many researchers [97, 416]. These models are usually three-dimensional and implemented with the help of powerful graphics libraries so that the required visualizations and physical embodiment can be realized.

It has been found in [40] that animats having a bilaterally symmetric body plan can generate more locomotive behaviors more efficiently. However, the neural controller is embedded into the model agent's morphology and only the connectivity of the actual morphology of the individual network could be changed. Note that encoding a network at the level of individual neuron positioning is essential to observe potential tendencies for certain types of neurons to aggregate, thereby potentially demonstrating the type of the central nervous system. In [91], an evolutionary developmental model of neurogenesis is developed to control the directional movements of a five-limbed starfish-like animat. The model presented here, however, does not consider the spatial positioning of the individual neurons either, focusing only on neurogenesis as a combined evolutionary developmental process.

In comparison to the above models, the model to be presented in this section [210] is interested in how central pattern-generating dynamics can yield directional movement. In addition, both neuroscientific principles, i.e., the types of network architectures that can be evolved and the type of movement behaviors that this evolution can shape are to be investigated. By evolving the individual neuron positions, the model aims to find out how a neural controller can become architecturally coupled with the symmetry of the body plan and the resulting behavior that can be brought about due to this coupling. Since the model is able to evolve the individual neuron positions, it becomes easier to examine the effect functional symmetry has on the self-organization of interconnected neurons.

In the following, the overall model, including the body plan, the neural controller as well as the environment the animat is situated in will be presented in detail.

6.2.2 A Physical Model for Undulatory Locomotion

Undulatory locomotion is a type of locomotion of bilaterally symmetric creatures requiring directional movement, in particular locomotive movement, e.g., an eel, a snake, or a fish. Detailed physics underlying locomotive movement can be found in [132]. In computational biology, the body of locomotive animals can be modeled by a spring-mass-damper system to simulate the mechanics [100, 167]. In addition, a friction model can be incorporated so that the animate can move within the simulated physical world. Finally, a control mechanism, usually a neural controller (the brain), is also embedded in the body to eventually achieve undulatory movements. In most existing models, continuous time recurrent neural networks (CTRNNs) [25, 26, 327] are employed in that they are able to serve as a central pattern generator (CPG) required for undulatory movements.

Many neural network-based CPGs for accomplishing undulatory locomotion are handcrafted [99, 100], based on findings in neurophysiology, which are used for simulating lamprey-like animats or leech-like swimming [476]. In addition, evolutionary algorithms are widely used to automatically design the parameters in a predefined neural network architecture [177]. A biologically more plausible model in [375] replicates some of the biological movement data observed from the *Anguilla anguilla* eel. In the model, the eel-like animat navigates by incorporating sensory inputs from the head of the animat. A similar work in [24] presents a model of an undulating lamprey-like animat that can follow a moving object. By contrast, in the physical model to be presented below [212], no neural architecture will be predefined and the connectivity will be evolved from scratch. The model consists of three main components, an animat body, a neural controller, and a simulated environment.

6.2.2.1 Body of the Animat

The body is also modeled by a mass-spring-damper system, a segment of which is illustrated in Fig. 6.1. In the figure, not all springs have been presented, since in reality, each block in the animat has a 'crane-like' structure of springs to prevent it from collapsing in on itself. Since the animat is three-dimensional, it is possible for it to undulate in multiple directions and/or demonstrate other types of movement depending on the output neurons of the neural network model. The equations governing the springs apply Hooke's Law with dampening dynamics. Given a spring with mass points p_1 and p_2 on either end, it is compressed by forcing p_1 toward p_2 and vice versa. Using p_1 as an example, the force exerted upon it by the internal dynamics of the spring is computed as follows:

$$F_{p_1} = -rV_{p_1} + kd, \tag{6.1}$$

where r is a dampening factor, V_{p_1} is the velocity of p_1, k is a spring constant defining spring torque, and d is the displacement of the spring from the resting length. A change in the mass point's velocity, V_{p_1}, is governed by a change in its acceleration, A_{p_1}:

$$A_{p_1}(t + \Delta t) = A_{p_1}(t) + \frac{F_{p_1} + F_{p_1}^E + F_{p_1}^W}{m_{p_1}}, \tag{6.2}$$

$$V_{p_1}(t + \Delta t) = V_{p_1}(t) + A_{p_1}(t + \Delta t) \cdot dt, \tag{6.3}$$

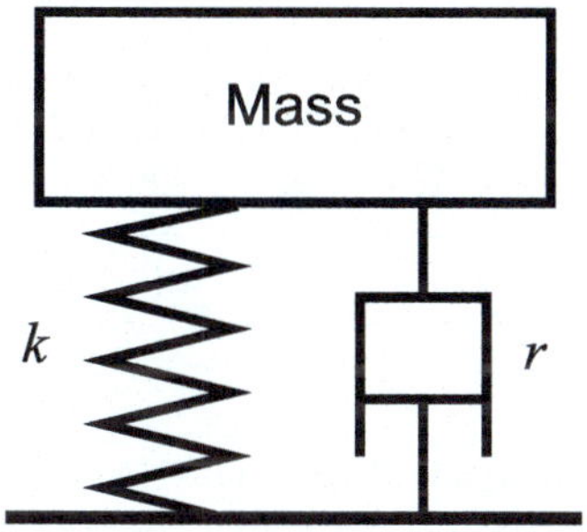

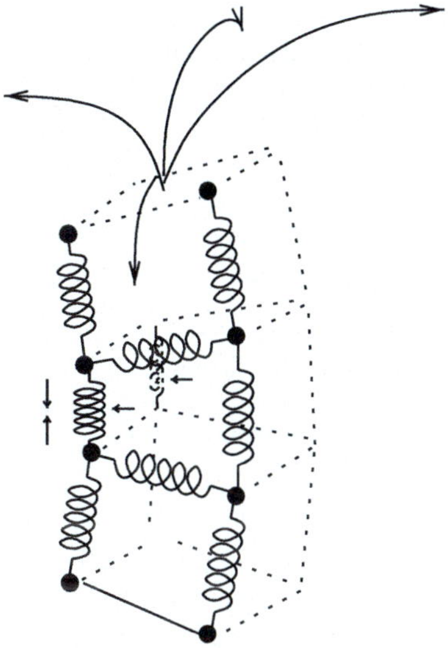

Fig. 6.1 The body of the animat. **a** Mass-spring-damper systems. **b** An illustration of the body plan, where each spring represents in mass-spring-damper system (adapted from [210])

where m_{p_1} is the mass of p_1 and dt is the time-step used during the integration process. Note that $F^E_{p_1}$ is an external force applied to the mass point whenever the output neuron controlling its associated spring becomes activated and $F^W_{p_1}$ represents the current environmental force yielded by the surrounding 'water'. Finally, the position of the mass point, and hence the length of the spring, is updated as follows:

$$P_{p_1}(t + \Delta t) = P_{p_1}(t) + V_{p_1}(t + \Delta t) \cdot dt. \tag{6.4}$$

6.2.3 *Neural Controller*

A continuous time recurrent neural network (CTRNN) is adopted to control the compression or relaxation of the springs in the body of the animat so that swimming-like behaviors can be generated by evolving the structure and parameters of the CTRNN. The activation of a motor neuron is used to calculate the spring constant of an associated spring with a preset force, although this force is only applied if the activation is between zero and one. Each block has its own subneural network architecture although the architectures of these sub-networks are equivalent and connected. Therefore, $A \longrightarrow B \longrightarrow C \longrightarrow D \longrightarrow E$ is a valid representation, while $A \longrightarrow D \longrightarrow B \longrightarrow C \longrightarrow E \longrightarrow A$ is not, where A, B, C, D, E are sub-networks and an arrow represents a connection. They are architecturally equivalent,

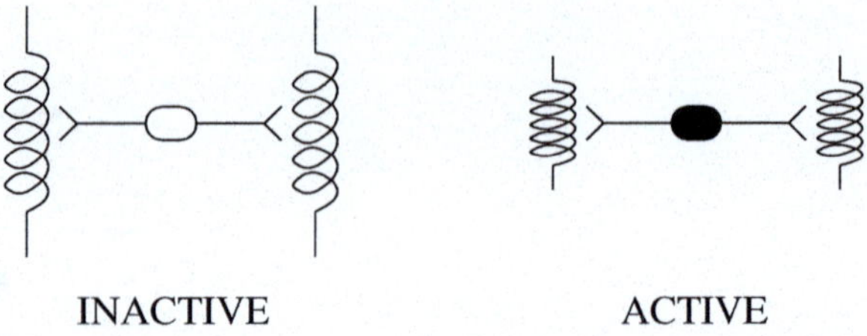

Fig. 6.2 An illustration of how a motor neuron controls the movement of springs

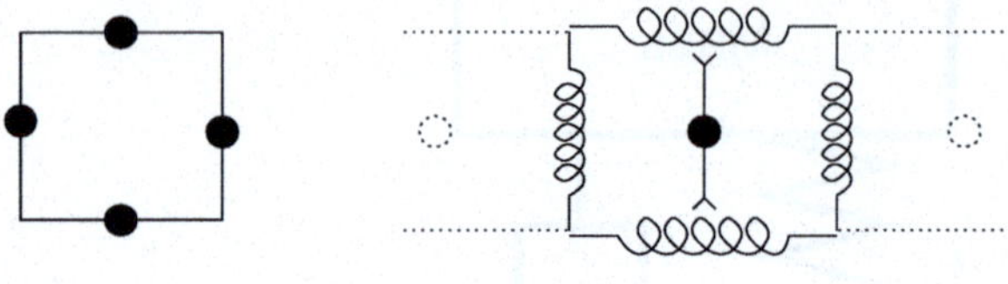

Fig. 6.3 A diagram showing the heads-on and side-on views of the location of the motor neurons in the body of the animat [210]

since only one set of architectural parameters is used to encode all sub-network structures in the evolutionary algorithm. Such simplified encoding can reduce the size of the search space, making the evolutionary search easier.

Figure 6.2 illustrates the inactive and active states of two springs controlled by a motor neuron. The springs will contract when a motor neuron is activated, and they will relax when the neuron is inactive.

The weight values and interconnectivity between the neurons fully depend on the neuron positions. Since their positions are encoded in the evolutionary algorithm, these neurons change their position during the evolutionary process. However, it is hypothesized that the motor neurons always reside within the centers of the block faces, as shown in Fig. 6.3.

The motor neurons can either bring about movement activity, or just serve as inter-neurons. Accordingly, different active motor configurations will have different impacts on the range of possible movements, so they are evolved together with the neural network architecture. Possible configurations of the motor neurons are illustrated in Fig. 6.4.

The membrane potential of a CTRNN neuron is computed according to its incoming presynaptic activity. In discrete time-steps, the activity (μ_i) of the i-th neuron can be modeled as follows [39]:

$$\mu_i(n+1) = \mu_i(n) + \frac{\left(-\mu_i(n) + \sum_{j=1}^{N} w_{ij} A_j(n) + I\right)}{\tau_i}, \tag{6.5}$$

where n is a discrete time-step and τ_i is the time constant for neuron i. A_j is the current output activity of presynaptic neuron j and I represents an external input current, which is always set to 1.0 for the first two neurons at time-step 0 and to 0 for every other time-step. A neuron might also be inhibitory, in which case the signs of all outgoing weights are flipped.

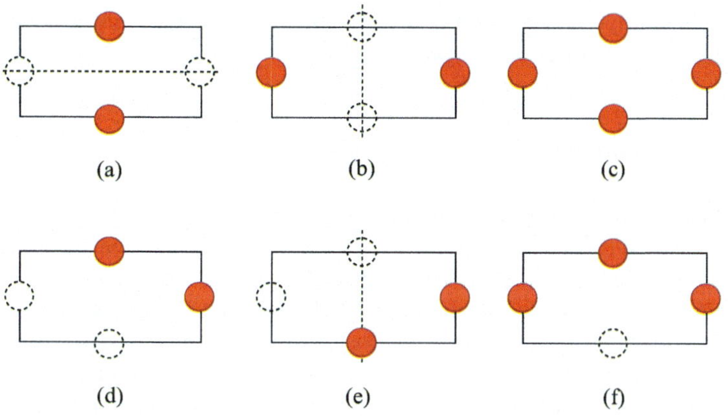

Fig. 6.4 Different active motor configurations (denoted by red filled circles) from the heads-on view, where the configurations in **a** and **b** are bilaterally symmetric

A weight value from neuron i to neuron j is derived according to the Euclidean distance between them, the impact of which is controlled by a parameter ξ:

$$\lambda_{ij} = \frac{\xi}{d_{ij}}, \tag{6.6}$$

$$w_{ij} = \begin{cases} w^{max} & \lambda_{ij} \geq w^{max}, \\ w^{min} & \lambda_{ij} \leq w^{min}, \\ \lambda_{ij} & \text{otherwise.} \end{cases} \tag{6.7}$$

6.2.3.1 Connectivity

Connectivity between a pair of neurons is established according to a minimum distance requirement. Hence, three threshold parameters are introduced. The first one decides inter-neuron-inter-neuron connectivity, the second one decides inter-neuron-effector connectivity, and the third one decides connectivity between neurons from a pair of contiguous subneural architectures. Since there are no sensory neurons currently employed in the model, there are no additional parameters as might be expected for more advanced architectures. A connection is formally decided as follows:

$$C_{ij} = \begin{cases} 1 & d_{ij} \leq \Gamma_q, \\ 0 & \text{otherwise,} \end{cases} \tag{6.8}$$

where Γ_q is a threshold value. For connectivity between sub-network architectures s_p and s_q, neuron i from s_p is allowed only to make one inter-sub-network connection, which is explicitly chosen to be neuron i from s_q. Note that motor neurons never make inter-sub-network connections.

6.2.3.2 Environment

To enable the animat to generate locomotive movements, it is necessary to provide the animat with a physical environment, typically through simulating the physics of water. For the sake of simplicity, many physical factors such as lift, vortex shedding, and buoyancy are neglected in the model, and only the environmental water force is considered, which depends on the velocity of the animat. To calculate this force ($F^W_{b_i,f_n}$), the summation of all mass point velocities of an individual block face, f_n of an animat block, b_i, is calculated, V_{b_i,f_n}. Then the force exerted on each of the four mass points is calculated as follows:

$$F^W_{b_i,f_n} = -\frac{1}{2}\upsilon\,\delta\,\alpha\ \mathrm{sign}(V_{b_i,f_n})\ \left(V_{b_i,f_n}\right)^2, \tag{6.9}$$

where υ and δ are the viscosity and drag coefficients, respectively, and α is the area of the face of the animat block. This process repeats for all other faces of the block, and for all blocks of the overall animat.

6.2.4 *Evolving Motor Configuration and Neural Controller*

6.2.4.1 Encoding

An evolutionary algorithm is used to optimize the parameters determining the architecture of the CTRNN network together with motor neuron configurations. A hybrid encoding is adopted since both integer and real-valued parameters are to be optimized. The details of the genotype are given in Fig. 6.5, which consists of eight chromosomes, each having a different color coding. All parameters specifying both the neural architecture and the body plan of the animat are encoded.

In principle, any evolutionary algorithm can be adopted for evolving the encoded parameters in Fig. 6.5. In [210], evolutionary programming with self-adaptive mutations in [258] is adopted.

6.2.4.2 Mutation

All real-valued genes are mutated with values drawn from a normal distribution of a zero mean and a variance that is self-adapted, which occurs for every gene with a predefined probability, Φ. For a real-valued y-position, the mutation is as follows:

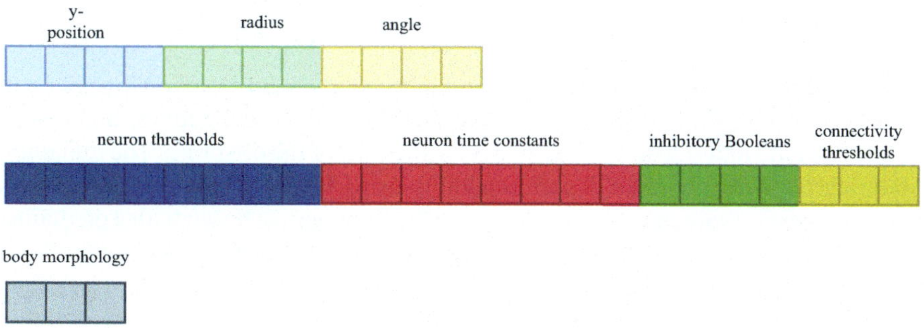

Fig. 6.5 The genotype for evolving the model. In this example, an individual has 8 neurons in each block. Note that there are only parameters for the positions because the positions of the four motor neurons remain fixed. Each group of genes (chromosome) is denoted with a different color

$$y_i = \begin{cases} y_i + N(0, \sigma_i) & \text{rand}() < \Phi, \\ y_i & \text{otherwise.} \end{cases} \tag{6.10}$$

For binary genes encoding the inhibitory or motor activity, the following mutation is implemented:

$$m_i = \begin{cases} !m_i & \text{rand}() < \Phi, \\ m_i & \text{otherwise.} \end{cases} \tag{6.11}$$

The adaptation of the mutation parameters relies on the setting of two strategy parameters, $\tau_o = 1.0/\sqrt{2D}$ and $\tau_1 = 1.0/\sqrt{2\sqrt{D}}$, where D is the dimension of the corresponding chromosome, which are similar to those in evolution strategies [19]. For example, for the genotype in Fig. 6.5, D=4 for the position parameters; D=8 for the thresholds and time constants, and D=3 for the connectivity thresholds. The strategy parameter σ are then self-adapted as follows:

$$\sigma_i \longleftarrow \sigma_i * \exp\left(N(0, \tau_0) + N(0, \tau_1)\right). \tag{6.12}$$

6.2.4.3 Crossover

A 1-point crossover is implemented at a predefined probability of χ between two randomly chosen parents. For both real-valued and binary genes, the crossover operator can be described as follows:

$$\langle y_i, y_j \rangle = \begin{cases} \langle y_j, y_i \rangle & \text{rand}() < \chi, \\ \langle y_i, y_j \rangle & \text{otherwise.} \end{cases} \tag{6.13}$$

6.2.4.4 Selection

The evolutionary programming [258] uses binary tournament selection together with an elitist strategy. First, the population is ranked according to the fitness and a given number of elite individuals is chosen to form the start of the offspring. The remaining offspring population is then chosen randomly using binary tournament selection in which two parents are picked at random, and the fitter one is chosen with a predefined probability. This repeats until the offspring population is filled up. The mutation and crossover operators are applied to all individuals except for the elitist.

6.2.4.5 Fitness Function

The fitness function is the distance that the animat can move forwards within a given time interval, which is to be maximized, meaning that the farther the animat can swim, the better the neural controller, the body, and their coupling. In case the mass-spring-damper system becomes unstable, a very large negative fitness value can be assigned to penalize such settings.

6.2.4.6 Simulation Results

From the simulation results [210], two most important observations can be made. First, the bilaterally symmetric motor configuration (the first one in Fig. 6.2) has emerged from the evolution, implying that for locomotive behaviors, this motor configuration is the most energy-efficient, as shown in Fig. 6.6. This optimal motor configuration has emerged in all six independent runs.

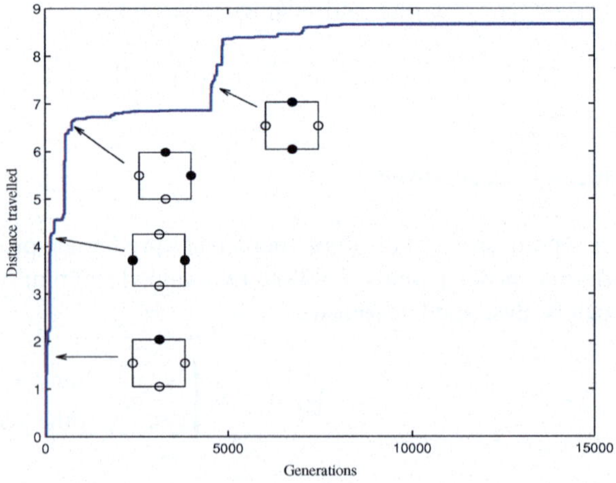

Fig. 6.6 The fitness profile of the best individual over the generations with the active motor configurations being shown [210]. It can be seen that the bilaterally symmetric configuration is favored in the evolution

The second observation that can be made from the results in [210] is that neural dynamics of a CPG has been successfully evolved without a predefined CPG architecture. The neural dynamics of the active motor neurons are plotted in Fig. 6.7. From these results, it can be seen that all blocks (segments) demonstrate various neural dynamics close to the CPG dynamics. Among them, both motor neurons (on two sides of the body) of the first, second, and sixth blocks show CPG dynamics. These dynamics directly contribute to the undulatory movements of the animat. This is very impressive; the architecture of the neural network is evolved from random initialization.

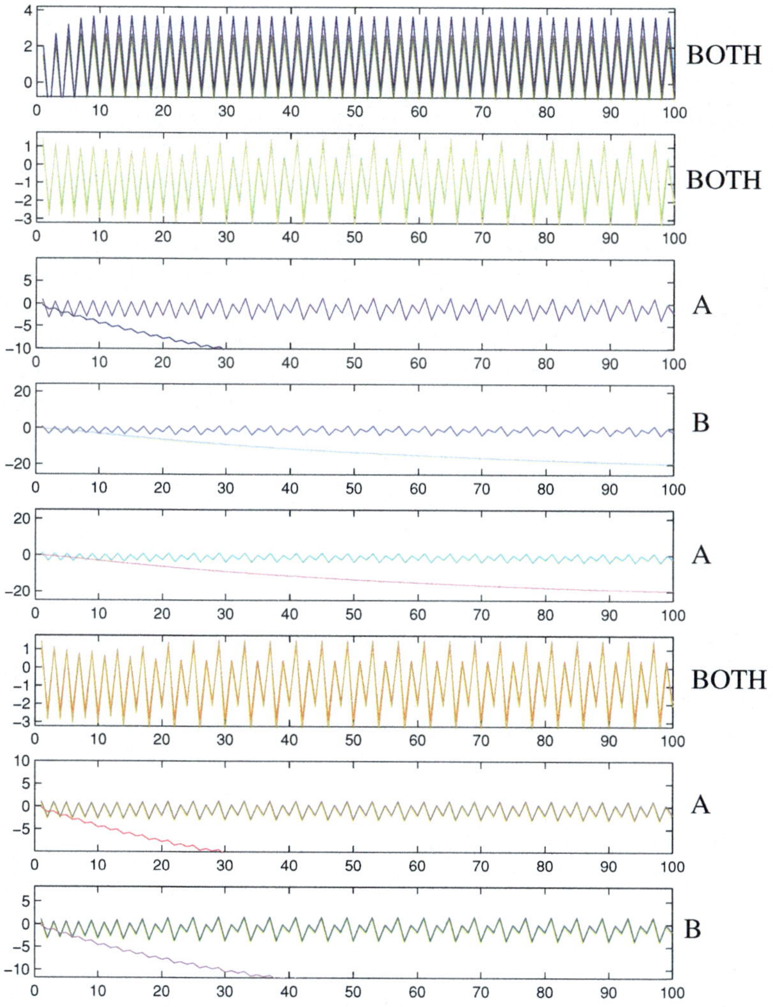

Fig. 6.7 The evolved neural dynamics of the active motor neurons in each of the eight body segments [210]. In the figure, 'BOTH', 'A', and 'B' denote both active motor neurons or only one of them (A or B) demonstrates CPG dynamics

6.3 Emerged Coupling in Brain-Body Co-Evolution

6.3.1 Motivation

In biological evolution, the neural controller (the brain) and the morphology (the body) are co-evolved, meaning that a morphological change in the body will very likely result in a change in the brain, or vice versa. This kind of coupling can happen either during the evolution, or during the development. Consequently, major evolutionary transitions in the nervous system (e.g., from radially symmetric to bilaterally symmetric) have resulted in drastic shifts in the body. More specifically, the coupling may happen in various components of the nervous system, e.g., the nervous system, the motor system, and the sensory system. Apparently, the brain-body coupling must also be considered, simply because the nervous system is embedded in the body. One good example is the jellyfish, which has a radially symmetric body plan with a large surface area and a diffused neural system, making it more easily to be carried by water currents. On the other hand, the flatworm has a bilaterally symmetric body plan and nervous structure. In biology, the close coupling between the brain and the body may result from the environment and the developmental process.

In the model described in the previous section, only the motor configuration is subject to evolution together with the neural controller, demonstrating the brain-body coupling effect at a very high abstraction level. In the next section, we will present one extension of the model presented in this section by further showing the brain-body coupling in computational evolution, where more components of the body, such as the number of segments (blocks) and the length of each segment, are subject to evolution.

6.3.2 The Model

The body, neural controller, and the environment are nearly the same as in the model presented in the previous section. The main difference lies in the encoding, meaning that more parameters of the body plan will be subject to evolution here, including the number of segments and the length of each segment, as shown in Fig. 6.8.

6.3.3 Results and Discussions

With the additional flexibility in the evolution of the body, the brain-body coupling can be more clearly shown by comparing the following settings [210]:

1. **evo-ALL** This is the full version where all parameters in the genotype in Fig. 6.8 will be evolved.
2. **evo-NOBC** All parameters except for the number of body segments are evolved.

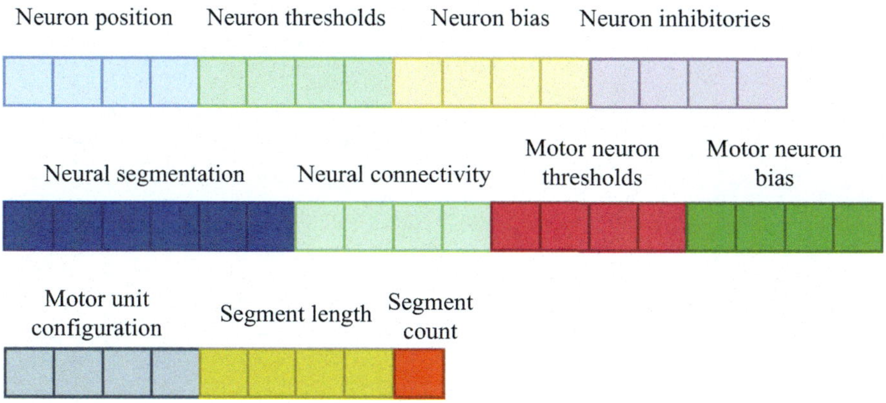

Fig. 6.8 The extended genotype, where several additional parameters of the body such as the number of segments and the length of the segments, in addition to the motor configuration, are subject to evolution

3. **evo-NOBL** All parameters except for the lengths of each segment are evolved.
4. **evo-NOBCBL** In this variant, all parameters except for the number of body segments and the lengths of each segment are evolved.

From the above settings, we can see that **evo-All** has the highest flexibility in evolving the body plan, while **evo-NOBCBL** has the lowest degree of flexibility. The motivation of the empirical study is to examine how these different settings may influence the performance of the evolved animats (the capability to swim), and the distribution of the workload between the controller in terms of the strength of the neural activity and the body plan.

The mean fitness profiles over 50 independent runs of the best individuals for the four settings are given in Fig. 6.9. From these results, we can see that the best average performance is obtained by **evo-NOBC**, in which the number of segments in the body is fixed to five, and the length of each segment is evolved together with the neural controller. This is not completely surprising because the change in the number of segments may introduce big perturbations to the evolutionary process, making it more challenging to find the global optimum. On the other hand, five segments might be fairly optimized for generating swimming behaviors, and focusing on evolving the length of each segment is more efficient.

It is also interesting to analyze the evolved lengths of the body segments. The median lengths of the best individuals in the four settings are plotted in Fig. 6.10. Two basic observations can be made from these results [209]. First, the second segment from the head always has the largest length. Second, the segments closer to the tail tend to have shorter lengths. These results appear to make sense, considering what we can observe from biological organisms. Note, however, that the segment lengths may impact the overall behaviors via different mechanisms, including an effect on the kinematics of the swimming animat due to its interaction with the

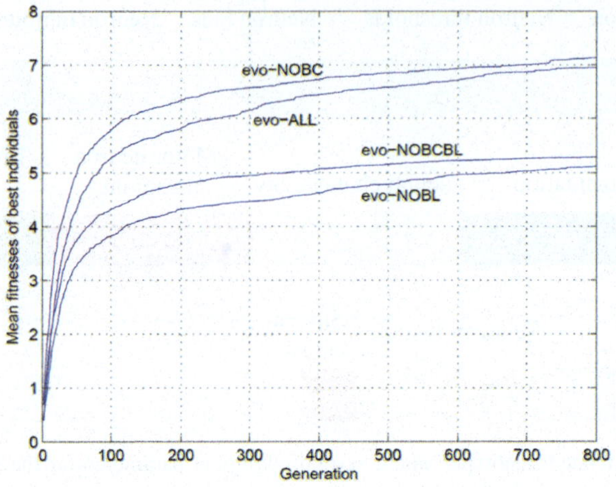

Fig. 6.9 The mean performance of the best individual from the four different settings [210]

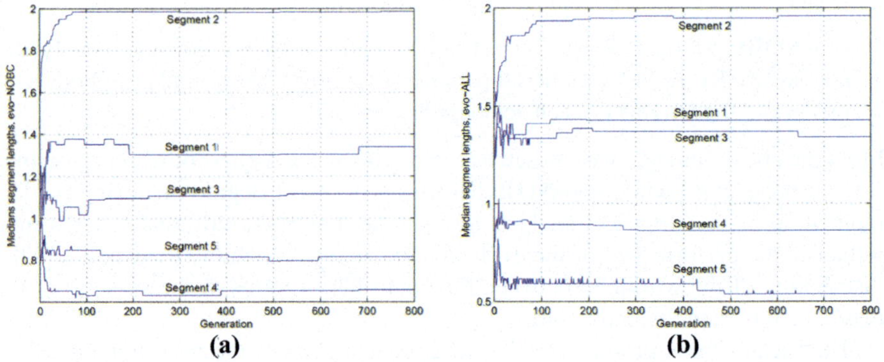

Fig. 6.10 The median segment lengths of the best individual from the two settings evolving the segment lengths [209]. **a evo-All**, and **b evo-NOBC**

environment, an effect on the kinematics of the animat due to the fact that springs having different lengths will have varying tensile and compressional properties, and an effect on the distribution of the neurons in the segment leading different neural dynamics. Interesting results on the neural controllers have also been obtained in the four settings. The first interesting observation is that the neuronal activities tend to decrease as the flexibility of the body segments increases, meaning that less control effort is needed when the body of the animat is better fitted to the environment. The median inter-neuronal and motor neuronal activities of the best individuals from the four different settings are plotted in Fig. 6.11, from which we can see that the neural activities are the largest when the body is completely fixed, and the second largest when the number of segment lengths is fixed. Second, the neural networks in different segments are disconnected if the animats are artificially equipped with proprioceptive

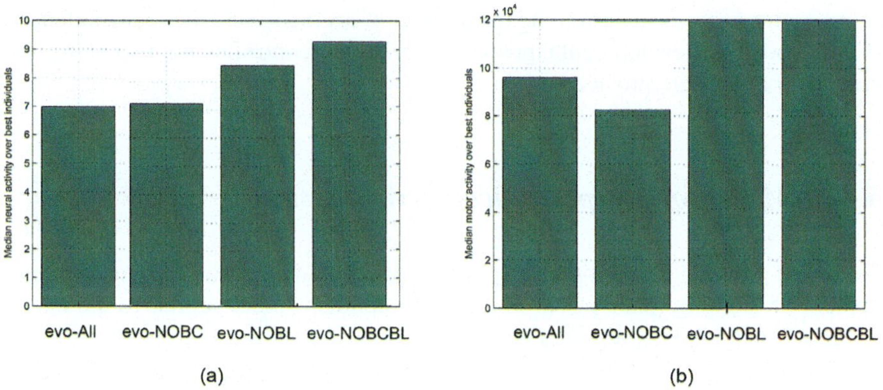

Fig. 6.11 The median activities of the inter-neurons and motor neurons of the best individuals obtained by the four settings [209]

feedback (an ability to sense the states of the body), implying that evolution is able to shape individual independent CPGs coordinating with neighboring segments. However, the sensory information from the head of the animat becomes indispensable if the proprioceptive feedback is unavailable, resulting in interconnected subneural networks in adjacent segments.

Although this extended model reveals several interesting findings, it is still limited for the following reasons. First, more realistic scenarios could be considered, e.g., following a moving object. This has been investigated in [211], and the results indicate that a higher selection pressure toward rich swimming behaviors results in a higher level of symmetry in the evolved neural controller, and asymmetry in the body plan can be compensated to a certain degree by the neural controller. Second, the developmental process of the nervous system and the body has not been included in the model. Finally, the strength of neural activities may not fully reflect the energy efficiency, which could be more precisely measured if a spiking neural network is used as a controller.

In the next section, we will introduce a model that includes the developmental process of both the neural controller and the morphology in the evolution based on the developmental models introduced in Sect. 1.4.

6.4 Coupling Between Evolution of Neural Controller and Morphological Development

The brain-body coupling effects shown in the previous section are based on an evolutionary model containing both a neural controller and a body plan. However, the body plan is represented by an abstract two-dimensional shape in which neurons are randomly positioned and then evolved together with the parameters configuring the active motor neurons, determining the number of segments and the segment lengths. In that model, no developmental process of the neural controller, nor that of the body plan, has been taken into account.

6.4.1 Model for Morphological Development

To fill the above gap, a biologically more plausible model considering the development of the body plan based on the multi-cellular model similar to the ones in Sects. 1.3 and 4.2.2 is presented below.

6.4.2 Neural Controller

To enable swimming behaviors, a motor controller is embedded into the developed morphology, in which the cells are all connected to form a whole body plan. Similarly, a mass-spring-damper system is used to simulate the body of the animat. To maintain the stability of the body plan, all motor cells are connected to at least three muscle neurons outside of the morphology. The springs connecting the neurons at the outside of the body contract if the motor neuron is activated; refer to Fig. 6.12. The movement is defined by changing the natural length of the springs connecting the lateral muscle cells. Therefore, there should be sufficient space between the cells for them to move. The natural length l of the springs switches between l_n and l_s within one period T, which is encoded in the chromosome and will be evolved. The overall morphology is split into 24 predefined segments, and all springs in the same segment have the same phase shift $\rho \in [0, T]$. From the above description, we can see that this is a relatively abstract representation of the neural controller, although the morphology is developed based on the multi-cellular model.

6.4.3 The Fitness

By evolving the gene regulatory network governing the growth of the body plan in a similar way in Sect. 4.2.2, an elongated body will be evolved. On top of this, the lengths between the muscle cells on the outside of the body are co-evolved using the following fitness function:

$$\text{minimize } f = f_{\text{swim}} + f_{\text{shape}}, \tag{6.14}$$

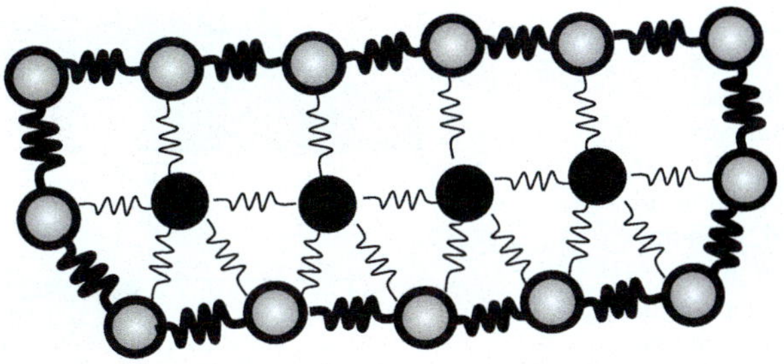

Fig. 6.12 An illustration of the body [360] of the animats containing muscle cells at the outside of the body (denoted by gray circles) and motor neurons (denoted by black circles) inside the body. If a motor neuron is activated, the spring of the connected muscle cells will contract, eventually resulting in deformation of the body. Such coordinated deformations of the body plan in a water-like environment will result in locomotive movement

where f_{swim} defines the displacement of the animate between a given time interval:

$$f_{\text{swim}} = -\left|\left(\sum_{i=0}^{n} \vec{x}^{i}(t=0)\right) - \left(\sum_{i=0}^{n} \vec{x}^{i}(t_{\text{end}})\right)\right| \tag{6.15}$$

and f_{shape} awards elongated shapes:

$$\begin{aligned} f_{\text{shape}} = \max\left\{\min_i \left\{\vec{x}^{i}(0)\right\}, -30\right\} - \min\left\{\max_i \left\{\vec{x}^{i}(0)\right\}, 30\right\} \ldots \\ \ldots - \min\left\{\min_i \left\{\vec{x}^{i}(1)\right\}, -5\right\} + \max\left\{\max_i \left\{\vec{x}^{i}(1)\right\}, 5\right\}. \end{aligned} \tag{6.16}$$

The fitness function contains two terms. The first term is to minimize the negative of f_{swim}, aiming to maximize the distance the animat swims during the given time. The second term f_{shape} makes sure that an elongated shape will be evolved so that the animat can make locomotive movements. The maximum value of f_{shape} is -50.

6.4.4 Coupling Between the Body and Controller

The most interesting observation from the experiments in [360] is that two different swimming behaviors emerge from the evolution. The first is a snake-like lateral undulatory locomotion, which is observed in two of the three body plans and the second is an earthworm-like peristaltic rectilinear locomotion [93] observed in one evolved body. The activation patterns of the three evolved animats are illustrated in Fig. 6.13, resulting in (a) and (c) undulatory, and (b) peristaltic locomotion. It is interesting to note that the two different swimming behaviors, both can be found in nature, are evolved without any manual definition of the precise desired control

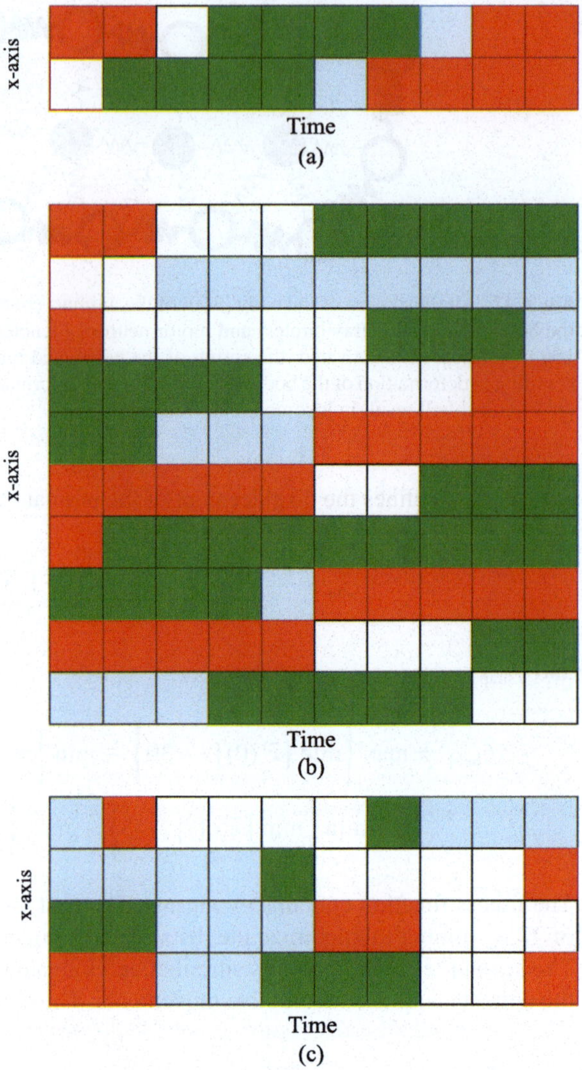

Fig. 6.13 Evolved activation patterns in the three evolved individuals, resulting in an undulatory movement in (**a**) and (**c**) and a peristaltic movement in (**b**) (adapted from [362]). In the figure, a red, green, light blue, and white square mean an activation state in the left, right, both, or none of the two-side muscle cells in the time instance

behavior or body plan. The settings for the different evolutionary runs are the same, and diversity in the morphology and swimming behavior occurs simply due to the randomness in the evolution and development, with the interaction with the physical environment. These results may partly explain the huge degree of biodiversity in nature.

6.5 Evolution of Brain-Body Co-Development

In the model introduced in the previous section, the neural controller is included in the evolution at a very high abstraction level without considering its developmental process. To make a step further, a simplified brain-body co-development model has been developed [364]. In the following, we briefly discuss the model, the controller, and the results.

6.5.1 GRN-Regulated Brain-Body Co-Development

Again, the multi-cellular model in Sect. 1.3 is adopted for modeling the development of the neural controller and the morphology. To this end, a minor extension is added to the genotype of the multi-cellular model, where a new structural unit for differentiating a muscle cell into a neuron SU^{neuron} is included. In addition, whether two neurons should be connected is also encoded in the chromosome of SU^{neuron} based on the following equation:

$$\varphi_{ij} = \frac{c_1}{1 + \exp^{c_2(d_{ij} - 10c_3)}}, \tag{6.17}$$

where c_1, c_2, and c_3 are parameters to be evolved, and d_{ij} is the distance between the i-th and j-th neurons. If a random number $p < \phi_{ij}$, where p is a random number drawn from a normal distribution of zero mean and standard devision one, then a connection will be established.

6.5.2 Neural Controller

Although the generation of the neurons is developed based on the status of the gene regulatory network that is also responsible for the morphological development, and the formation of the neural network is subject to evolution, the neuronal dynamics is more or less predefined with a CPG, unlike in the model in Sect. 6.2.3, where the structure of the neural controller is fully evolved. Here, the muscle cells contract with the output of one of the neurons of the closest CPG. When the distance to the closest CPG is higher than a predefined threshold, the muscle cell will be passive. When a muscle cell contracts, the associated spring at the outside of the individual will change its length.

There are multiple models that can generate CPG dynamics. One general challenge for these CPG models is to produce stable oscillatory dynamics, which depends on the connectivity and strengths of the connections. In [364], the coupled Hopf-Kuramoto oscillator consisting of two neurons is adopted [62]:

Fig. 6.14 Illustration of a CPG consisting of two interconnected neurons (adapted from [364])

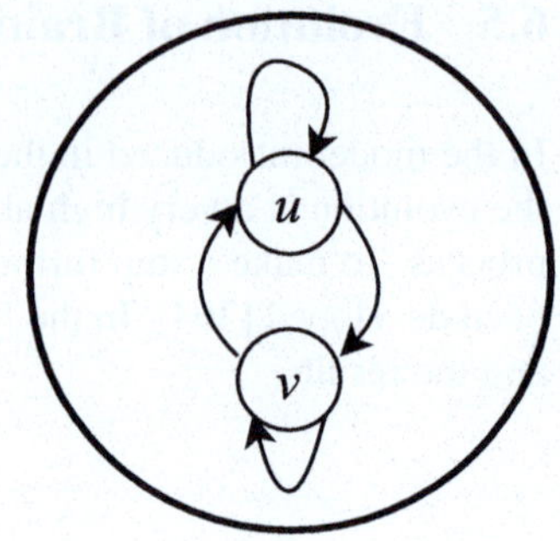

$$\dot{\mathbf{x}} = \mathbf{f}(\mathbf{x}_i, \rho_i) - k \sum_{j \in N_i}^{m_i} \left(\mathbf{x}_i - \frac{\rho_i}{\rho_j} R(\phi_{ij}) \mathbf{x}_j \right), \tag{6.18}$$

where $\mathbf{x}_i(t) = (u_i(t), v_i(t))^T$ and

$$\mathbf{f}(\mathbf{x}, \rho) = \begin{pmatrix} -\lambda/\rho^2(u^2 + v^2 - \rho^2\sigma)\, u - \omega(t)\, v \\ \omega(t)\, u - \lambda/\rho^2(u^2 + v^2 - \rho^2\sigma)\, v \end{pmatrix}. \tag{6.19}$$

The parameters $k, \omega, \rho, \lambda, \sigma$ need to be properly defined to generate CPG neural dynamics. In the CPG, an orientation is introduced to define to which neuron a cell is connected. The orientation of the CPG itself is defined by the gradient of a transcription factor (TF) defined in SU^{neuron} for neuron formation. Cells connecting to the CPG with an angle from 0 to 180 degrees are connected to the neuron u and those with an angle from 180 to 360 degrees are connected to the neuron v of the CPG. The configuration of a CPG neuron composed of two connected neurons (u and v) is shown in Fig. 6.14. Refer to [359] for more details of the parameter settings of the genotype of SU^{neuron}.

6.5.3 *Fitness Function*

The fitness function for evolving the brain-body co-development is the distance an animat swims during a predefined time period. In addition, constraints are applied to control the complexity of the evolved animat using the number of cells and requirements on the connectivity of the developed neurons.

6.5.4 *Discussion*

The empirical study in [364] shows that the model has successfully evolved swimming behavior. However, a close coupling between the morphology and neural controller has not been clearly shown. This might be due to the fact that the neural architecture of the CPG is mostly predefined, and the complexity of the evolved animat is constrained due to computational complexity. In addition, the evolved organisms do not exhibit a clear symmetry in morphology and therefore, it would be interesting to find out under which constraints such a symmetry may emerge.

6.6 Summary

Several models of various abstraction levels are presented in this chapter to examine the coupling between the evolution and development of the neural controller and the morphology. Overall, a close coupling between the body and brain is clearly shown in evolution and/or development. Several interesting findings have been revealed in the studies, which are not only of interest to understand the evolution and development of the neural controller and body plan but are also of biological significance.

Chapter 7
Evolutionary Morphogenetic Self-organization of Swarm Robots

7.1 Morphological Development and Self-organization

7.1.1 Biological Morphogenesis

In biology, morphogenesis is a general term denoting the biological process that causes the creation of the form of multi-cellular organisms [459], including the body plan and the nervous systems. Morphogenesis of animals can be divided into early embryonic development and later embryonic development [131]. Early embryonic development involves typically cleavage, gastrulation, and axis formation, while later embryonic development is mainly responsible for the development of the nervous systems, starting with the segregation of neural and glial cells from the ectoderm germ layer [352].

Metamorphosis is another interesting stage of biological development during which both the shape and size of the organisms change [36]. There are two types of metamorphosis, namely incomplete and complete metamorphosis. For organisms underlying incomplete metamorphosis, there are three developmental stages, in which nymphs look similar to adults. By contrast, organisms that undergo complete metamorphosis have four developmental stages, in which the shape of the organisms changes drastically.

Both multi-cellular morphogenesis and metamorphosis are under the control of gene regulatory networks. When a gene is expressed, information stored in the genome is transcribed into mRNA and then translated into proteins. Some of these proteins are transcription factors that can regulate the expression of their own or other genes, thus resulting in a complex network of interacting genes termed as a gene regulatory network (GRN). In the following, we will discuss the computational modeling of GRNs and show how these models can be used for understanding biology and solving engineering problems.

Y. Jin, *Computational Evolution of Neural and Morphological Development*, Natural Computing Series, https://doi.org/10.1007/978-981-99-1854-6_7

7.1.2 Morphogenetic Self-organization

It has been suggested that morphogenesis can be regarded as a self-organizing and self-assembling process through cellular and molecular interactions under the genetic and environmental control [28, 386]. In addition, biological morphogenesis has also shown a surprising degree of robustness [30]. A recent overview of computational modeling of biological development can be seen in [408].

Due to the attractive properties that biological morphogenesis exhibits, much attention has been paid to employ genetic and cellular mechanisms for designing robotic systems, in particular for self-organizing swarm robotic systems and self-reconfigurable modular robots. Meanwhile, a large body of research has been performed in artificial life and robotics to design simultaneously the body plan and neural controller of robots or animats using the evolutionary developmental approach [387, 399, 415].

The term of *morphogenetic robotics* was first coined in [200], which was proposed as a new research field in robotics that concentrates on the application of biological morphogenetic mechanisms for designing self-organizing, self-configurable, and self-adaptive robots [310]. It is argued that morphogenetic robotics may include, although not limited to the following three main topics:

- Morphogenetic swarm robotic systems that deal with the self-organization of swarm robots using genetic and cellular mechanisms underlying the biological early morphogenesis [144, 267, 379, 383].
- Morphogenetic modular robots where modular robots adapt their configuration autonomously using morphogenetic principles [285, 286].
- Evolutionary developmental approaches to the design of the body plan of robots, including sensors and actuators, and/or the controller of robots [121, 170, 242].

Morphogenetic robotics differ from epigenetic robotics, also known as developmental robotics [266, 453], or developmental cognitive robotics [233]. The main difference between morphogenetic robotics and epigenetic robotics lies in the fact that the former focuses on self-organization, self-reconfiguration, self-assembly, and self-adaptive control of robots using genetic and cellular mechanisms inspired from biological morphogenesis, whereas the latter emphasizes the development of robots' cognitive capabilities, such as language, emotion, and social skills, through experience during their lifetime. Morphogenetic robotics is closely connected to developmental biology and systems biology, while epigenetic robotics is related to cognitive science, developmental psychology, and neuroscience. As a whole, morphogenetic robotics and epigenetic robotics together lay the foundation of developmental robotics in a more general sense. Meanwhile, morphogenetic robotics can be seen as a sub-field of morphogenetic engineering [89].

Other closely related notations are evolutionary robotics [345], morphological computation, also known as morphological robotics [325], and morphogenetic systems [133]. Traditionally, evolutionary robotics is concerned with the design of robot controllers using evolutionary algorithms. Complementary to evolutionary robotics, where the role of robot morphology in realizing an intelligent behavior is neglected to a large degree, morphological computation was targeted for connecting brain, body, and environment in robot design. Unfortunately, it turned out that morphological computation has overly stressed the role of the morphology of robots in shaping intelligent robotic behaviors and has not paid sufficient attention to the developmental aspects.

7.2 Morphogenetic Fixed Pattern Generation

In this section, we first present the basic idea of applying gene regulatory and cell-cell signaling mechanisms for self-organizing swarm robots that form a desired fixed pattern. This is realized by establishing a metaphor between multi-cellular systems and multi-robot systems. Then, the gene regulatory network model presented in [144] is introduced in detail.

7.2.1 A Metaphor Between Multi-cellular and Multi-robot Systems

To employ the genetic and cellular mechanisms that govern biological morphogenesis for self-organized multi-robot pattern formation, it is necessary to establish a metaphor between multi-cellular organisms and multi-robot systems. For example, a cell is mapped to a robot, where protein concentrations in a cell correspond to internal states or the physical location of the robot. The proteins that represent the location of the robots can diffuse out of the cell to generate cell-cell signaling. To simulate cell-cell interactions through the diffusion of proteins, the distance between the neighboring robots is taken into account in the model to influence the robot movement dynamics. Finally, morphogen gradients in multi-cellular organisms are used to define the target pattern to be formed by the multi-robot system.

There is, however, an important difference between a multi-cellular organism and a multi-robot system. In biological morphogenesis, the morphogen gradients are distributed in the form of biochemicals in the embryo, which are available to all cells inside the embryo. However, in a multi-robot system, the target pattern cannot be distributed in terms of chemical concentrations in the environment, unless additional information is added to the environment, which is not always realistic. In Sect. 7.3, we will describe a model that can adaptively generate patterns according to the information collected from the environment, e.g., moving targets to be surrounded.

7.2.2 Problem Statement

In general, the problem we are addressing here is to stationary and/or mobile targets using a few mobile robots. The entrapping task consists of two steps, namely pattern generation and pattern formation. During pattern generation, the robots are expected to generate an appropriate pattern according to the number and location of the targets. In pattern formation, the robots should deploy themselves onto the generated pattern to entrap the targets without a centralized control.

In order to employ GRNs for multi-robot pattern formation, the following assumptions are made:

1. The robots can localize themselves at anytime with their on-board sensors, such as encoders and sonar sensors. This assumption can be lifted by building up a local coordinate system via local communications among the robots, as reported in [200]. A Kalman filter-based algorithm has also been developed in our previous work [197] to improve the accuracy of localization.
2. There is a base station containing a sufficient number of robots. If the robots in mission cannot accomplish the pattern formation task, e.g., the perimeter of the generated target pattern is too long to be fully covered by the robots in mission, they can call for additional robots from the base station. On the other hand, if some robots are no longer needed in entrapping the existing targets, these robots can find their way back to the base station.
3. All robots have a limited sensing range, and therefore, they can only detect targets and other robots that are within their sensing range.
4. The communication range between robots is limited. The robots can communicate information such as targets' location and velocity with their immediate neighbors. By immediate neighbors, we mean that the distance between the two robots is within the communication range. However, note that the communication range between robots and the base station is not limited. In other words, each robot can always communicate with the base station.
5. All targets in the concerned region can be detected by at least one robot. To satisfy this assumption, the coverage algorithm proposed in [172] can be adopted to deploy the robots in the region during initialization.

7.2.3 Gene Regulatory Network Models

Many gene regulatory network models have been proposed, as presented in Sect. 1.2. Here, the following ODE-based gene regulatory network model is used [349]:

$$\frac{d\mathbf{g}_i}{dt} = -a\mathbf{z}_i + m\mathbf{p}_i, \tag{7.1}$$

$$\frac{d\mathbf{p}_i}{dt} = -c\mathbf{p}_i + kf(\mathbf{z}_i) + b\mathbf{d}_i, \tag{7.2}$$

where $i = 1, 2, ..., n$, where n is the total number of robots in the system, $\mathbf{g}_i$ is the concentration of the protein representing the position of the i-th robot in a two-dimensional space, and $\mathbf{p}_i$ is the concentration of the protein representing the internal states (sort of velocity, but not mathematically equivalent) of the i-th robot. a, c are the decay rates, and k, b are the synthesis rates. In Eqs. (7.1) and (7.2), $\mathbf{d}_i$ represents the concentration of protein $\mathbf{g}_i$ diffused out of the cell, indicating the density of robots and obstacles in its neighborhood:

$$\mathbf{d}_i = \sum_{j=1}^{n_i} \mathbf{d}_i^j, \tag{7.3}$$

where n_i is the number of robots in the immediate neighborhood of the i-th robot, and $\mathbf{d}_i^j$ denotes the diffused protein concentration vector from the j-th robot:

$$\mathbf{d}_i^j = \frac{\mathbf{g}_i - \mathbf{g}_j}{\| \mathbf{g}_i - \mathbf{g}_j \|}. \tag{7.4}$$

Note that the diffused protein will be activated only if its concentration is above a given threshold. The diffusion term in the regulatory model simulates the cell-cell signaling in multi-cellular systems. In swarm robotic systems, however, this term will produce a repulsive force to maintain a predefined minimum distance between two neighboring robots.

In Eqs. (7.1) and (7.2), $f(\mathbf{z}_i)$ is an activation function known as the Hill coefficient, which is a sigmoid function:

$$f(\mathbf{z}_i) = \frac{1 - \exp^{\alpha \mathbf{z}_i}}{1 + \exp^{\alpha \mathbf{z}_i}}, \tag{7.5}$$

where α is a constant, and $\mathbf{z}_i$ represents the maternal morphogen gradient at the i-th robot's current position:

$$\mathbf{z}_i = \frac{dh_i(\mathbf{g}_i)}{d\mathbf{g}_i}, \tag{7.6}$$

where $h_i(\mathbf{g}_i)$ is a function that defines the boundary of the pattern to be generated by the robots. For example, we can use the following function to define a circle as the target pattern in a two-dimensional space:

$$h_i = [g_{i,x} - c_x)^2 + (g_{i,y} - c_y)^2 - r^2]^2, \tag{7.7}$$

where $g_{i,x}$ and $g_{i,y}$ are the protein concentrations corresponding to the i-th robot's x- and y-coordinates, (c_x, c_y) is the center of the circle, and r is the radius of the circle.

Theoretically, any closed two-dimensional shape can be defined by function $f(\cdot)$. For example, a unit square whose four corners are (−0.5, −0.5), (−0.5, 0.5), (0.5, −0.5), and (0.5, 0.5) can be defined in the following steps. First, the following circle centered at (0,0) of a radium of $\sqrt{1/2}$ is defined:

$$s_1(g_{i,x}, g_{i,y}) = g_{i,x}^2 + g_{i,y}^2 - \frac{1}{2} = 0, \tag{7.8}$$

$$f_1 = (s_1(g_{i,x}, g_{i,y}))^2. \tag{7.9}$$

Next, the following function $s_2(g_{i,x}, g_{i,y})$ can be defined:

$$s_2(g_{i,x}, g_{i,y}) = \begin{cases} g_{i,x} + 0.5; & \text{if } g_{i,x} \leq 0 \text{ and } -0.5 \leq g_{i,y} \leq 0.5, \\ g_{i,x} - 0.5; & \text{if } g_{i,x} \geq 0 \text{ and } -0.5 \leq g_{i,y} \leq 0.5, \\ g_{i,y} + 0.5; & \text{if } g_{i,y} \leq 0 \text{ and } -0.5 \leq g_{i,x} \leq 0.5, \\ g_{i,y} - 0.5; & \text{if } g_{i,y} \geq 0 \text{ and } -0.5 \leq g_{i,x} \leq 0.5. \end{cases} \tag{7.10}$$

Based on this, we have

$$f_2 = (s_2(g_{i,x}, g_{i,y}))^2. \tag{7.11}$$

Although we are able to define many different shapes using the above approach, they are limited to closed form shapes. Thus, different geometry representation methods must be introduced to form more complex and non-closed shapes. For example, non-uniform rational B-Splines [326] was adopted in [144], and radial-basis functions were used in [323, 472]. Theoretically, any other geometry representation methods can also be used, and not only two-dimensional, but also three-dimensional shapes can also be generated.

To achieve self-organized generation of predefined patterns using the gene regulatory network model in the previous section, the following issues should be resolved:

- The parameters in the gene regulatory network in Eqs. (7.1) and (7.2) must be properly defined. In [144], it is mathematically proved that the following condition must be satisfied to ensure that the states (positions of the robots) will converge to an arbitrary point on the function (7.7):

$$a\,c \geq m\,k, \text{ and } a, c, m, k > 0. \tag{7.12}$$

- In addition to the condition in (7.7), parameters a, c, m, k can be further optimized using a multi-objective evolutionary algorithm to minimize the time for the robots to converge to the desired pattern, and the overall distance traveled by all robots from their initial position to a position on the desired pattern.
- The final position a robot converged to depends on its initial position and the interaction between the robots if the distance between them is smaller than a predefined threshold when converging to the desired pattern. That is to say, each robot moves toward the desired pattern independently before robot-robot interaction (the repulsive force) occurs. The repulsive force makes sure that the robots distribute on the desired patterns evenly. However, the robots may slightly deviate from the defined shape because of the repulsive force between the robots. This, however, does not prevent the robots from roughly forming the desired shape. Note that it is not critical to have an exact circle in multi-robot pattern formation.

- The neighborhood size that controls robot-robot interaction can be adapted according to the number of robots available and the perimeter of the shape to be covered to make sure that the robots evenly distribute on the boundary of the shape [144].
- If there are stationary or moving obstacles in the field, the robots are able to avoid colliding with them if these obstacles are detected. Once the obstacles move away, the robots will reformulate the desired shape, although they may return to a different position on the shape.

7.2.4 Evolutionary Optimization of the GRN

Five parameters in the GRN, a, m, c, k, and b in Eqs. (7.1) and (7.2), need to be optimized using the NSGA-II [83]. The goal of the optimization is to minimize the robots' travel distance and the convergence time, while ensuring that the stability of the system is maintained. In the evolutionary optimization, the number of robots is set to 20 and the population size is set to 100. The crossover probability is 0.9 and the distribution index for the simulated crossover in NSGA-II is set to 20. The mutation probability is set to 0.2 and the distribution index for mutation is 20. The evolutionary optimization runs for 50 generations. In the first generation, the parameters k, c, a, and m are randomly initialized between 1 and 100, and b is randomly initialized between 200 and 1000. A large value for b allows the robots to move far away from each other such that they can converge to different points on the target shape. Note that the computational complexity of the algorithm is nearly insensitive to the number of robots [197]. From the achieved Pareto optimal solutions, one solution having a balanced performance with respect to travel distance and convergence time is picked out for experiments, where $\mathrm{a} = 62.65$, $\mathrm{m} = 63.95$, $\mathrm{c} = 70.27$, $\mathrm{k} = 45.52$, and $\mathrm{b} = 380.42$. Clearly, the stability condition in Eq. (7.12) is satisfied.

7.2.5 Empirical Results

Two examples of multi-robot pattern formation are plotted in Fig. 7.1. Eight robots are randomly deployed in a field of 4 by 4, indicated by star symbols. The robots move to a circular shape finally, where the trajectories are indicated by the curves. The final positions are denoted by the circles, from which we can see that a circle and a square are formed in Fig. 7.1a, b, respectively. We also note that the eight robots are evenly distributed on the shapes, forming a regular coverage of the boundary of the desired shapes.

The formation of a more complex non-closed shape by 20 robots, a letter 'N', which is represented by B-splines, is shown in Fig. 7.2.

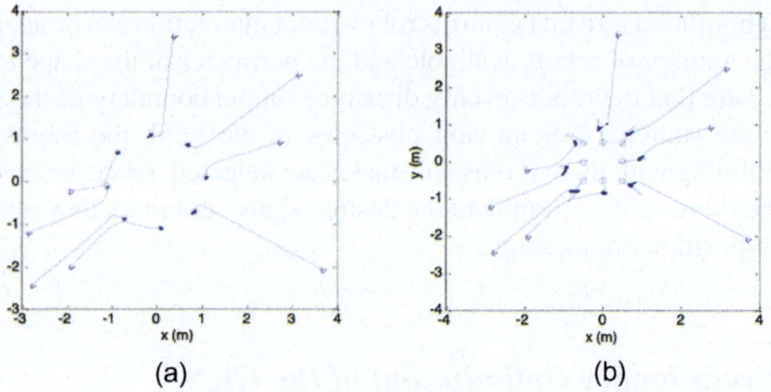

Fig. 7.1 Formation of two predefined shapes by eight robots. **a** A circle and **b** A square. The initial positions of the robots are denoted by a '*' and the final positions are denoted by a 'o'

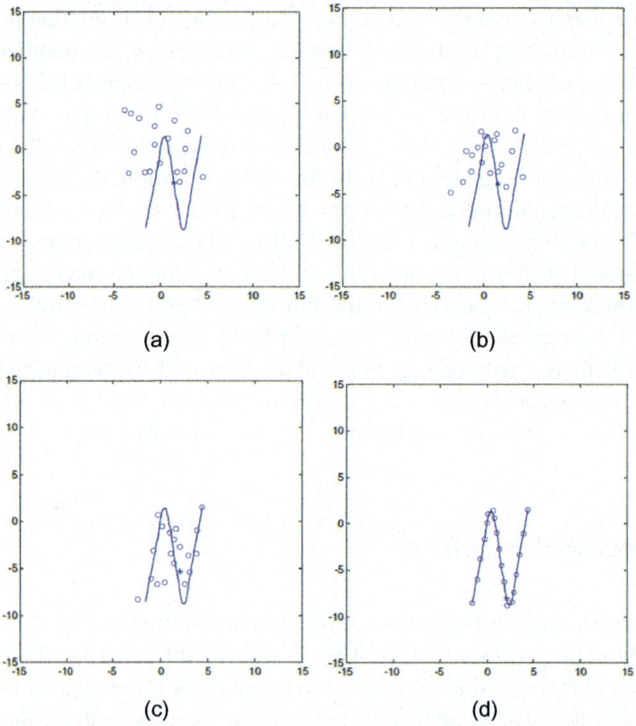

Fig. 7.2 Formation of a shape 'N' with 20 robots. **a** Initial positions; **b** and **c** Intermediate positions; **d** Final positions forming a shape 'N', on which the robots are evenly distributed

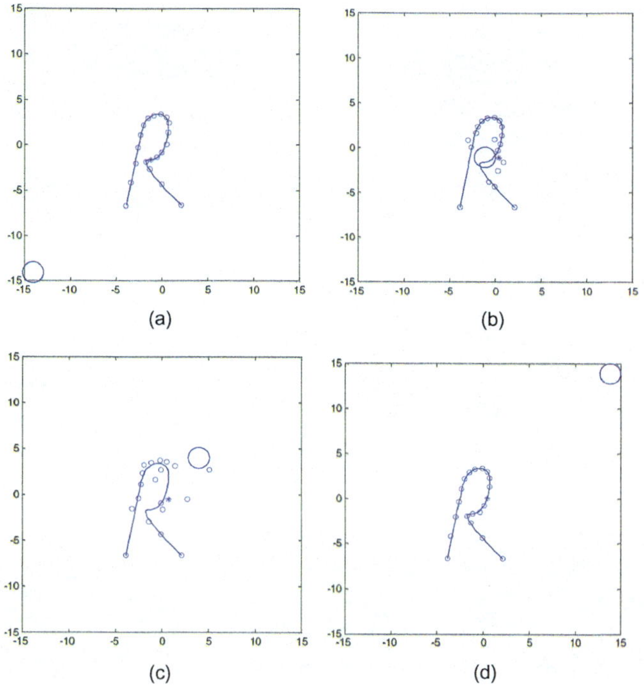

Fig. 7.3 Snapshots of an obstacle avoidance process after a shape 'R' is formulated by 20 robots denoted by small circles. **a** Formation of shape 'R'. **b** The obstacle (denoted by the big circle) moves through the shape 'R' and robots move away to avoid collision. **c** The obstacle moves away, while the robots return to the shape 'R'. **d** Once the obstacle moves away, the robots distribute on letter 'R' evenly again

In the next, we show one example of avoidance of a moving obstacle after a shape 'R' has been formed, as shown in Fig. 7.3. From the figure, we can verify that the robots are able to avoid a moving obstacle denoted by a big circle automatically and self-reorganize themselves again into the shape 'R' with an even distribution. Note that the positions of the robots may have changed even if they form the shape 'R' again after the obstacle moves away, since their positions on the shape are not pre-specified. Instead, the robots find their position in a self-organized way through the genetic dynamics and robot-robot interaction.

Proof of concept physical experiments using eight e-puck robots, as shown in Fig. 7.4, have been performed. Each e-puck robot (http://www.e-puck.org/) is about six centimeters in diameter with a circumferential ring of eight infrared proximity sensors, a pair of step motors in a differential drive configuration, three microphones, and a ZigBee wireless communication card. Infrared proximity sensors are used for measuring distances. Step motors are used for odometer measurement and microphones are used to trigger the start of the experiment. Snapshots of the formation process generating a letter 'R' are plotted in Fig. 7.5. From these results, we can see

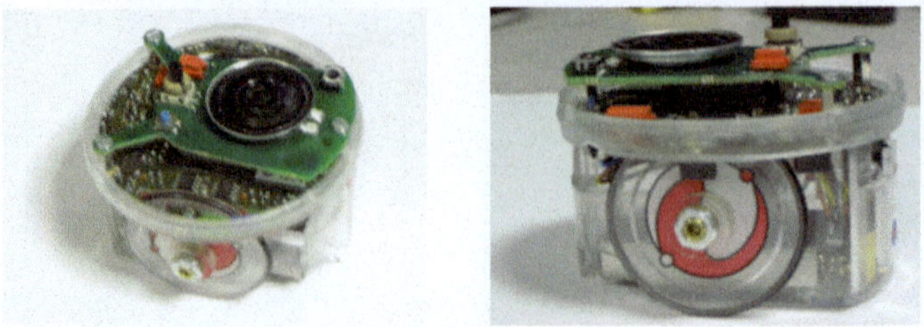

Fig. 7.4 The e-puck robot

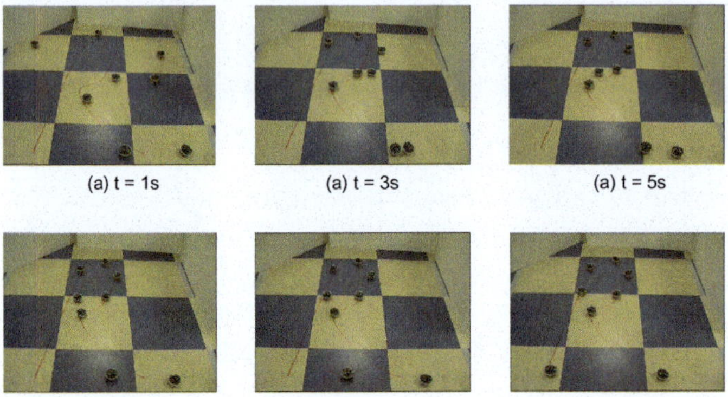

(a) t = 1s (a) t = 3s (a) t = 5s

(a) t = 6s (a) t = 7s (a) t = 11s

Fig. 7.5 Formation of the letter 'R' using eight e-puck robots

that the e-puck robots are able to form a 'R' shape, although it is not fully evenly covered due to the small number of robots.

The gene regulatory models described in this section are able to drive the robots to a pre-described shape in a self-organizing way so that the robots evenly cover the boundary of the shape. Additional functions such as avoiding moving obstacles and adapting to the changing number of robots or the changing length of the perimeter to be covered can also be achieved. However, the requirement to predefine the target shape makes the system unrealistic, since it is very likely in many application scenarios, that the target shape is unknown beforehand. For example, the desired shape depends on the locations of objects to be surrounded if the robots are supposed to follow and entrap the objects. In the following, we introduce a two-layer gene regulatory network that is able to generate the desired pattern dynamically according to the positions of the targets.

7.3 Adaptive Morphogenetic Boundary Coverage

7.3.1 Motivations

In the previous section, the concept of morphogenetic robotics [200], which employs genetic and cellular mechanisms inspired from biological morphogenesis, is introduced. The model presented in Guo et al. [144] that defines the target shape using a function and a variant of the algorithm that uses NURBS [326] to construct more complex 2D or 3D patterns [283] is also briefly introduced. Interestingly, this requires to establish a coordinate system so that the positions of the robots and obstacles will become relative, removing the dependence on the availability of a global coordinate system [201].

A substantial limitation of the above models is that the target shape must be predefined. As a result, the generated patterns are not adaptable to unknown environmental changes. This holds also true for many other bio-inspired multi-robot self-organizing systems [379]. An additional limitation of many control algorithms for multi-robot systems is that they can form only a small number of simple shapes. Actually, not much work has been reported that can adaptively change the desired shape based on the target objects to be surrounded.

In the following, a hierarchical gene regulatory network consisting of two gene regulatory sub-networks [200] will be presented, which enables the automatic design of the desired shape using a bio-inspired gene regulatory network, which is then used to self-organize the robots for pattern formation by a second gene regulatory network introduced in the previous section.

7.3.2 A Hierarchical Gene Regulatory Network

The biological development process is not fully flat, and there are many hierarchies in both temporal and spatial control of the genetic and molecular processes in biological development [348, 440]. It has been found that hierarchies in gene regulatory networks play a central role in the evolution of biological development [102]. Findings in developmental biology suggest that a hierarchical gene regulatory network (GRN) is responsible for the patterning strategies in Drosophila [65]. In computational modeling of biological GRNs for pattern formation, hierarchical GRNs have shown to be powerful in generating stable complex patterns [86]. Although the exact reasons behind the importance of a hierarchical GRNs remain elusive, it is hypothesized that hierarchies make it possible for GRNs to produce a modular functional structure, resulting in better adaptability and evolvability.

Motivated from the above findings, a hierarchical gene regulatory network (H-GRN) was proposed in [200] for designing a decentralized control mechanism for adaptive multi-robot pattern generation in changing environments. The proposed H-GRN for adaptive multi-robot pattern generation consists of two layers, as

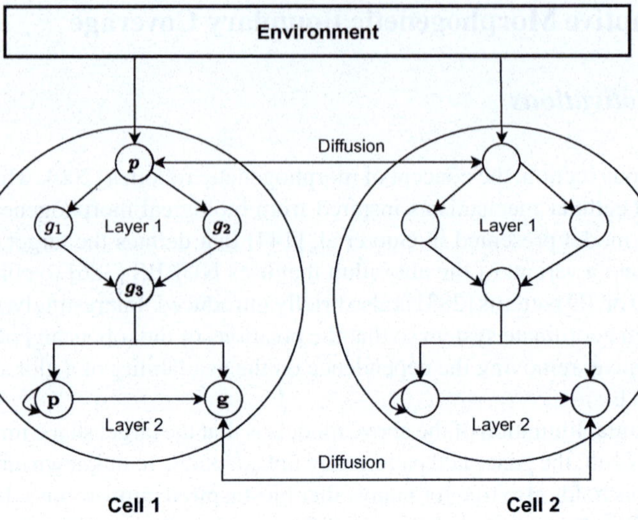

Fig. 7.6 A diagram of the hierarchical gene regulatory network for adaptive self-organized multi-robot pattern generation, showing two cells (robots) in the environment. Each cell contains a hierarchical gene regulatory network containing two layers

illustrated in Fig. 7.6. The basic idea is that the first layer defines the desired pattern according to the positions of the targets and the second layer is the same as the GRN described in Eqs. 7.1 and 7.2. Layer 1 generates patterns in terms of protein concentrations depending on the position of the targets. The concentration of one particular protein (g_3) plays the role of morphogen and its gradients can be read in by receptors in the GRN in Layer 2, which is responsible for controlling the robot's movement dynamics. In Layer 2, protein types **g** and **p** represent the position and internal states of the robots, respectively. If the patterns are generated in a two-dimensional (2D) space, the vector length of G and P is two. In a 3D space, the dimension of both position and velocity vectors is three. Note that only protein type **g** can diffuse into other cells and influence the motion dynamics of other robots.

The dynamics of GRN in Layer 1 are activated only in the organizing robots, which can detect at least one target. Based on the location of the detected targets, a target pattern will be generated by the GRN in Layer 1. The generated pattern will be sensed by the GRN in Layer 2 to drive the organizing robots to the target pattern. So the target pattern may change as the organizing robots are moving toward the target pattern, since the number of targets the organizing robots can detect may change. Meanwhile, if a robot does not detect any target, it will follow the movement of any neighboring organizing robots. Once the targets are within their sensing range, they will become an organizing robot and move toward the targets.

The main advantage of having a hierarchical GRN in the system is that the desired pattern no longer needs to be predefined, making the model more practical. If the positions of the targets change, the pattern generated by Layer 1 will change automatically.

7.3.2.1 Layer 1: Pattern Generation

In the previous section, the target pattern to be formed by the robots is predefined explicitly using either an analytic function [144] or a freeform representation [283], such as the NURBS model [326]. In the H-GRN, the target pattern is generated online using the GRN in Layer 1.

It can happen that a robot detects only a subset of the targets. Therefore, the target patterns (in terms of protein concentration simulated in each robot) generated by different robots may be different when they are approaching the targets. We assume that when the robots are close to the targets, they are able to know the position of all targets either detected by the robots themselves or informed of by other organizing robots through local communication. In this way, the pattern generated by all robots will be the same after convergence.

During pattern generation, each organizing robot will run the following genetic system to generate chemical concentrations that define the target pattern. The input of these equations is γ_j, which corresponds to the position of the detected targets. Here, the organizing robots can either detect the target position through their on-board sensors or acquire the target position through neighboring organizing robots that know the target positions.

$$\frac{dp_j}{dt} = \nabla^2 p_j + \gamma_j - p_j, \tag{7.13}$$

$$\frac{dg_1}{dt} = -g_1 + s(p_j, \theta_1, k), \tag{7.14}$$

$$\frac{dg_2}{dt} = -g_2 + [1 - s(p_j, \theta_2, k)], \tag{7.15}$$

$$\frac{dg_3}{dt} = -g_3 + s(g_1 + g_2, \theta_3, k), \tag{7.16}$$

where $s(x, z, k) = \frac{1}{1+e^{-k(x-z)}}$, where x equals p_j in Eqs. (7.14) and (7.15), a protein concentration that is produced from the environmental input (γ_j) resulting from the j-th target, and $g_1 + g_2$ in Eq. (7.16) p_j, z is the threshold of the sigmoid function, which equals to θ_1, θ_2, and θ_3 in Eqs. (7.14), (7.15), and (7.15), respectively, ∇^2 is the Laplacian operator, which is defined as the second-order derivative of p_j in the spatial domain. g_1, g_2, and g_3 are protein concentrations, where g_3 is the morphogen gradient that defines the target pattern. In the model, g_3 is regulated by both g_1 and g_2, which are regulated by p_j. This setup can be easily implemented by the robotic system, as the targets can be detected by the on-board sensors.

To explain how g_3 can generate a stable pattern in the space based on different environmental inputs, it is assumed that there is only one target in the environment.

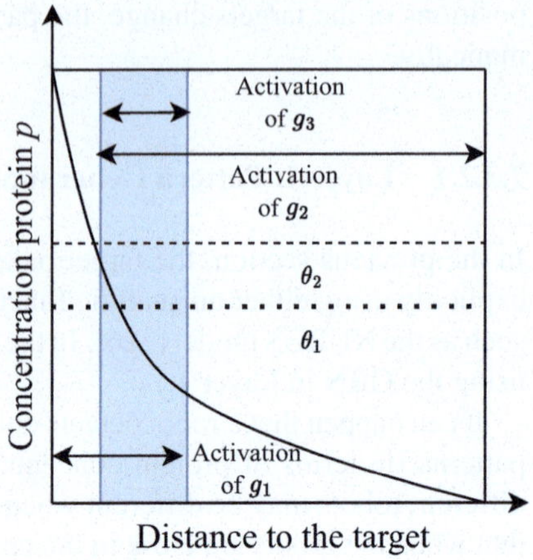

Fig. 7.7 The concentration distribution of protein p and the resulting expression profiles of proteins g_1, g_2, and g_3 in one-dimensional space

At first, a constant input γ at the target position will be generated. Then, protein p will build up a concentration distribution that decays as the distance to the target increases, as shown in Fig. 7.7. From the figure, we can see that proteins g_1, g_2, and g_3 can be regulated by protein p only within a particular range of its concentration, refer to Eqs. (7.13)–(7.16). Particularly, protein g_3 can be regulated by protein p only when the concentration of p is between θ_1 and θ_2. Therefore, the activation pattern of g_3 resulting from one detected target in a two-dimensional space will be a band of circle.

In order for the generated target pattern to be used by the GRN of Layer 2, it is necessary to extract the contour of the generated pattern and represent it using a NURBS model. This can be done by choosing a few representative points on the pattern and using these data points to generate control points of the NURBS representation. In the next, we assume that the generated target pattern has already been represented by a number of control points of the NURBS curve.

7.3.2.2 Layer 2: Pattern Formation

Once the target pattern is generated by Layer 1 of the H-GRN model, it will function as an input of Layer 2 to trigger its dynamics. The dynamics of the GRN in Layer 2 guide the robots to the target pattern. It is assumed that each robot has a sensing range of r. The GRN in Layer 2 is in principle the same as the model used in the previous section:

$$\frac{dg_{i,x}}{dt} = -a\mathbf{z}_{i,x} + mp_{i,x}, \tag{7.17}$$

$$\frac{dg_{i,y}}{dt} = -a\mathbf{z}_{i,y} + mp_{i,y}, \tag{7.18}$$

$$\frac{dp_{i,x}}{dt} = -cp_{i,x} + rf(\mathbf{z}_{i,x}) + bD_{i,x}, \tag{7.19}$$

$$\frac{dp_{i,y}}{dt} = -cp_{i,y} + rf(\mathbf{z}_{i,y}) + bD_{i,y}, \tag{7.20}$$

where $i = 1, 2, \ldots, n$, and n is the total number of organizing robots. $g_{i,x}$ and $g_{i,y}$ represent the x- and y-positions of the i-th organizing robot, which correspond to concentrations of two proteins of type **g** within the cells. $p_{i,x}$ and $p_{i,y}$ are two internal states of the robots, which are the concentration of two proteins of type **p** in the cells. D_i can be seen as the concentration of protein **g** that is diffused out of the cell, which represents the number of robots and obstacles in the neighborhood in a swarm robotic system. The size of the neighborhood is predefined and should be smaller than the sensing range. It can be seen that simulating protein diffusion in multi-robot systems in this way is practical, as it requires only the detection of neighboring robots and obstacles. More specifically,

$$D_{i,x} = \sum_{j=1}^{N_i} D^j_{i,x}, \tag{7.21}$$

$$D_{i,y} = \sum_{j=1}^{N_i} D^j_{i,y}, \tag{7.22}$$

where N_i denotes the number of robots in the neighborhood of robot i, $D^j_{i,x}$ and $D^j_{i,y}$ represent the diffused protein concentrations along x- and y-axis received by robot i from robot j, respectively, which is defined as follows:

$$D^j_{i,x} = \frac{(g_{i,x} - g_{j,x})}{\sqrt{(g_{i,x} - g_{j,x})^2 + (g_{i,y} - g_{j,y})^2}}, \tag{7.23}$$

$$D^j_{i,y} = \frac{(g_{i,y} - g_{j,y})}{\sqrt{(g_{i,x} - g_{j,x})^2 + (g_{i,y} - g_{j,y})^2}}. \tag{7.24}$$

Here, $f(\mathbf{z}_i)$ is defined to be the following sigmoid functions:

$$\begin{aligned} f(\mathbf{z}_{i,x}) &= \tfrac{1-e^{-\mathbf{z}_{i,x}}}{1+e^{-\mathbf{z}_{i,x}}} \\ f(\mathbf{z}_{i,y}) &= \tfrac{1-e^{-\mathbf{z}_{i,y}}}{1+e^{-\mathbf{z}_{i,y}}} \end{aligned} \tag{7.25}$$

where $\mathbf{z}_{i,x}$ and $\mathbf{z}_{i,y}$ are defined by the following:

$$
\begin{aligned}
\mathbf{z_{i,x}} &= (\mathbf{g_{i,x}} - \mathbf{g_{i,x}}(\mathbf{u})), \\
\mathbf{z_{i,y}} &= (\mathbf{g_{i,y}} - \mathbf{g_{i,y}}(\mathbf{u})),
\end{aligned}
\tag{7.26}
$$

where $g_{i,x}(u)$ and $g_{i,y}(u)$ are $x-$ and $y-$ coordinates of the target pattern represented by the NURBS model [326]. Here, u is a parameter in the NURBS model ranging from 0 to 1. Refer to [200] for more details about geometry representation using NURBS.

It has also been proved in [283] that the dynamics of the GRN in Layer 2 converge to the target pattern. Since no target position is predefined for the robots, a random value for $u \in [0, 1]$ will be chosen by each robot. As a result, it is possible that more than one robot goes to the same point on the target shape. However, due to the diffusion term in Eqs. (7.19) and (7.20), the robots will adjust their dynamics automatically to keep a certain distance from each other so that the robots distribute on the target shape evenly. These diffusion terms also contribute to obstacle avoidance. Once a robot detects an obstacle in its neighborhood, the diffusion terms will drive the robots away from the obstacle.

7.3.2.3 Following Dynamics

For those robots that do not detect any target, their movement behavior is governed by the following dynamics, assuming that a non-organizing robot has N neighbors within its sensing range:

$$
\frac{dx}{dt} = \sum_{j=1}^{N} \left(\frac{dx_j}{dt} - \frac{dx}{dt} \right),
\tag{7.27}
$$

where x denotes the current position of the non-organizing robot. $\frac{dx}{dt}$ is the velocity of the robot. Once a robot detects a target, it will become an organizing robot and its GRN dynamics will be activated.

7.3.3 Evolutionary Optimization of the H-GRN Model

7.3.3.1 Layer 1

Several parameters, including θ_1, θ_2, θ_3, and k in Layer 1 (Eqs. (7.13)–(7.16)) need to be specified. To achieve this, the covariance matrix adaptation evolution strategy (CMA-ES) [154] is described in Sect. 1.1.2 to optimize these parameters, as CMA-ES has shown to be efficient for continuous optimization with a small population size only. For evolutionary optimization of the H-GRN parameters, we first need to define a fitness function for pattern generation. Since we are considering pattern generation for entrapping multiple targets, the formulated pattern, to which the robots will converge, should neither be too far away from (may not be able to trap the targets) nor too close to the targets (the targets may pose danger to the robots if they are too close to the targets). Therefore, the following fitness function is adopted:

$$\text{fitness} = 1 - s(d_{\min}, d_0, k) + s(d_{\max}, d_0 + 1, k), \tag{7.28}$$

where d_0 is set to 1, $\dot{()}$ is a sigmoid function as defined in Eq. (7.25), $d_{\min}$ and $d_{\max}$ are the allowed minimum and maximum distances between the pattern and the targets, respectively, where we define $d_{\min} > d_0$ and $d_{\max} < d_0 + 1$. In other words, the distance between the generated pattern and the targets should be between d_0 and $d_0 + 1$. The above fitness function is to be minimized to find the optimal parameters.

In evolutionary optimization, θ_1 and θ_2 are initialized randomly between 0 to 1. The mean value of θ_3 is initialized between 1 to 2. The mean value of k is initialized between 1 to 100. The evolutionary process is terminated when the sum of the fitness changes within the most recent 20 generations is smaller than 0.005, or the number of generations exceeds 50. The population size is set to 100. These parameters are chosen empirically. The optimized parameters are as follows: $\theta_1 = 0.271$, $\theta_2 = 0.326$, $\theta_3 = 1.672$, and $k = 81.338$. According to the convergence condition for the GRN dynamics of Layer 1, γ is set to $\sqrt{\pi/2}$.

7.3.3.2 Layer 2

The robots are expected to quickly converge to the target pattern generated by the GRN in Layer 1 with a minimum travel distance. Therefore, optimization of the parameters of the GRN in Layer 2 has two objectives, namely to minimize the robots' travel distance and to minimize the convergence time. Similar to Sect. 7.3.3, NSGA-II [83] is adopted to solve this bi-objective optimization problem to optimize the GRN parameters in Layer 2.

One interesting observation from the Pareto optimal solutions one can make is that there is a trade-off between the minimization of the sum of the travel distances of all robots and the convergence time. Three representative Pareto solutions are presented in Table 7.1.

Table 7.1 Three representative Pareto solutions, where td refers to travel distance in meters and ct refers to the convergence time in seconds

Solution	r	c	b	a	m	td (m)	ct (s)
(a)	45.88	69.13	387.5	69.28	63.77	19.64	0.39
(b)	79.46	70.06	420.8	44.39	6.85	19.77	0.25
(c)	91.81	57.65	599.9	1.04	1.00	20.5	0.09

7.3.4 *Results and Discussions*

7.3.4.1 Adaptation to Environmental Changes

The main new feature of the proposed model is its ability to adapt the generated pattern to environmental changes. To demonstrate this ability, a test scenario in which two targets move in the simulated area in such a way that the desired target pattern must change in both the shape and the number of the patterns. In the considered scenario, two targets are close in the beginning and one shape can be generated to encircle the two targets. Then, the two targets start to move apart toward two different directions, meaning that the shape will change gradually and finally two separate shapes must be generated. Snapshots of the pattern generation process during the above scenario are provided in Fig. 7.8, showing the ability of the H-GRN model to adapt its generated pattern to environmental changes. Note that splitting or merging the patterns is fully automatically resulted from the dynamics of the GRN of Layer 1. The robots only need to detect the locations of the targets and update the input (γ_j) of the GRN dynamics of Layer 1 and no user interference is needed.

7.3.4.2 Scalability to Number of Robots

Then, the scalability of the model in terms of the time needed for all robots to converge to the target pattern is examined. The time to convergence includes the time needed for deploying the robots from the base to cover the whole area, the time for the organizing robots to generate a stable target pattern, and the time for all robots to converge to the target pattern. It is assumed that the deployment is complete if the velocity of all robots is lower than 1% of the maximum velocity during the employment. A pattern is regarded as stable when the concentration change is smaller than 1% of the maximum concentration change. In the pattern formation stage, a robot is considered to have converged to the target pattern once the position error (distance between the robots and a point on the target pattern closest to the robot) is smaller than 0.05, which is 5% of the allowed minimum distance between two robots entrapping targets. These criteria are empirically defined but a slight change in these criteria will not influence our conclusion.

The area to be covered is 20 by 20 m, and the targets are randomly put in a 6 by 6 m grid in the center of the area. All robots are initialized in a base of 5 by 5 m in bottom-left corner of the area to be covered. In the simulations, nine cases are considered in which the number of targets is 5, 10, and 15, and the number of robots is 25, 36, and 49, respectively.

The time elapses needed for each of the three stages are listed in Tables 7.2, 7.3, and 7.4, respectively. From these results, we can conclude that the convergence time is insensitive to the number of robots and the number of targets. However, the time for communication to get the position of the targets that the organizing robots cannot detect by themselves is not calculated.

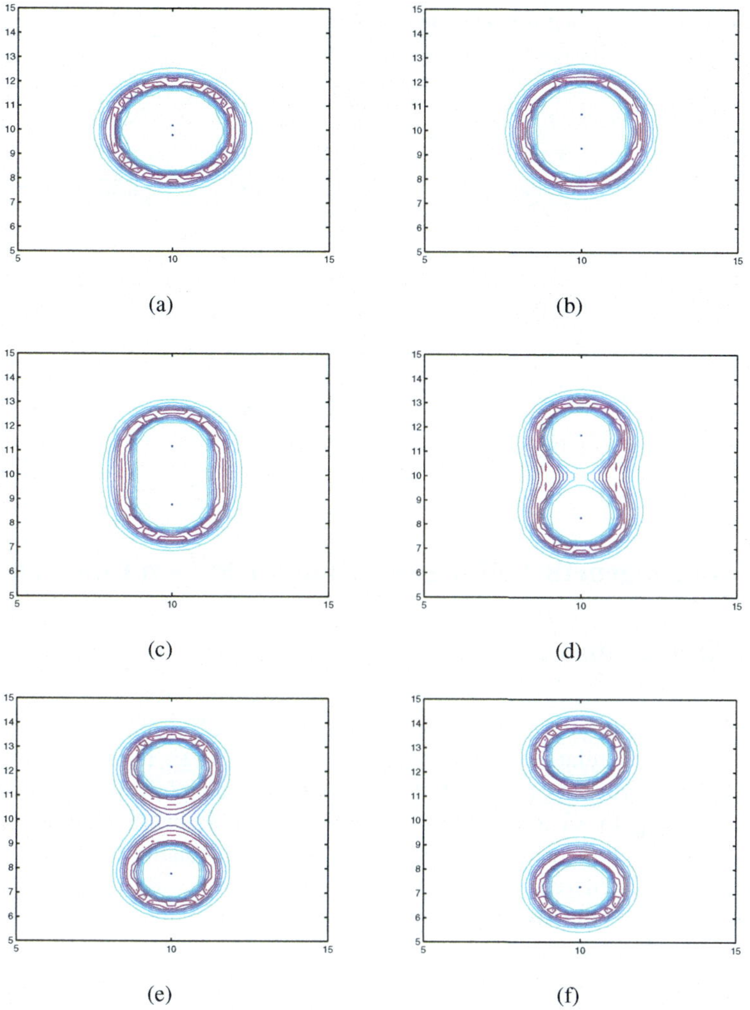

Fig. 7.8 Automatic change of the desired pattern when two closely located targets are moving away from each other

Table 7.2 Time for deployment (Mean ± Std) (s)

No. of targets	No. of robots		
	25	36	49
5	8.891±1.063	9.046±0.965	9.297±1.197
10	8.936±1.539	9.810±1.436	10.700±1.259
15	9.589±1.389	10.193±1.071	10.533±1.537

Table 7.3 Time for pattern generation (Mean ± Std) (s)

No. of targets	No. of robots		
	25	36	49
5	8.326±0.343	8.314±0.304	8.383±0.263
10	8.336±0.234	8.352±0.309	8.375±0.205
15	8.394±0.250	8.415±0.199	8.352±0.199

Table 7.4 Time for the robots to converge to the target pattern (Mean ± Std) (s)

No. of targets	No. of robots		
	25	36	49
5	1.198±0.031	1.203±0.025	1.203±0.031
10	1.192±0.029	1.191±0.025	1.200±0.031
15	1.195±0.030	1.202±0.028	1.202±0.031

7.4 Morphogenetic Self-organization for Region Coverage

7.4.1 Motivations

Although a large body of research on self-organized multi-robot pattern formation has been reported, relatively little work can control a large number of robots for generating adaptive shapes interacting with an unknown and dynamic environment. Swarm Chemistry [354], a computational model of particle swarms following certain kinetic rules, is a good example for a swarm pattern formation approach that can self-organize and self-repair. However, it requires the velocity of nearby particles, which is difficult to obtain, and it is not straightforward to design the kinetic rules producing the desired patterns. Mamei et al. [274] used the concept of morphogen gradient diffusion to achieve pattern formation, but patterns are limited to polygonal shapes, and the robots are assumed to be able to pass through each other without a physical size. The hierarchical two-layer GRN (H-GRN) introduced in the previous section is able to able to generate patterns in an adaptive way with the help of two gene regulatory sub-networks. However, only the boundary of the target pattern can be covered, which might not be robust enough for certain real-world scenarios.

A region-based shape control using a potential field was proposed in [173], where each robot in the group stays within a region as a group while maintaining a minimum distance from each other to achieve an even coverage of the pre-specified region. The algorithm in [173] provided a convergence analysis and considered various shapes of the desired region. However, the target shape needs to be predefined with a limited degree of complexity, which may become inadequate for handling unknown environmental changes.

By extending the H-GRN in the previous section, here we introduce a morphogenetic approach to region coverage using self-organized swarm robots [311]. In the H-GRN, the upper layer is responsible for pattern generation, providing a morphogen gradient to the lower layer. The lower layer, different from the one in the previous section, implements a region-based shape control strategy that guides all robots into the desired region from the upper layer and maintains a specified minimum distance between each robot and its neighboring robots as well. The minimum distance between robots is adapted to evenly distribute robots into the desired region. Using this region-based shape control for target entrapment instead of boundary coverage enables the robots to entrap targets more tightly and herd them into a desired shape or position by directly contacting and pushing the targets.

Similarly, entrapping stationary or moving targets using a swarm of robots consists of two tasks, region generation and region-based shape control. Based on the detected target locations, the entrapping region (desired pattern) is generated, and the robots are deployed into the generated region. Like in [200], we distinguish between *organizing robots* that can detect at least one target and are responsible for the region generation, and *non-organizing robots* that have not yet detected any target. The non-organizing robots will follow their neighboring robots until they detect a target or receive information on the target from organizing robots. It is assumed that global (or at least local) 2D position is available for the robots, and when the robots are close to the targets, they can obtain the position of targets either by direct sensing or from other organizing robots through local communications.

7.4.2 A H-GRN Model for Region Coverage

The structure of the two-layer hierarchical gene regulatory network (H-GRN) model is the same as the one in Fig. 7.6.

For region coverage, the organizing robots will utilize the following gene regulatory dynamics of the upper layer to generate the desired concentrations:

$$\frac{dp_j}{dt} = -p_j + \nabla^2 p_j + \gamma_j, \quad \mathbf{p} = \sum_{j=1}^{n_t} p_j, \tag{7.29}$$

$$\frac{dg_1}{dt} = -g_1 + \text{sig}(\mathbf{p}, \theta_1, k), \tag{7.30}$$

$$\frac{dg_2}{dt} = -g_2 + [1 - \text{sig}(\mathbf{p}, \theta_2, k)], \tag{7.31}$$

$$\frac{dg_3}{dt} = -g_3 + \text{sig}(g_1 + g_2, \theta_3, k), \tag{7.32}$$

$$\text{sig}(x, z, k) = \frac{1}{1 + e^{-k(x-z)}}, \tag{7.33}$$

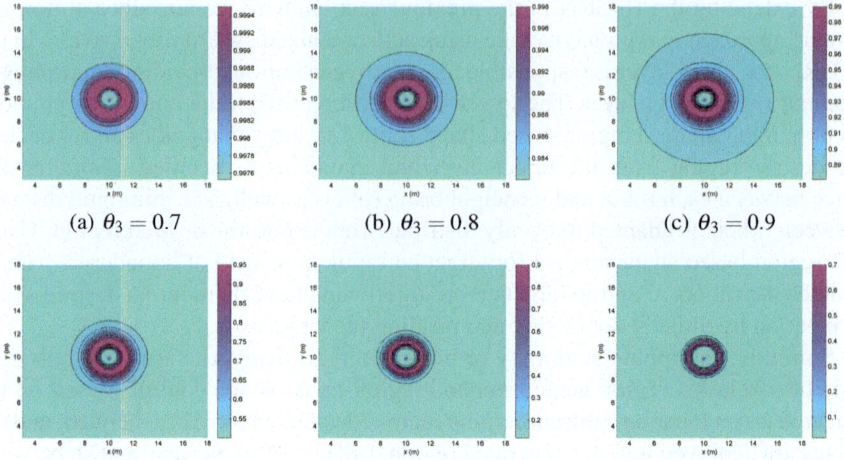

(a) $\theta_3 = 0.7$ (b) $\theta_3 = 0.8$ (c) $\theta_3 = 0.9$

(d) $\theta_3 = 1.0$ (e) $\theta_3 = 1.1$ (f) $\theta_3 = 1.2$

Fig. 7.9 Examples of the target pattern generated by the protein concentration g_3 of the upper layer of the H-GRN, when threshold θ_3 takes different values while $\theta_1 = 0.25$ and $\theta_2 = 0.3$ are fixed. The target is located at $(x, y) = (10, 10)$

where p_j represents the protein concentration produced by the j-th target input γ_j, and p the sum of concentrations of all detected n_t targets. The integrated protein p will activate the internal protein concentrations. k is a positive constant which determines the slope of the sigmoid function, and g_1, g_2 and g_3 are protein concentrations, where g_3 defines the target entrapping pattern. Note that g_3 is regulated by both g_1 and g_2 that are regulated by $\mathbf{p}$ only within a particular range of its concentration. In particular, protein g_3 can be regulated by protein $\mathbf{p}$ only when the concentration of p is between θ_1 and θ_2. This leads to an activating band of circle shapes depending on the thresholds of a sigmoid function θ_1, θ_2, and θ_3, as shown in Fig. 7.9.

For the generated pattern to be used by the lower layer of the H-GRN, the contour of the target pattern needs to be extracted. To this end, a subset of a class of object termed Splinegons [455] is used to create a set of vertices that are connected by line segments of a constant curvature, representing the curved nature of the target pattern as a set of arc segments. First, a set of representative points whose g_3 concentration is higher than a threshold value on the pattern are selected to describe the target pattern. Then, these points are connected by line segments of a constant curvature, which allows us to form shape constraints to be easily used for region coverage. More details for constructing the whole Splinegon can be found in [455]. Figure 7.10 shows the procedure for Splinegon representation from a given target pattern using selected representative points. The generated region by the upper layer of the H-GRN can then be modeled as a set of circular arc segments $s_l \in S = \{s_1, \ldots, s_{n_s}\}$. For each arc segment, only the center position and its curvature will be used for region coverage. The lower layer of the H-GRN aims to guide all robots into the desired region

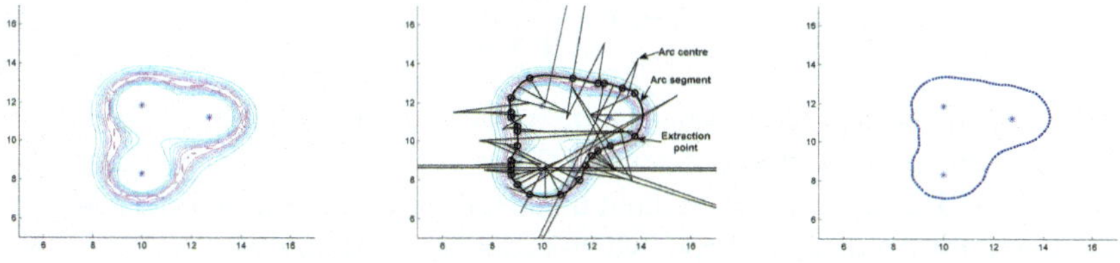

(a) Gradient value from g_3 (b) Curvature segments (c) Splinegon

Fig. 7.10 Extraction of Splinegon representation from a target pattern generated by the GRN in the upper layer

generated by the upper layer and to maintain a specified minimum distance between robots. In the following, we first describe the GRN dynamics for driving the robots to the target region and then present the region-based shape control logic. A few issues for implementing the region-based shape control within the GRN framework will then be discussed. The GRN in the lower layer is basically the same as the one in the previous section:

$$\frac{d\mathbf{g}_i}{dt} = -a\mathbf{z}_i + m\mathbf{p}_i, \tag{7.34}$$

$$\frac{d\mathbf{p}_i}{dt} = -c\mathbf{p}_i + rf(\mathbf{z}_i) + b\mathbf{D}_i, \tag{7.35}$$

where $i \in \{1, \ldots, N_o\}$ the i-th organizing robot, and protein types $\mathbf{g}_i$ and $\mathbf{p}_i$ correspond to a 2D position and internal state vector of robot i, respectively. a, m, c, r, and b are constants to be optimized depending on the objectives of the task. $\mathbf{D}_i$ represents the concentration of protein $\mathbf{g}$ diffused out of the cell, indicating the density of robots and obstacles in the neighborhood:

$$\mathbf{D}_i = \sum_{j=1}^{n_i} \mathbf{D}_i^j, \tag{7.36}$$

where n_i denotes the number of the robots in the neighborhood of robot i, and $\mathbf{D}_i^j$ represents the diffused protein concentration vector from robot j:

$$\mathbf{D}_i^j = \frac{\mathbf{g}_i - \mathbf{g}_j}{\|\mathbf{g}_i - \mathbf{g}_j\|}. \tag{7.37}$$

The diffusion process is activated only when the distance to the neighbor is less than a threshold r_n. In order to embed the desired shape into the regulatory dynamics, $f(\mathbf{z}_i)$ is defined as the following sigmoid function:

$$f(\mathbf{z}_i) = \frac{1 - e^{\alpha \mathbf{z}_i}}{1 + e^{\alpha \mathbf{z}_i}}, \tag{7.38}$$

where $\mathbf{z}_i$ represents a gradient value at the robot's current position, and $\alpha > 0$ determines the slope of the sigmoid function. This $\mathbf{z}_i$ regulates the concentration of both proteins $\mathbf{g}$ and $\mathbf{P}$ as a feedforward input in the GRN dynamics as in Eq. (7.34) so that robots could form a desired shape, which is defined as follows:

$$\mathbf{z}_i = \frac{dh_i}{d\mathbf{G}_i} \tag{7.39}$$

where h_i is the desired shape on which the robots need to be deployed defined by the upper layer. For instance, if the robots need to be deployed onto a circle of a radius R at a point $\mathbf{c} = [c_x, c_y]^T$, the shape function can be defined as follows:

$$h_i = [(g_{i,x} - c_x)^2 + (g_{i,y} - c_y)^2 - R^2]^2. \tag{7.40}$$

Note that if a circle radius R approaches to zero, then the robot will be deployed onto a point $\mathbf{c}$ rather than a circle. The region-based shape control is developed in which the desired region is specified by an inequality function as follows:

$$F(X) \leq 0, \tag{7.41}$$

where $F(X)$ is a continuous scalar function with continuous first partial derivatives. Here, the target pattern is represented as a set of arc segments:

$$S_l(\mathbf{g}_i) = (g_{i,x} - c_{l,x})^2 + (g_{i,y} - c_{l,y})^2 - \left(\frac{1}{\kappa_l}\right)^2 = 0, \tag{7.42}$$

where $[c_{l,x}, c_{l,y}]^T$ and κ_l are the center position and the curvature of the l-th arc segment, respectively. Thus, inequality functions for region shape control can be defined by $S_l(\mathbf{g}_i) \leq 0$ for keeping robots inside an arc segment or $-S_l(\mathbf{g}_i) \leq 0$ for pushing robots toward outside an arc. The inequality functions for the entire complex shape using the Splinegon representation can then be expressed as follows:

$$F(\mathbf{g}_i) = [\pm S_1(\mathbf{g}_i), \pm S_2(\mathbf{g}_i), \ldots, \pm S_{n_s}(\mathbf{g}_i)]^T \leq 0. \tag{7.43}$$

For an efficient region coverage, each robot finds its nearest arc segment $s_{l_i^*}$ at its current location rather than considering all arc segments consisting of the Splinegon:

$$s_{l_i^*} = arg \min_{s_l \in S} \left(\|\mathbf{g}_i - s_l^m\| \right), \tag{7.44}$$

where s_l^m is the middle point of the l-th arc segment.

Depending on the sign of the arc curvature and the position of targets, the shape control logic is divided into three cases with a gradient $\mathbf{z}_i$ and the inequality function of the nearest arc segment as follows:

- **Case 1:** $\kappa_{l_i^*} \leq 0$

$$\mathbf{z}_i = \begin{cases} 0, & \text{if } S_{l_i^*}(\mathbf{g}_i) \leq 0 \\ \frac{dh_{1,i}}{d\mathbf{g}_i}, & \text{otherwise} \end{cases} \tag{7.45}$$

where $h_{1,i} = \left[(g_{i,x} - c_{l_i^*,x})^2 + (g_{i,y} - c_{l_i^*,y})^2\right]^2$. This gradient drives a robot toward the center of its nearest arc segment, and once the robot enters a circular sector of a desired Splinegon, the gradient becomes zero, as illustrated in Fig. 7.11.

- **Case 2:** $\kappa_{l_i^*} > 0$

$$\mathbf{z}_i = \begin{cases} 0, & \text{if } -S_{l_i^*}(\mathbf{g}_i) \leq 0 \text{ and} \\ & \|\mathbf{g}_i - \bar{\mathbf{c}}_t\| \leq \|\mathbf{g}_i - \mathbf{c}_{l_i^*}\| \\ \frac{dh_{2,i}}{d\mathbf{g}_i}, & \text{otherwise} \end{cases} \tag{7.46}$$

where $h_{2,i} = \left[(g_{i,x} - \bar{c}_{t,x})^2 + (g_{i,y} - \bar{c}_{t,y})^2\right]^2$. Here, $\bar{\mathbf{c}}_t = [\bar{c}_{t,x}, \bar{c}_{t,y}]^T$ is the center position of a target group. This gradient brings a robot toward the center of a target group instead of an arc segment, which is outside the Splinegon. Only if a robot is outside a circular sector ($-S_{l_i^*}(\mathbf{g}_i) \leq 0$) as well as closer to the center of the target group than the one of an arc (i.e., $\|\mathbf{g}_i - \bar{\mathbf{c}}_t\| \leq \|\mathbf{g}_i - \mathbf{c}_{l_i^*}\|$), the gradient is zero.

- **Case 3:** $\|\mathbf{g}_i - \mathbf{g}_t^j\| < R_d$ (for all $j \in \{1, \ldots, n_t\}$, regardless of the sign of $\kappa_{l_i^*}$)

$$\mathbf{z}_i = \frac{dh_{3,i}}{d\mathbf{g}_i} \tag{7.47}$$

where $h_{3,i} = [(1+\epsilon)R_d^2 - (g_{i,x} - c_{t,x}^j)^2 - (g_{i,y} - c_{t,y}^j)^2]^2$. Here, ϵ is a positive constant. If a robot is inside a certain circular boundary of R_d around the target,

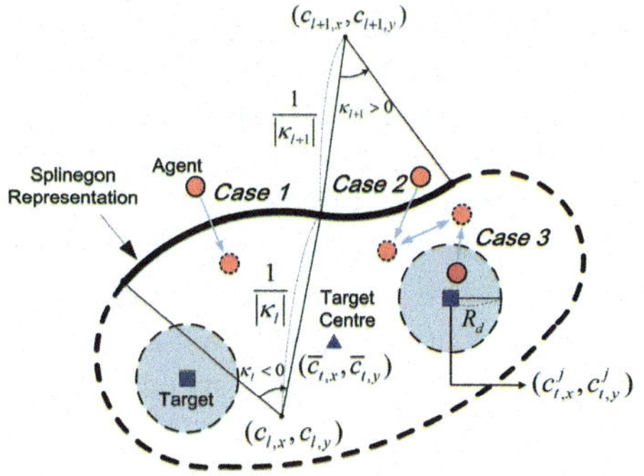

Fig. 7.11 Geometric relationship for region coverage

the robot is pushed out of that area, which is larger than R_d, $(1+\epsilon)R_d$, to avoid any collision with the targets.

7.4.2.1 Discussions

There are several issues to be addressed to implement the proposed algorithm in real physical robots in a changing environment.

- The robots should be able to localize themselves with their own on-board sensors such as encoders or an inertial navigation system. Note that this localization can be done in a local coordinate system by choosing a reference point as the origin with the help robot-robot communications. For example, the position of the robot that first detects a target can be chosen as the reference point, although many other methods can also be employed depending on the task objectives.
- An optimal neighborhood size r_n for the diffusion process in Eq. (7.36) cannot be predefined in practice to ensure the convergence of the system or to evenly distribute the robots in the region to be covered, as the total number of robots and targets is unknown to each robot. Here, the neighborhood adaptation method is used in [144], which adjusts the initial guess of the distance with two physical constraints of the robot, namely the bumper range d_{min} and the sensor range d_{max}. If the current neighbor distance d is found to be small, then it is updated with a half-sum of d and d_{max}, resulting in a distance increase. By contrast, it is updated with d_{min} if the distance is large. Whether the current distance is close to optimal or not is determined by calculating the average number of neighbors after a sufficiently large time elapse for the robots to stabilize inside the region for several iterations. Lastly, the effect of communication/sensing noise and delay should be carefully considered, since this may lead to performance degradation or even instability of the shape control.

7.4.3 Simulation Results

In this section, we will present some simulation results for surrounding stationary or moving targets to validate the feasibility and benefit of the morphogenetic region coverage algorithm based on the H-GRN. The number of robots used in the simulation is 100. To set up the parameters, an evolutionary algorithm [200] is run to optimize them, which are further fine-tuned for stabilizing the robots inside the desired region within a reasonable time. Finally, parameters for the upper layer of the H-GRN are set as follows: $\theta_1 = 0.25$, $\theta_2 = 0.3$, $\theta_3 = 1.2$, and $k = 20$. For the lower layer, the parameters are set as follows: $a = 6.5$, $m = 4.2$, $c = 9.9$, $r = 4.3$, and $b = 3.5$. In addition, the maximum speed of the robot is bounded by 0.8 m/s considering the robots' physical capability.

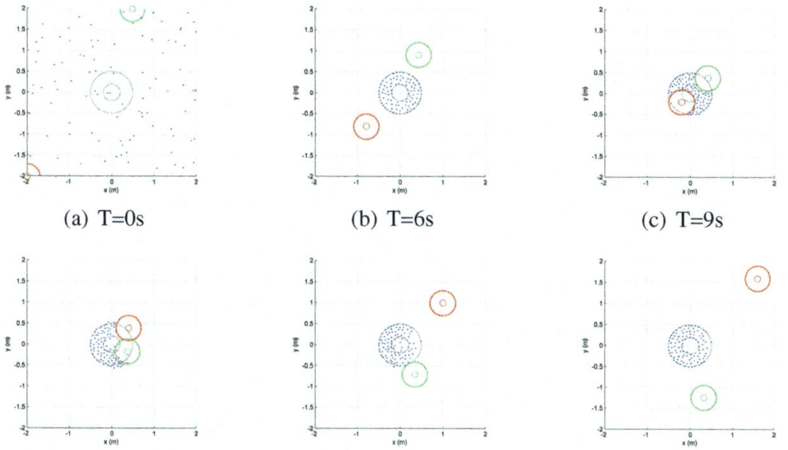

(a) T=0s (b) T=6s (c) T=9s

(d) T=12s (e) T=15s (f) T=18s

Fig. 7.12 100 robots (blue points) forming a target shape denoted by a ring and avoiding two moving obstacles, which are denoted by red and green circles

7.4.3.1 Forming a Predefined Target Region

To entrap a stationary target, a predefined simple ring shape is first employed, assuming that the target is at the center of a circle. The inner and out radii are 0.2 and 0.5 m, respectively. A desired distance between the robots is 0.065 m, and the one between robots and obstacles is 0.3 m. Six snapshots of a situation where 100 robots are randomly distributed in the space, forming the target region driven by the H-GRN and avoiding two moving obstacles, are shown in Fig. 7.12. Note that for obstacle avoidance, a diffusion term **D** between robots and obstacles as in Eqs. (7.34)–(7.37) is used so that the GRN dynamics inherently adapt itself to the two moving obstacles.

Next, we consider tracking of a single moving target while maintaining a ring shape. If the trajectory of the target is known or can be estimated, it can be embedded in the GRN dynamics by modifying Eq. (7.34) as follows:

$$\frac{d\mathbf{g}_i}{dt} = -a\mathbf{z}_i + m\mathbf{p}_i + \mathbf{v}_t, \tag{7.48}$$

where $\mathbf{v}_t$ is the velocity of the target to be followed. The simulation results in tracking the moving target while maintaining a ring shape with or without information about the target movement. The trajectory of the moving target is denoted by a red line, which is specified by $\mathbf{c}_t = [c_x, c_y]^T = 0.1[t \;\; 3\sin(t/3)]^T$, where t represents time in seconds. It is assumed that each robot is equipped with a sensor that can sense the distance and orientation of the target. Sensory noise is set to be white Gaussian noise with a standard deviation of $\sigma_r = 0.2$ m for the range and $\sigma_\phi = 2°$ for the direction. To obtain accurate estimates of the target motion, the decentralized extended information

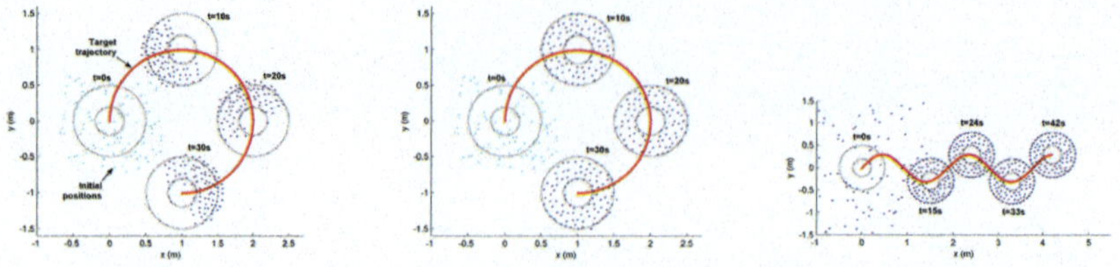

(a) (b) (c)

Fig. 7.13 Forming a ring shape while tracking a target when **a** the target position is known but the velocity is unknown; **b** the target position and the velocity are estimated; and **c** both the position and velocity of the target are known

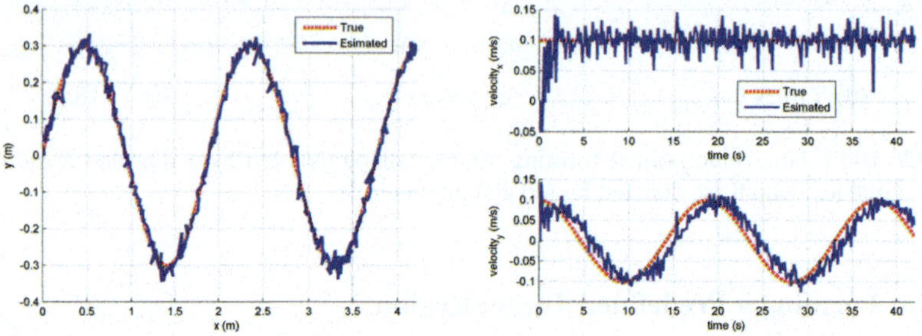

(a) Position (b) Velocity

Fig. 7.14 Position and velocity estimation results using the DEIF

filter (DEIF) [302] is applied with a constant velocity target model. It will be difficult for the swarm of robots to closely follow the moving target while forming the desired ring shape, if the target velocity is unknown, as shown in Fig. 7.13a. By contrast, the robots can successfully surround the moving target if the target position and velocity can be estimated by the DEIF filter, as shown in Fig. 7.13b and the estimation results are plotted in Fig. 7.14. The tracking results when both the position and velocity of the target are known are shown in Fig. 7.13c.

7.4.3.2 Adaptive Formation of Complex Regions

To entrap multiple targets, the desired region generated by the upper layer of the H-GRN may become very complex and needs to be represented using Splinegon. Figure 7.15a shows a case of multiple stationary targets with an initial guess of a neighborhood size of 0.55 m, which fails to stabilize the motion inside the region and maintain the given distance between robots. By adjusting the neighborhood size with $d_{max} = 1.0$ m and $d_{min} = 0.1$ m through iterations of adaptation as shown in Fig. 7.16, the final neighborhood size of 0.4234 m is obtained resulting in an even distribution of the robots, as shown in Fig. 7.15b.

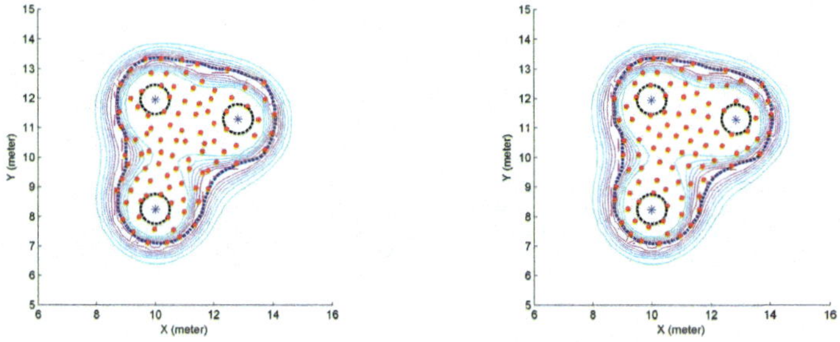

(a) Iteration = 1 (b) Iteration = 6

Fig. 7.15 Entrapping of multiple static targets with neighborhood size adaptation. By adjusting the neighborhood size through iterations using the bumper range d_{min} and the maximum sensor range d_{max}, the final size of 0.4234 m is obtained at iteration 6

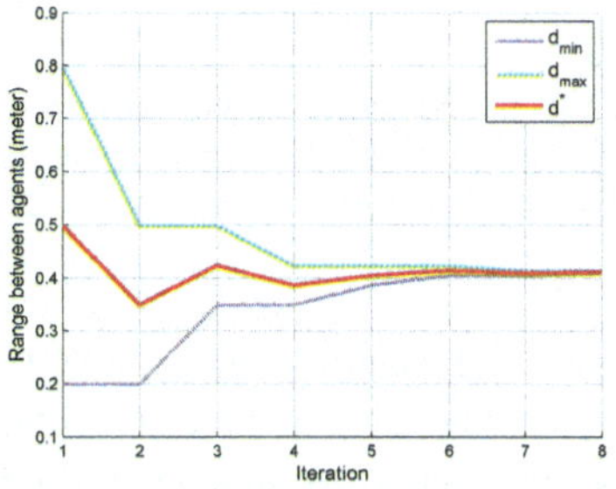

Fig. 7.16 Neighborhood distance adaptation over iterations

Finally, the entrapping of multiple moving targets with a changing complex shape is shown in Fig. 7.17. As the targets move away from each other, the desired shape is dynamically changing thanks to the upper layer of the H-GRN. After 20 s, the targets stop moving, and the robots are able to organize themselves into a region that surrounds the targets, as shown in Fig. 7.17.

7.5 Evolving Genetic Motifs for Morphogenetic Pattern Generation

7.5.1 Motivations

A significant limitation of most existing pattern formation algorithms for multi-robot systems is that the target shape needs to be predefined and forms only a small number of simple geometric shapes, which may become inadequate for dealing with unknown complex environments. To address this problem, Beta and Kumar [29] proposed a formation control method for a large number of robots based on an abstraction of

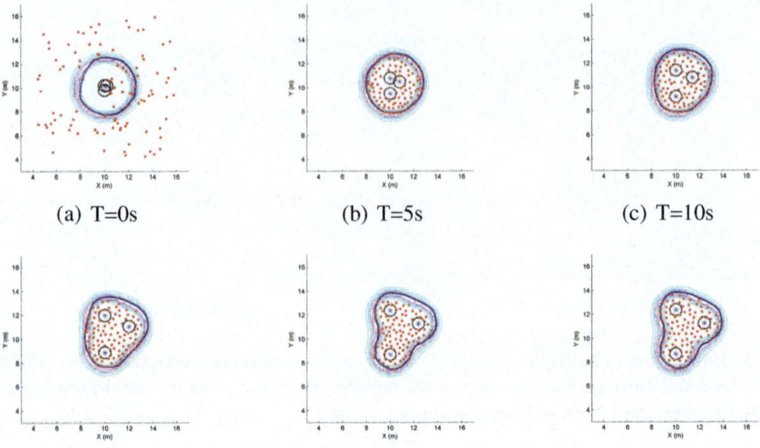

(a) T=0s (b) T=5s (c) T=10s

(d) T=15s (e) T=20s (f) T=35s

Fig. 7.17 Entrapping of multiple moving targets with a changing complex region. As the targets move away from each other, an entrapping shape is changing accordingly, and consequently, the robots are organizing themselves inside a shape

the team to a small dimensional manifold independent of the number of robots. Although target shapes can be adaptable to environmental changes by changing the shape size (as demonstrated with a narrow tunnel passing scenario), only two simple shapes are considered and the final formation results in the loose distribution of robots inside the shape. Cheah et al. have introduced a region-based shape control for robot swarms based on potential field, where each robot in the group stays within a moving region while maintaining a minimum distance from each other [54, 171]. However, the complexity of the target shapes is still limited, and the target shape must be predefined.

The H-GRN-based morphogenetic self-organizing systems introduced in the previous sections are able to generate patterns that are adaptive to the positions of the targets to be trapped. Nevertheless, the architecture of the upper layer in the H-GRN must be manually defined, and its capability to generate diverse patterns may be limited. In addition, the algorithm must rely on a repulsive avoidance scheme using a diffusion term of robots and obstacles to avoid obstacles. To fix these weaknesses, we introduce an algorithm [311] that can automatically generate the GRN by evolving a number of widely seen network motifs described in Sect. 2.2 as the building blocks. The evolved GRN shall be able to generate different yet suitable shapes for robots to traverse an unknown dynamic environment. This is termed an evolving hierarchical gene regulatory network (EH-GRN).

The EH-GRN also consists of two layers. The upper layer is for adaptive pattern generation where the GRN model is evolved by network motifs with genes and environmental inputs, and the lower GRN layer is for driving robots to the target pattern generated by the upper layer. This framework enables us to generate patterns adaptive and robust to environmental changes. Particularly, by introducing obstacles as one of the environmental inputs along with targets, the weakness of the patterns generated by the H-GRN can be mitigated because it allows all the robots in the swarm to simultaneously accomplish both missions, namely entrapping targets and avoiding obstacles. In addition, the EH-GRN framework improves the flexibility of the pattern generation to be applied for various tasks as it has an open structure of gene regulations in a GRN model to some extent. Separability (splitting or merging) of multiple patterns resulting from target movement is addressed by the inherent feature of the upper layer and the k-means clustering algorithm [37]. More recent research along the same line can be found in [254, 310].

7.5.2 Network Motifs

Five basic network motifs, including the positive, negative, AND, OR, and XOR regulatory motifs, are used here as the building blocks to construct the GRN in the upper layer.

Firstly, a positive (or negative) regulation is defined as gene X activates (or inhibits) or poses a positive (or negative) feedback to gene Y. The mathematical description of the positive regulation from X to Y can be represented as follows:

$$\frac{dy}{dt} = -y + \text{sig}(x, \theta), \tag{7.49}$$

$$\text{sig}(x, \theta) = \frac{1}{1 + e^{-k(x-\theta)}}, \tag{7.50}$$

where x and y represent the expression level of gene X and Y, respectively, and θ is a regulatory parameter for the gene expression. Similarly, the negative regulation is represented as follows:

$$\frac{dy}{dt} = -y + (1 - \text{sig}(x, \theta)). \tag{7.51}$$

In the AND regulation, gene Y expresses only if both genes X_1 and X_2 express, as defined by,

$$\frac{dy}{dt} = -y + \text{sig}(g_1 \cdot g_2, \theta), \tag{7.52}$$

where g_1 and g_2 are the expression levels of gene X_1 and X_2, respectively. The expression level of each gene is governed by a positive or negative regulation with a regulatory parameter θ_1 or θ_2. In the OR regulation logic, gene Y expresses if either gene X_1 or X_2 expresses:

$$\frac{dy}{dt} = -y + \text{sig}(g_1 + g_2, \theta). \tag{7.53}$$

Finally, in the XOR regulation logic, gene Y expresses only if genes X_1 and X_2 have disparate expression levels, as defined by

$$\frac{dy}{dt} = -y + \text{sig}(g_1 \cdot (1 - g_2), \theta) + \text{sig}((1 - g_1) \cdot g_2, \theta). \tag{7.54}$$

Thus, if gene X_1 expresses and gene X_2 does not express (or vice versa), gene Y expresses. These regulation motifs will be the basic building blocks of the GRN structure, and a regulatory parameter θ will be optimized by an evolutionary algorithm.

7.5.3 Evolving H-GRN Based on Network Motifs

7.5.3.1 Overall Framework

The general structure of the proposed EH-GRN using network motifs to be evolved is illustrated in Fig. 7.18. Firstly, the upper layer of the EH-GRN is for adaptive pattern generation where the GRN model is evolved by the combination of basic network motifs with genes and environmental inputs. In Fig. 7.18, p_1 and p_2 represent the protein concentrations produced by the environmental inputs, i.e., the target and obstacle if any, and diffused spatially. These will serve as the inputs to the upper GRN layer and activate genes g_1, g_2, and g_3. In particular, the concentration of M plays the role of morphogen to form the desired shapes around targets, which will

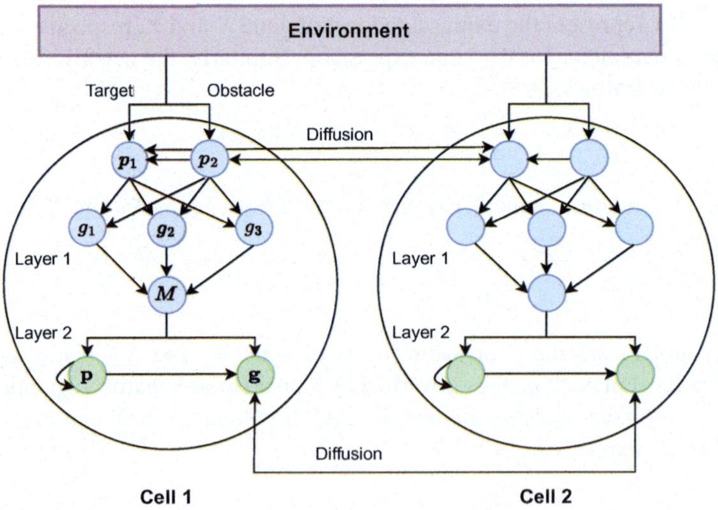

Fig. 7.18 An EH-GRN consisting of two layers for target entrapping

be transmitted to the lower layer. Note that the dynamics of the GRN in the upper layer are activated only in the organizing robots that are able to detect targets, while the non-organizing robots simply follow the movement of neighboring organizing robots. By introducing obstacles as one of the environmental inputs along with targets, a target entrapping pattern itself changes as an obstacle approaches for all the robots to entrap targets and avoid obstacles at the same time. Here, only the upper layer is evolved with the fixed number of genes. However, the entire structure including the lower layer and a varying number of genes can be used with the same methodology exploiting basic network motifs.

Once the target pattern is generated by the upper layer, it will function as the input of the lower layer to trigger its dynamics. The lower layer is responsible for pattern formation control guiding all robots on to the generated pattern and maintaining a minimum distance between the robots at all times. These functions are realized by another GRN dynamics where proteins $\mathbf{g}$ and $\mathbf{p}$ are used to represent the current position and internal state of the robot. With the diffusion of p_1, p_2, and $\mathbf{g}$, the robots can share the information on the target and the obstacle and maintain the desired minimum distance among robots in order to avoid robot-to-robot collisions.

7.5.3.2 Upper Layer of the EH-GRN

The generation process of an adaptive target entrapping pattern consists of two steps, forming a morphogen gradient space according to the evolving GRN model and extracting candidate points whose gradient value is higher than a threshold and representing candidate points with a Non-Uniform Rational B-Splines (NURBS) [326] to generate target pattern for robots to follow, as illustrated in Fig. 7.19. A morphogen gradient space is first generated by the regulation parameters and environment inputs using the upper layer as shown in Fig. 7.19a, and the points whose gradient value is higher than a certain percentage of the maximum value are selected as candidate points in Fig. 7.19b. The desired morphogen gradient space for target entrapping requires as many candidate points as possible that are not too close (d_{min}) to or far away (d_{max}) from targets and maintains a minimum distance (d^{obs}_{min}) to the obstacle. Thus, the following fitness function for the evolutionary algorithm can be defined:

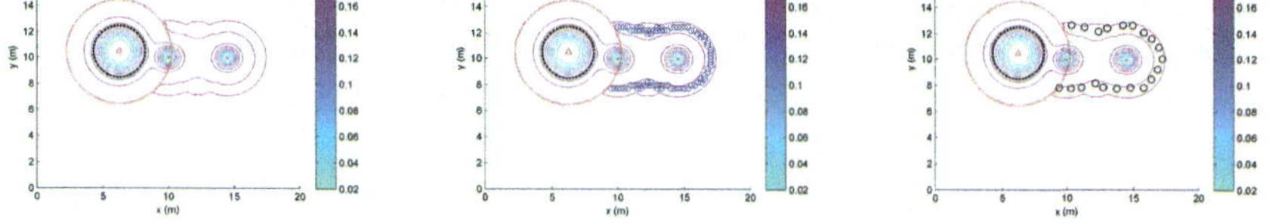

(a) Morphogen gradient sapce (b) Highest concentration (c) NURBS points

Fig. 7.19 NURBS representation of a target entrapping pattern

$$f = \sum_{i=1}^{N_p}\sum_{j=1}^{N_t} \frac{\mathrm{sig}(d_{pt}^{ij}, d_{\max}, k_1) + \mathrm{sig}(d_{\min}, d_{pt}^{ij}, k_2) - A_{ij}}{N_p N_t} + \sum_{i=1}^{N_p}\sum_{k=1}^{N_o} \frac{\mathrm{sig}(d_{po}^{ik}, d_{\min}^{obs}, k_3)}{N_p N_o} \tag{7.55}$$

where

$$A_{ij} = \begin{cases} 1, & \text{if } d_{\min} \le d_{pt}^{ij} \le d_{\max} \\ 0, & \text{otherwise} \end{cases} \tag{7.56}$$

and N_p, N_t, and N_o represent the number of candidate points, targets, and obstacles, respectively. d_{pt}^{ij} denotes a distance between i-th candidate point and j-th target, and d_{po}^{ik} denotes a distance between i-th candidate point and k-th obstacle. A_{ij} is the candidate point satisfies the restriction as:

For the upper layer, there are 26 parameters to be optimized: θ_i $(i = 1, \ldots, 13)$ for a regulatory parameter between environmental inputs p_i $(i = 1, 2)$ and genes g_i $(i = 1, 2, 3)$, and genes and morphogen gradient M as shown in Fig. 7.18; t_i $(i = 1, \ldots, 9)$ for a positive or negative regulation; and c_i $(i = 1, \ldots, 4)$ for AND or OR regulation. XOR regulation is not used here for simplicity. t_i can be 0 (positive regulation) or 1 (negative regulation), and also c_i can be 0 (AND regulation) and 1 (OR regulation). For instance, if $t_1 = 1, t_2 = 0$, and $c_1 = 0$, then the partial dynamics between target p_1, obstacle (p_2), and gene 1 (g_1) can be described as follows:

$$\frac{dg_{11}}{dt} = -g_{11} + \mathrm{sig}(p_1, \theta_1), \tag{7.57}$$

$$\frac{dg_{12}}{dt} = -g_{12} + (1 - \mathrm{sig}(p_2, \theta_2)), \tag{7.58}$$

$$\frac{dg_1}{dt} = -g_1 + \mathrm{sig}(g_{11} \cdot g_{12}, \theta_7), \tag{7.59}$$

where g_{11} and g_{12} are internal genes between environmental inputs and g_1. The entire upper layer dynamics of the EH-GRN can be constructed in a similar manner using the defined 26 parameters.

The covariance matrix adaptation evolution strategy (CMA-ES) presented in Sect. 1.1.2 is used to optimize the regulatory parameters. Since CMA-ES performs optimization in a continuous domain, discrete parameters such as t (positive/negative regulation) and c (AND/OR logic) need to be transformed into continuous values. For example, if the parameter is less than 0.5, it is considered as 0, and otherwise 1. The population size is set to 400, and the CMA-ES was run for 50 generations. θ_7, θ_8, θ_9, and θ_{13} are assigned to be real numbers from 0 to 2 and the rest of the parameters to be real numbers from 0 to 1. The resulting evolved structure of the upper layer is shown in Fig. 7.20 with the following parameter values: $\theta_1 = 0.0891$, $\theta_2 = 0.2764$, $\theta_3 = 0.0783$, $\theta_4 = 0.8363$, $\theta_5 = 0$, $\theta_6 = 0$, $\theta_7 = 1.1956$, $\theta_8 = 0.2957$, $\theta_9 = 0.8352$,

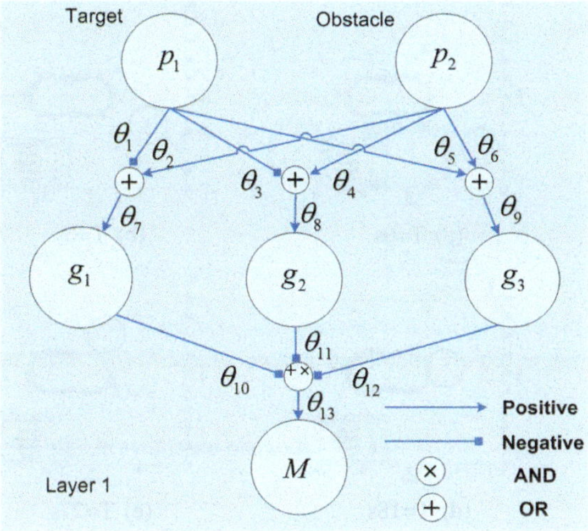

Fig. 7.20 Illustration of an evolved two-layer H-GRN structure for target entrapping

$\theta_{10} = 0.4230$, $\theta_{11} = 0.3256$, $\theta_{12} = 0.3226$, and $\theta_{13} = 1.4299$. In Fig. 7.20, morphogen M that defines a gradient space for the target pattern is regulated by (i) g_1 activated by both p_1 and p_2, which finally inhibits M, (ii) g_2 inhibited by p_1 and p_2, which activates M, and (iii) g_3 activated by p_1 and inhibited by p_2. Thus, gene g_1 can be considered making the target pattern being not too close to both targets and obstacles and g_2 is for the target pattern not far away from both environmental inputs. The combinatorial regulation of gene g_1 with a regulatory parameter $\theta_{10} = 0.4230$ and g_2 with $\theta_{11} = 0.3256$ to M makes the target pattern a band of circle as shown in Fig. 7.19b. Lastly, gene g_3 works as activating a morphogen gradient (M) value when there is target expression and inhibiting it when there is obstacle expression, adding one more design dimension to the target shape.

Note that as the targets move away from or get closer to each other, the number of entrapping patterns can be varied. Thus, this pattern separability aspect is addressed using the k-means clustering algorithm to group candidate points based on attributes into predefined k number of groups [37]. The grouping is done by minimizing the sum of squares of distances J between points and the corresponding cluster centroid. As there is no initial knowledge on the number of patterns, the number of cluster k starts from one and increases by one while comparing the cost J in order to find the best k.

The lower layer GRN is the same as that in the previous sections.

7.5.4 Numerical Simulations

Below we show some illustrative simulation results using scenarios containing static/moving targets as well as a moving obstacle to validate the feasibility and

7.6 Morphogen Diffusion Self-organizing Minimalist Robots

7.6.1 Motivations

The morphogenetic self-organizing algorithms presented in the previous sections are able to form complex patterns and self-organize swarm robots to cover the boundary or region of the desired patterns, even in the presence of dynamics, e.g., when the targets to be surrounded are moving. However, these algorithms require that the robots are able to self-localize themselves and are able to detect the distance and orientation of other objects in the environment, including other robots, obstacles, or targets. These requirements are not always feasible, making it less practical to construct large-scale swarm systems consisting of hundreds or even thousands of robots. In general, the functional simplicity of agents is essential for swarm robotic systems [277, 377]. The functionally simpler a robot is, the more affordable it will be and the smaller its size can be. Moreover, a smaller and simpler robot consumes less energy and can potentially operate for a longer period of time. However, control of a swarm of minimalist robots is also challenging [297].

In collective flocking, a large number of robots move toward a target, where swarm connectivity becomes more challenging because of limited sensing ability and possible disruptions in local interactions [346]. Lack of directional sensing, non-holonomic motion, and no motion feedback together highly increases the chance of collisions as the number of robots in the swarm increases, which has a destructive impact on motion planning.

7.6.2 Kilobots: A Minimalist Robot Platform

In the following, three algorithms for self-organizing minimalist robots using morphogen diffusion will be presented. The algorithms were designed based on the settings of kilobots, the swarm robot testbed developed by Rubenstein et al. [341]. For locomotion, kilobots use two sealed coin-shaped vibration motors with three sticks, adopting the slip-stick principle [438], as shown in Fig. 7.24. Kilobots can make both clockwise and counter-clockwise rotations as well as straight movements by differentially controlling the magnitude of vibration of two motors. The slip-stick locomotion significantly reduces the hardware complexity, robot size as well as cost compared to typical wheel drive robotic systems. For communication, an infrared (IR) emitter and transmitter are used while allowing digital message transmission between robots at rates up to 30 kb/s within a range of 10 cm, and a distance to a neighboring robot is estimated by measuring the intensity of the incoming IR light. Current setting is sending a message at about every 0.1 s by interrupt and receiving process is triggered at the end of each byte coming. The kilobot also has an ambient light intensity sensor. Furthermore, an overhead infrared transmitter is developed so

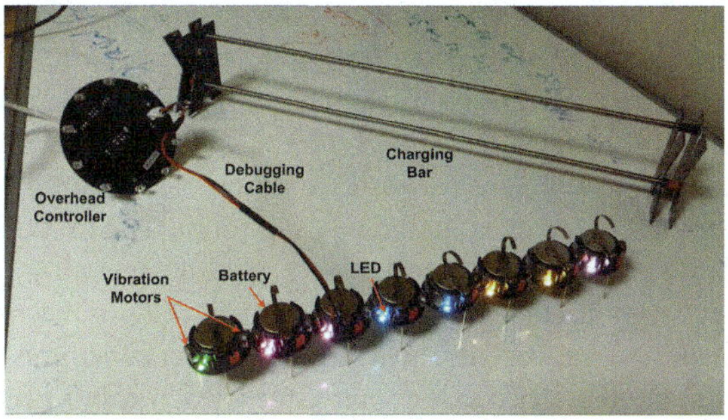

Fig. 7.24 Kilobot platform including overhead controller and battery charging bar

that the power can be switched on and off and the program can be uploaded remotely to all robots in the collective at the same time regardless of the number of robots, making the system scalable to a large number of robots. Battery charging is also scalable by hanging the robots over a charging bar as shown in Fig. 7.24. More details of the kilobots can be found in [341].

It should be pointed out that the functions of the kilobot testbed are highly constrained. Its IR communication is up to 10 cm, distance and ambient light sensing do not provide orientation, and there is no self-localization capability, meaning that kilobots know neither their position nor orientation as there is no encoder/sensor to trace the movement or dead-reckoning. In addition, their locomotion made by vibrating motors is sensitive to the surface and calibration settings and therefore inaccurate. Besides, on starting a vibration motor, a spin up period is required by applying a high vibration power to overcome the friction and inertia, causing unpredictable slipping, bouncing off the surface, and control delays.

For prototyping patterning algorithms on a swarm of kilobots before testing them on real robots, a kilobot simulator called *kbsim* is used, which is designed for simulating the kilobots in Python code with Pygame library for a user interface by Halme [1]. The *kbsim* simulates the most important functionality of the kilobots in terms of sensing, communication, and movement; however, noise and disturbance from sensing, motor control, and physical interactions between robots are not explicitly considered.

7.6.3 Morphogen Diffusion for Tracking and Herding

7.6.3.1 Maintaining Connectivity

It is challenging yet interesting to design algorithms that can make movement while maintaining the connectivity of the robots with highly limited sensing, communication, and motion capabilities. The tasks of the algorithm to be presented below [309] are tracking and herding, which are basic swarm behaviors.

Assuming that an initial swarm group is formed and part of the robots in the group is in the effective range of a light source, the objective to be achieved here is to approach to the light source, which can be considered tracking and surrounding a stationary target. Meanwhile, the robots should maintain group aggregation and the network connectivity with the group [457]. For this purpose, spatiotemporal decision parameters are evaluated by each robot. The first one is $\alpha = N_k - N_{k-1}$ as a temporal network connectivity measure at time instant k, where N represents the number of neighbors. The second one is $\beta_k = N_k - \bar{N}_k$ as a spatial connectivity measure, where $\bar{N}$ represents the average number of neighbors' neighbors. By analyzing these parameters, the system can infer if each robot behaves correctly to maintain swarm coherence at the right place during the tracking. For instance, if α is less than zero, it means that the robot is losing some of its network connections, or if β is less than zero, it means that the robot is either at the edge or sparse area in the group. In those cases, the robot is about to lose swarm coherence, and therefore, a necessary action needs to be taken. To regain swarm coherence from this situation during an object tracking process, an approach similar to the one in [457] can be taken, in which a lost robot will turn backwards (180 degrees) immediately, and otherwise go toward a light source (target). The object tracking can be achieved by three behaviors depending on the network connectivity:

$$\begin{cases} \text{Turn backwards,} & \text{if } N < N_{th},\ \alpha \leq 0,\ \beta < \beta_{th} \\ \text{Straight,} & \text{if } N < N_{th},\ \alpha > 0,\ \beta < \beta_{th} \\ \text{Toward target,} & \text{otherwise,} \end{cases} \tag{7.60}$$

where the values N_{th} and β_{th} are positive threshold values for the number of neighbors and the spatial connectivity measure, respectively.

When a robot does not have enough neighbors ($N < N_{th}$) and is at the edge of the group ($\beta < \beta_{th}$) while losing network connections ($\alpha \leq 0$), the robot should turn backwards to avoid losing swarm coherence. When the robot is in an unfavorable situation when $N < N_{th}$ and $\beta < \beta_{th}$, but $\alpha > 0$, the robot should go straight for a few time-steps to gain more network connections and secure swarm coherence. Otherwise, the robot should move toward a target at the center of a light source. The direction of travel toward a target is obtained by performing a few random movements until the right direction is identified. The robot will make random movements if it is out of the light source range. The benefit of this algorithm is that it requires the minimal exchange of information between robots, which is the number of neighbors N to accomplish the mission. Considering the locomotion of kilobots, the 'turning backwards' behavior is achieved by performing one direction turning for a certain number of time-steps continuously. Note that the 'turning backwards' behavior does not necessarily need to be a precise 180-degree turn; instead, it is sufficient to make a turn in the opposite direction to maintain swarm coherence.

Figure 7.25 shows the sample process of target (light source) tracking. An initial robot group is given and the part of the group is within the effective range of the light source represented as a solid circle line. The robots outside the light source range initially move randomly, but as the remaining robots move toward the light source,

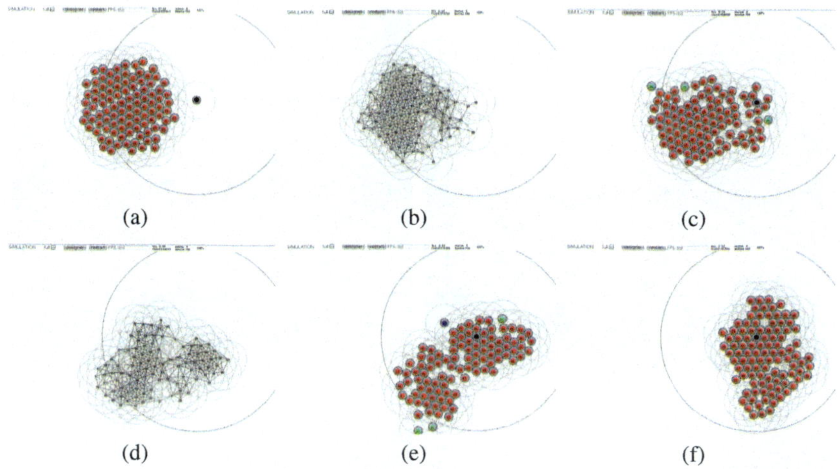

Fig. 7.25 Light source tracking process while maintaining network connectivity using the kilobot simulator. Parameter settings: $(N_{th}, \beta_{th}) = (3, 2)$. In the figure, large circles denote the effective range of a light source, and small circles the communication range. A robot in red means the robot is moving toward the target, a blue robot means it is going straight, and a green robot is turning backwards

they were attracted to the swarm by 'turning backwards' and 'go straight' behavior. Throughout the process of Fig. 7.25a–d, the entire robot swarm successfully moves toward the light source (the robot with a black dot) to surround it while maintaining swarm coherence. Moreover, a proof-of-concept experiment is performed using 11 kilobots as shown in Fig. 7.26, which demonstrates swarm movement toward the light source in spite of the ambient light sensing noise and inaccuracy.

7.6.3.2 Morphogen Gradient Following

Next, morphogen diffusion is used for object tracking. To maintain swarm coherence while tracking a moving object, each robot moves in a way that maintains its desired gradient concentration values. A morphogen gradient is initiated by a target object (or origin) by sending its neighbors a message with an initial value of $h = 0$. This gradient value h is forwarded outwards from the origin (target) incremented by one within a communication range of r. Each robot maintains the minimum gradient value (termed as concentration hereafter) among received ones and ignores messages containing larger values to prevent the gradient from backwardly diffusing. This gradient diffusion forms concentric circular rings of a width of approximately r and of different colors with corresponding concentration values around a target as shown in Fig. 7.27. It is worth mentioning that the concentration value can provide an estimation of the shortest distance d of the robot from the target by the relation $d \leq h \cdot r$, which can then be improved by using an average neighbor density and smoothing algorithms, as explained in [303].

(a)

(b)

(c)

(d)

Fig. 7.26 Light source tracking using kilobots with parameters $(N_{th}, \beta_{th}) = (3, 2)$

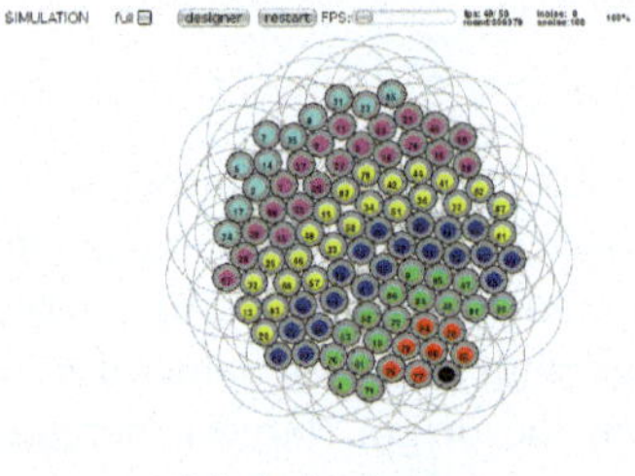

(a) A single target

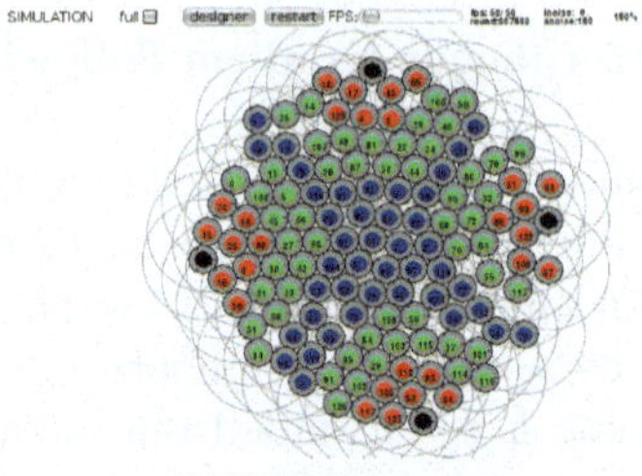

(b) Multiple targets

Fig. 7.27 Morphogen diffusion from the target (the robot with a black dot). Different colors represent different classes of concentration values

To make the robot follow other robots with the corresponding concentration value, a distance-based attraction/repulsion controller is used. Assuming that the robot at the center is attracted to the homogeneous robot (blue) and repelled to the heterogeneous robot (red), the net direction of travel is supposed to be where the green arrow points. However, since the kilobot has no directional sensing and no self-localization capabilities, the movement direction is obtained indirectly by defining and analyzing the following objective function for each robot:

$$f = w_a sig(\Delta d_a, k_a) + (1 - w_r sig(\Delta d_r, k_r)), \tag{7.61}$$

where

$$\Delta d_a = \sum_{j=1}^{N_a} \frac{d_j - d_{a,th}}{N_a}, \quad \Delta d_r = \sum_{j=1}^{N_r} \frac{d_j - d_{r,th}}{N_r}, \tag{7.62}$$

$$sig(x, k) = \frac{1}{1 + e^{-kx}}, \tag{7.63}$$

where k_a and k_r represent the slope of a sigmoid function, w_a and w_r are positive weighting factors, N_a and N_r are the number of robots to be attracted and repelled, respectively, and d_j is the distance between the robot in concern and its neighbor robots. $d_{a,th}$ and $d_{r,th}$ are the threshold distance that determines an effective sensing range. From the above equation, it can be shown that the less the value f is, the better the robot is located for the attraction and repulsion mission. Thus, the robot keeps its previous movement direction if the time derivative of the objective function is less than zero ($df/dt < 0$). Otherwise, the robot will move randomly. If the objective of the robot is to be either attracted or repelled to the others, then only the corresponding term in the function f is used. In the following, two gradient following algorithms for object tracking are proposed depending on the concentration values.

7.6.3.3 Following the Morphogen Gradient

To follow an object while maintaining swarm coherence, the robot is programmed to follow its preceding neighbors whose morphogen concentration is lower than its own using the attraction term in Eq. (7.61). By doing this, robots are attracted to each other as well as going toward the target with only local communication of the morphogen gradient and distance between them. This allows a robot swarm to aggregate in one group around a target as long as the robot is connected to a target via multiple communication hops, and consequently to follow a moving target as a group. Figure 7.28 shows the process of tracking and surrounding a randomly moving target using morphogen diffusion and distance-based attraction controller with a kilobot simulator. In the beginning, the robots are randomly distributed, and a gradient value is propagated from a target (colored as black) in the center to the other robots. The robots in white, which are not connected to the target through

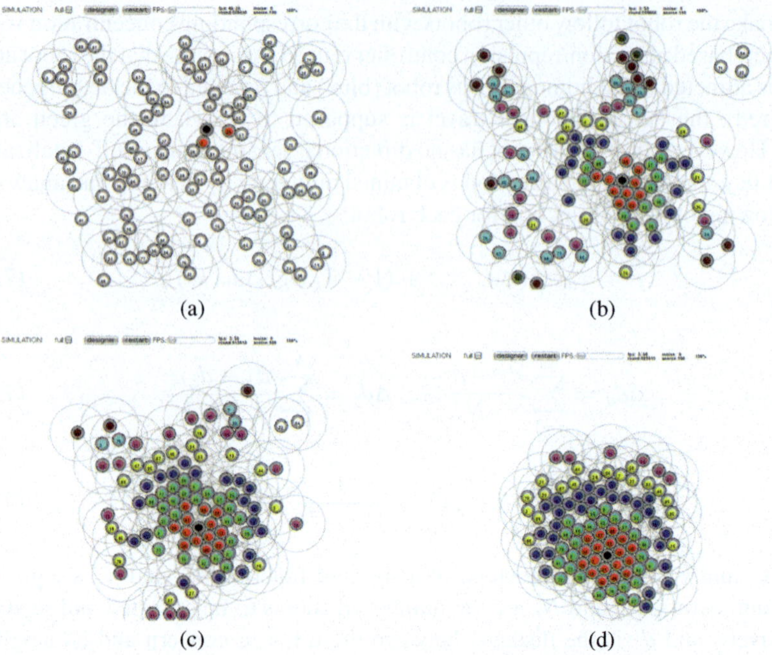

Fig. 7.28 Process of surrounding a moving target by following robots having a lower concentration of morphogen with a distance-based attraction controller with parameters $(d_{a,th}, k_a, w_a) = (69, 0.05, 128)$

morphogen diffusion, randomly move until they obtain a gradient value from their neighbors. The robots are progressively attracted toward the target by following their neighbors whose concentration is lower than their own, leading to surrounding the moving target successfully. Figure 7.29 shows a preliminary experiment result of the same process using 23 kilobots.

7.6.3.4 Maintaining Morphogen Gradients

To make robots aggregate around a target as well as maintain a desired formation initially given, the difference between the initial concentration value and current value is included in the objective equation along with the attraction term as follows:

$$f = w_a sig(\Delta d_a, k_a) + w_h |\Delta h|, \tag{7.64}$$

where $|\Delta h| = h_{ini} - h_{current}$ is the difference between initial and current concentration value, and w_h is a positive weighting factor. The robots should maintain their initial morphogen concentration so that they can be positioned at the same relative location from the target as the initial configuration. This is similar to the previous

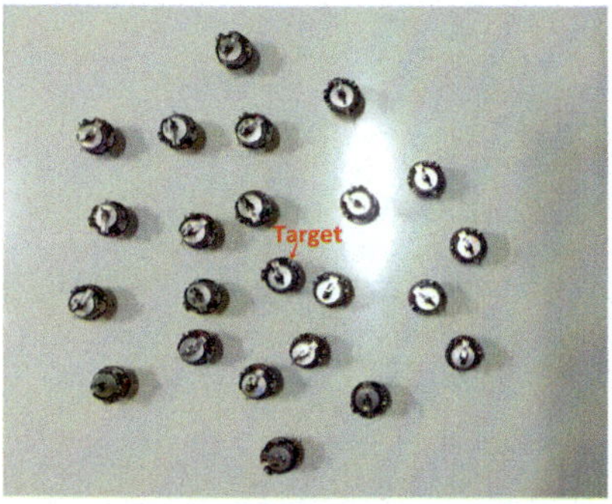

(a)

(b)

(c)

(d)

Fig. 7.29 Experiment result of surrounding a moving target by following the robots having a lower concentration with a distance-based attraction controller. Parameter settings: $(d_{a,th}, k_a, w_a) = (80, 0.05, 128)$

algorithm if the morphogen concentration remains the same during the tracking process. However, the robot should follow other robots whose concentration is higher or lower than itself to restore their initial value, if the concentration of the robot is changed. As a result, the initial spatial arrangement during the tracking process can be loosely maintained, as shown in Fig. 7.30. An initial group spatial arrangement remains similar to the initial one in terms of the number of the same colored robots throughout Fig. 7.30. Figure 7.31 shows some preliminary results for nine kilobots tracking a moving object.

7.6.3.5 Morphogen Diffusion-Based Herding

In the following, an algorithm for object tracking and herding is presented, which is essentially a combination of the light source tracking and gradient following behaviors. To achieve this, the following cost on the distance from a light source is considered along with the objective function in Eq. (7.64):

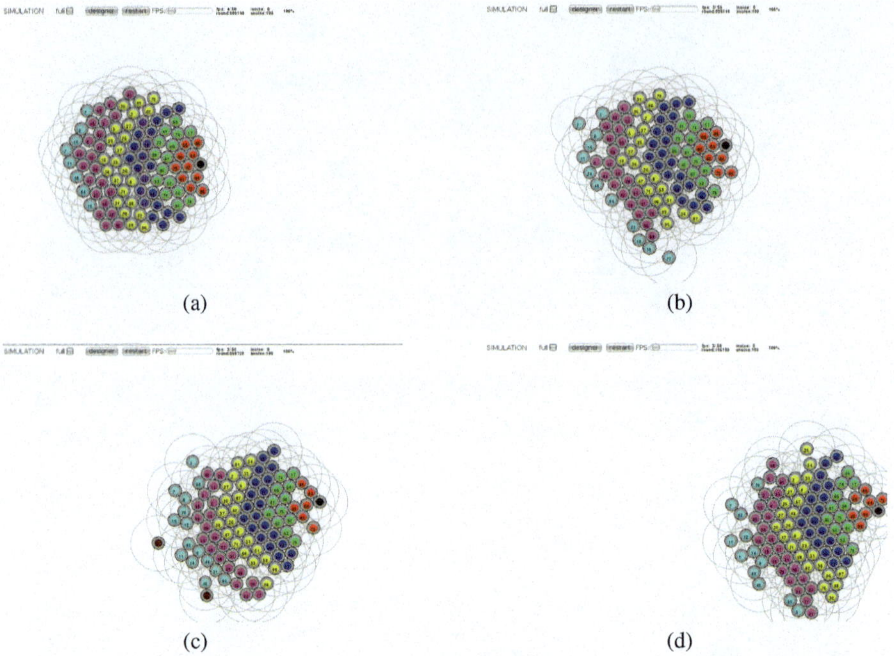

Fig. 7.30 Object tracking by maintaining the initial morphogen concentration for formation keeping. Parameter settings: $(d_{a,th}, k_a, w_a, w_h) = (69, 0.05, 128, 20)$

$$f = w_a sig(\Delta d_a, k_a) + w_h |\Delta h| + w_l sig(\Delta d_l, k_l), \tag{7.65}$$

where w_l is a positive weighting factor. $\Delta d_l = d_l - d_{l,th}$, where d_l is the distance between the robot and the light source, and $d_{l,th}$ is the threshold distance. Thus, the above objective function ensures the robots maintain their initial morphogen concentration and move toward the light source.

Consider a scenario in which one stationary target at the center of the swarm group, as shown in Fig. 7.32a, is already surrounded by the robots, and the part of the group is within the effective range (denoted by solid black circles) of the second target. By applying Eq. (7.65), herding of the first target to the second target can be accomplished, as shown in Fig. 7.32. Since robots should maintain their initial concentration values, the target in the group is pushed toward the second target, which is a light source, by the surrounding robots as they are moving toward the second target. Once some robots get close enough to have a concentration value diffused from the second target, they maintain the received minimum concentration value. Then, the robots at the edge of the group having a high concentration value will become an inner robot with respect to the second target with the lowest concentration value, as shown in Fig. 7.32b. Therefore, they keep moving outwards until they restore their initial concentration values. After a sufficient period of time, both targets are surrounded by the robots forming one swarm group with a similar spatial arrangement to the

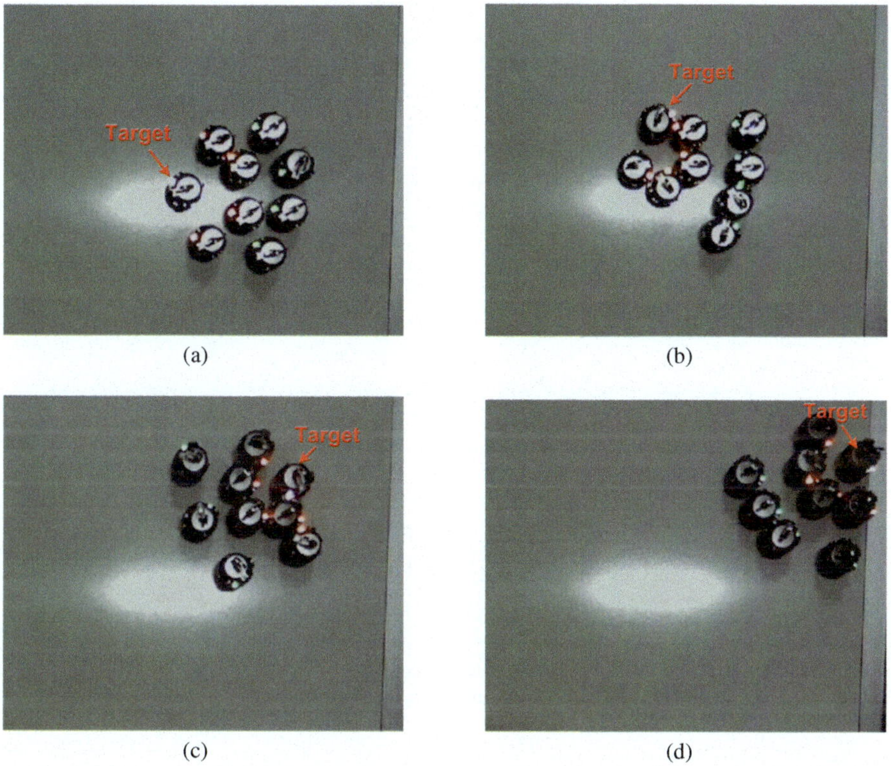

Fig. 7.31 Object tracking by maintaining the initial morphogen concentration. Parameter settings: $(d_{a,th}, k_a, w_a, w_h) = (75, 0.05, 128, 35)$

initial formation, as presented in Fig. 7.32d. The above behaviors can be considered as herding of the first target to the other target.

7.6.4 Self-organized Coordinated Motion of Minimalist Robots

7.6.4.1 Motivations

As discussed in the previous section, many minimalist robots, such as kilobots, do not have self-localization capability or directional sensing. In addition, kilobots move with three pins, which very easily leads to collisions and entanglements. To solve this problem, coordinated movement of the robots becomes particularly important and the algorithm developed in [382] will be introduced below.

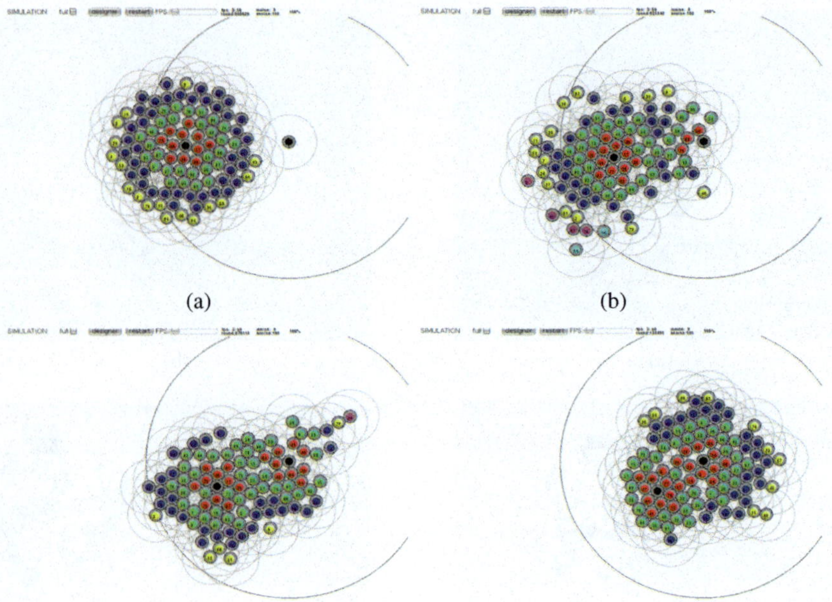

(a) (b) (c) (d)

Fig. 7.32 Herding one target to the other by combining the gradient following and light source tracking behaviors. Parameter settings: $(d_{a,th}, d_{l,th}, k_a, k_l, w_a, w_h, w_l) = (65, 30, 0.02, 0.01, 128, 50, 100)$

To achieve this, one indispensable step is to establish a local coordinate system via local robot-robot communications. Consider a swarm of robots, where each robot is able to measure its distance to the robots inside its communication range. Each robot in a swarm can be seen as a node in a graph, which is termed a communication graph. Node A in the graph is linked to node B if node B is in the communication range of node A. Based on the assumption that the communication range of all robots is equal, the communication graph's links are bidirectional. If a node has three localized nodes with known distances in its neighborhood, then it is able to calculate its own relative position using trilateration. To this end, one assumption must be made that the robots in the swarm are properly distributed so that the communication graph is connected, i.e., each node has more than two links to other nodes. In this way, positional information can propagate through the swarm by trilateration.

To establish a coordinate system for localization, the first step is to determine the center of the local coordinate system to be established. In [332], a mechanism for identification of the robot at the center of a swarm is proposed, termed center-bot, based on artificial morphogen diffusion [200, 309]. Here, it is assumed that all robots are initially aggregated so that they are able to communicate with at least their immediate neighbors. Consequently, the algorithm starts with the differentiation of the agents at the boundary of the swarm, termed edge-bots, based on the comparison of the number of neighboring robots of this robot's neighbors. Then, all edge-bots act as a source of morphogen and concentration gradient forms. Considering a convex shape swarm, this gradient will be the minimum at the center. Hence, it is possible to identify the robot that is at the center of the swarm by comparing the concentration level of a robot with its neighbors. Refer to [332] for more details. Once the center-bot is identified, its localization status is set to *determined*, and its position to (0, 0).

At first, the center-bot determines the center of the coordinate system at first in setting up the coordinate system. Next, the center-bot propagates the ID of its closest neighbor. This closest neighbor, named x-bot, determines the x-axes of the swarm coordinate system. Once x-bot has received a message from center-bot and identified itself as the x-bot, it sets its position to $(d_{cx}, 0)$, where d_{cx} is the distance between the x-bot and center-bot. Then, it changes its localization status to be determined and propagates its position.

Finally, a third bot, termed h-bot, needs to be identified, which will determine the handedness of the coordinate system. The h-bot should be within the communication range of both the x-bot and center-bot and should not be close to the line passing through the x-bot and the center-bot. This is because other robots may fail to properly localize themselves because of noisy measurements otherwise.

Once the x-bot has identified itself, it sets one dedicated bit in the status byte to label itself as the x-bot. Consequently, those robots that can communicate with the x-bot set another bit in the status byte to indicate that '*I see the x-bot*', and send a message containing their distance from the x-bot. Once this is done, one robot can be chosen as the h-bot by the center-bot from those that can see the x-bot. The robot that minimizes the following criterion is eventually chosen as the h-bot, and its ID is propagated by the center-bot.

$$\min_i(|d_{cx} - d_{ix}| + |d_{cx} - d_{ci}|), \tag{7.66}$$

where d_{cx} and d_{ci} are the distances from the center-bot to the x-bot and robot i, respectively, and d_{ix} is the distance between robot i and the x-bot, which is measured and sent by robot i.

Because of omnidirectional sensing and the lack of global information, the determination of the handedness of the coordinate system is impossible without movement. Note that at this stage, the flip ambiguity of the coordinate system is ignored at first. Once the robots start to move, this flip ambiguity will be detected and corrected. Therefore, it is assumed that the h-bot is on the top of the x-axis, and the h-bot can calculate its relative position as $[d_{hc}\cos(\alpha), d_{hc}\sqrt{(1-\cos(\alpha)}]$, where d_{hc} is the distance between the h-bot and the center-bot measured by the h-bot, and $\cos(\alpha)$ is equal to

$$\cos(\alpha) = \frac{d_{cx}^2 + d_{hc}^2 - d_{hx}^2}{2d_{cx}d_{hc}}, \tag{7.67}$$

where d_{hx} is the distance from the h-bot to the x-bot, measured by the h-bot, and $d_{c}x$ is the distance between the center-bot and the x-bot, which can be calculated by the h-bot after they propagate their relative positions.

Until now, three robots in the swarm are aware of their position, which will enable all robots that can communicate with them to figure out their relative positions using trilateration. Once each robot calculates its relative position, it propagates its estimated position to its neighbors, and eventually the position information is propagated through the whole swarm. Whenever one of the x-bot or h-bot leaves the neighborhood of the center-bot, the center-bot will determine a new pair of x-bot and h-bot and propagates their IDs. Then a new procedure for calculating and propagating the positional information will be triggered.

7.6.4.2 Motion Coordination

For localization and coordination of the movement of minimalist robots, some robots must stay still for a while to serve as beacons for others for the following reasons. First, it takes time to calculate the position and propagate it to neighboring robots because of limited computational and communication capacity. If all robots are moving, the displacements of the robots during this time will significantly impact the accuracy of the estimated positions based on trilateration. Moreover, due to the lack of the capability to observe other robots' velocity and heading, an increased number of collisions between the robots will occur when all robots are moving. Finally, it is difficult to approach a moving target robot based on the distance measurement when the velocity of the target robot is unknown.

Consequently, there will be a number of stationary robots surrounding each moving robot to avoid collisions. This way, the mobile robot will also have a sufficient number of immobile neighbors to localize itself.

An illustrative example is given in Fig. 7.33 of stationary and moving robots, in which a moving robot in blue is surrounded by a 'polygon' of immobile robots in green inside a region within its communication range R_c. These neighboring robots are termed the first-order natural neighbors. To maximize the number of mobile robots, it is desired that only the first-order natural neighbors of a mobile robot stay still. Thus, a robot shall be able to distinguish between its first-order natural neighbors with other neighboring robots. Therefore, the next step is to identify the first-order natural neighbors for each robot. In the following, we implement the Delaunay triangulation method to generate a graph that links each robot to its first-order natural neighbors. For a planar straight-line graph $G = (V, E)$, Delaunay triangulation returns a subgraph $G_D = (V, E')$, where $E' \subseteq E$ such that each node $v_i \subseteq V$ is linked to its first-order natural neighbors but not high order neighbors [241]. Here, triangulation is adopted to design a motion schedule that divides the robots into two groups, one mobile and the other immobile, where all mobile robots are surrounded by immobile ones.

Each robot carries out triangulation once each of its neighbors has localized itself and broadcasts a message containing its estimated position. It can happen, however,

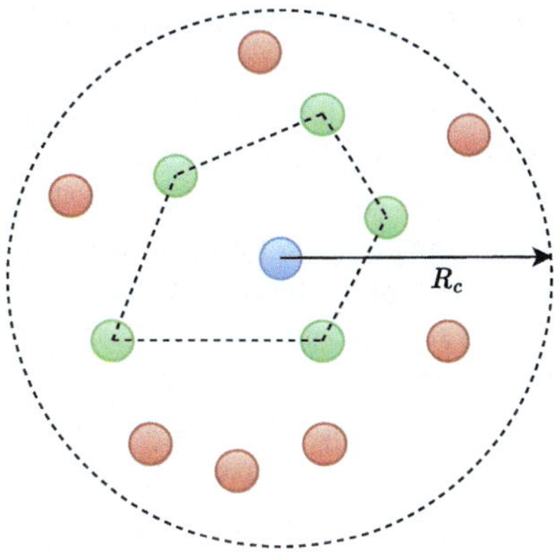

Fig. 7.33 First-order natural neighbors of an agent (in green) form the operational polygon for movement coordination

that a robot fails to localize itself due to the lack of three position-aware robots in its neighborhood or missing messages in communication. In this case, the robot will wait for a certain period of time before triangulation. Therefore, a robot performs triangulation either when all its neighbors are localized or when a certain period of time is passed.

The Delaunay triangulation algorithm in [241] is adopted here. Assume $\mathbf{T}_v$ is initialized using the positional information received from the neighboring nodes of node v. Thus, $\mathbf{T}_v^{i,pos}$ stores the i-th neighbor's estimated position, $\mathbf{P}_{u_i}$. Likewise, $\mathbf{T}_v^{i,id}$ stores the associated ID of the i-th neighbor, ID_{u_i}. Let $\mathbf{S}_v \subset \mathbf{T}_v$ be the nodes that are linked to node v in the Delaunay graph G_D. At first, the algorithm sorts the elements in $\mathbf{T}_v$ based on the increasing angle. It has been proved that the link to the nearest neighbor, $\min(\overline{vu_i})$, is a member of E'. Therefore, the array $\mathbf{T}_v$ is shifted till the nearest neighbor locates at the beginning of the array. Then, a local optimization procedure (LOP) is performed on every three radially consecutive neighboring nodes starting from the nearest neighbor to remove the links that are not in the Delaunay graph from $\mathbf{T}_v$. Function $LOP\left(\mathbf{P}_v, \mathbf{P}_{u_i}, \mathbf{P}_{u_{i+1}}, \mathbf{P}_{u_{i+2}}\right)$ denotes the results of the optimization, where $\mathbf{P}_v$ is the position of node v, and $\mathbf{P}_{u_i}$, $\mathbf{P}_{u_i+1}$, and $\mathbf{P}_{u_{i+2}}$ are the positions of three radially consecutive neighboring nodes of node v. The optimization procedure returns $true$ if $\mathbf{u_{i+1}}$ is a member of $\mathbf{S}_v$, and $false$ if it is not.

After performing localization and triangulation, a robot is able to distinguish its immediate neighbors. In the following, a motion scheduling is required to make sure that when a robot becomes mobile, its immediate neighbors must remain immobile. The motion scheduling strategy also ensures that an immobile robot will be switched to mobile once all its first-order natural neighbors have become mobile. Details of motion scheduling can be found in [382].

7.6.4.3 Motion Planning and Collision Avoidance

A motion planning method is then designed to achieve the following objectives. First, it should be able to preserve the swarm connectivity. Second, it should minimize the number of collisions. Third, it should allow for the maximum progress toward the target during each motion cycle.

To avoid collisions with other mobile robots, each mobile robot must remain inside the operational polygon defined by its immediate immobile robots. To this end, a minimum distance d_{min} to its immediate neighbors needs to be defined. In addition, we define a maximum distance d_{max} from a robot's immediate neighbors for preserving swarm connectivity. To maximize the progress toward the target during each motion cycle, each robot will calculate an alternative direction along which the robot can still progress without violating the constraints. Let $\overrightarrow{\mathbf{u}_i\mathbf{v}}$ be a member of the vector set $\mathcal{O}$ if it satisfies one of the two conditions:

$$\mathcal{O} = \begin{cases} \textbf{If } |\overline{\mathbf{vu}_i}| < d_{\min} \vee \angle\mathbf{Gvu}_i < \frac{\pi}{2} \textbf{ Then } \overline{\mathbf{vu}_i} \in \mathcal{O}, \\ \textbf{If } |\overline{\mathbf{vu}_i}| > d_{max} \vee \angle\mathbf{Gvu}_i > \frac{\pi}{2} \textbf{ Then } \overline{\mathbf{vu}_i} \in \mathcal{O}. \end{cases} \tag{7.68}$$

Assuming $\widehat{\mathbf{G}}$ is the unique vector pointing the target direction from the robot, the following equation calculates the alternative direction $\overrightarrow{\mathcal{H}}$ in which the robot can move:

$$\vec{\mathbf{O}} = \sum_{\overline{\mathbf{vu}_i} \in \mathcal{O}} \frac{\overline{\mathbf{vu}_i}}{|\overline{\mathbf{vu}_i}|} \tag{7.69}$$

$$\begin{cases} \textbf{If } \angle\mathbf{GvO} >= 2\theta_T \textbf{ Then } \overrightarrow{\mathcal{H}} = \widehat{\mathbf{G}} - \frac{\vec{\mathbf{O}}}{|\vec{\mathbf{O}}|}, \\ \textbf{If } \angle\mathbf{GvO} < 2\theta_T \textbf{ Then } state_v \leftarrow immobile. \end{cases} \tag{7.70}$$

In the above equation, θ_T specifies the maximum difference between the alternative direction and the goal direction that is allowed, $\mathcal{O}$ is a set containing all directions that violates the constraints, and $\vec{\mathbf{O}}$ is the resulting directions. The direction of the vectors engaged in motion planning is illustrated in Fig. 7.34. To estimate the goal direction, $\widehat{\mathbf{G}}$, a robot uses information from their immediate neighbors only. This information can be either some environmental variables, such as the intensity of a chemical, heat, and light, or some internal variables that are calculated using the information of neighbors' neighbors.

To define an additional term for Eq. (7.70) that provides more control over preserving connectivity, a vector $\vec{\mathbf{F}}$, termed coherence force, is defined as follows:

$$\vec{\mathbf{F}} = \frac{\sum_{u_i \in \mathbf{S}_v} \overrightarrow{vu_i}}{\sum_{u_i \in \mathbf{S}_v} |\overrightarrow{vu_i}|}. \tag{7.71}$$

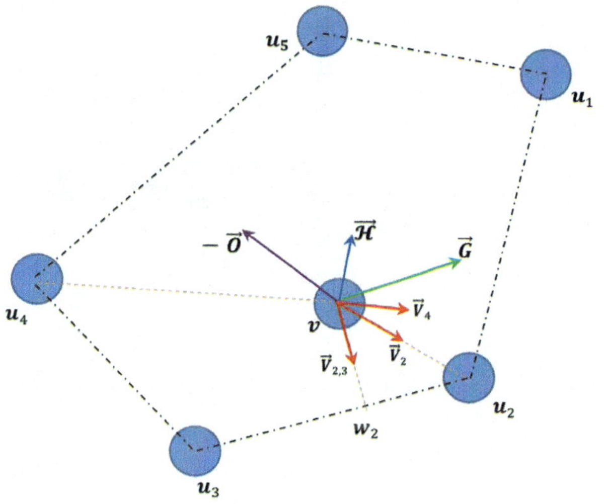

Fig. 7.34 Vectors engage in motion planning of robot v. The dash-dot lines represent the polygon defined by the immediate neighbors of robot v

Taking the coherence force into account, the alternative direction $\overrightarrow{\mathcal{H}}$ is updated as follows:

$$\overrightarrow{\mathcal{H}'} = \overrightarrow{\mathcal{H}} + \beta \overrightarrow{\mathbf{F}} . \tag{7.72}$$

As the coherence force is divided by the sum of the immediate neighbors' distances, its magnitude depends on the number of the immediate neighbors and their distances. Hence, this force becomes larger at the boundary of the swarm where there are fewer immediate neighbors, or in an area in which there are a large number of robots, resulting in a much shorter distance to their immediate neighbors. An illustrative example of the magnitude and direction of the coherence force for a swarm with an irregular shape is shown in Fig. 7.35, from which we can see that the direction of the coherence force points toward the center of the swarm when the robot is at the boundary and to some vacant space when it is inside the polygon.

A positive value for β can result in a tendency of moving toward the center of the polygon defined by the immediate neighbors. For the edge robots, which are outside of this polygon, the coherence force is toward the center of the swarm, which can promote connectivity. Note that the value of β will influence the shape of the swarm during flocking and it is found that negative β values will prevent the swarm from becoming elongated.

Details about the communication of information between robots can be found in [382].

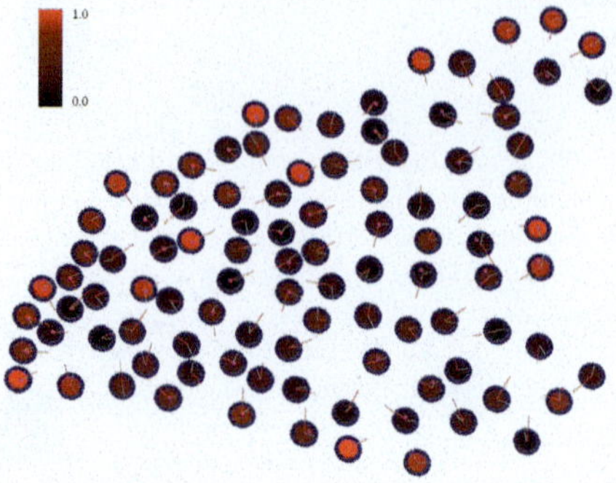

Fig. 7.35 The magnitude and direction of the coherence force. The red lines indicate the direction of the coherence force

7.6.4.4 Results and Discussions

To evaluate the effects of the coordination algorithm, computer simulations are performed in the kilobot simulator at first. Table 7.5 lists the parameter settings used in the simulations. For simplicity, a robot stops moving once it cannot see any localized robot in its neighborhood.

For minimalist robots, it is important that they stay connected in communication as aggregation is difficult once a robot is disconnected from the swarm in the absence of global positional information. To quantitatively assess the performance of the coordination algorithm in maintaining the swarm, a metric that is sensitive to every single loss of communication is proposed, which measures the connectivity of a swarm with a limited communication range defined as follows:

$$\text{connectivity} = \frac{|E_{cr}|}{|E_{\infty}|}, \tag{7.73}$$

where E_{cr} represents the number of links in the Delaunay graph when the communication range is equal to cr, while E_{∞} is the number of links in the Delaunay

Table 7.5 Parameter settings used in the simulations

Parameter	Description	Chosen value
$d_{\min}$	Equation (7.68)	47 mm
$d_{\max}$	Equation (7.68)	80 mm
θ_T	Equation (7.70)	$\pi/5$
β	Coherence force coefficient Eq. (7.72)	−0.2
δ_{MAX}	Maximum allowed time to be in the mobile state	30 s

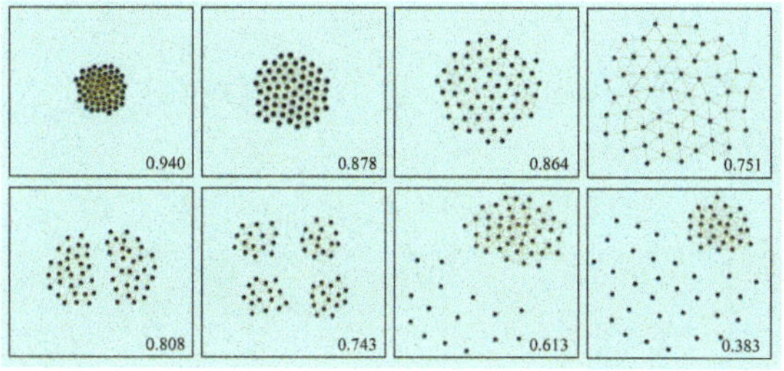

Fig. 7.36 Connectivity metric for different swarm arrangements. The robots within their communication range are connected by a line

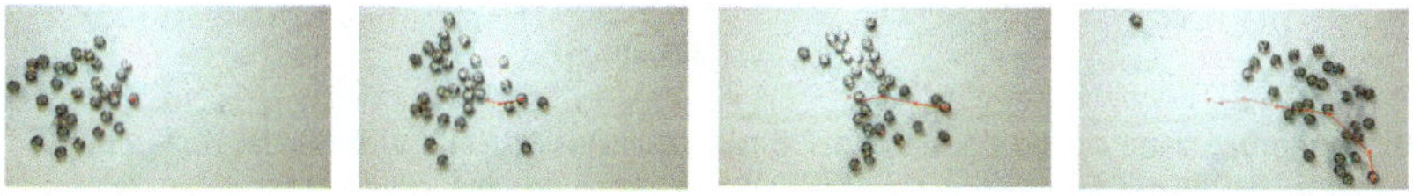

(a)

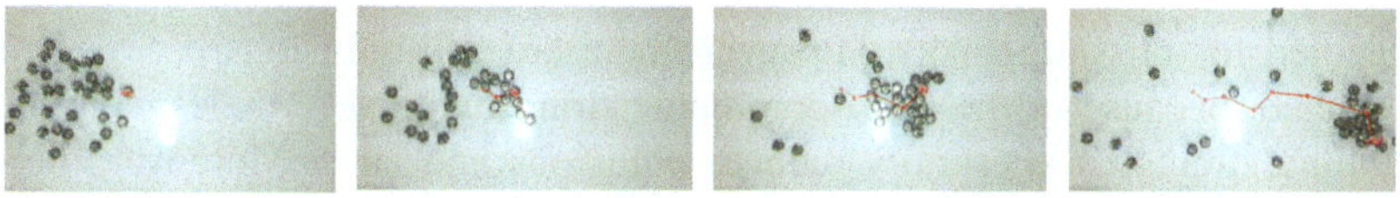

(b)

Fig. 7.37 Morphogen diffusion-based target following (**a**) with and (**b**) without the coordinated motion scheme. The red dots in each snapshot represent the location of the target robot in the current and previous snapshots

graph when the communication range is equal to infinity. The value of the proposed connectivity metric for different swarm arrangements is plotted in Fig. 7.36. From these results, it can be seen that this metric is sensitive to the loss of communication between robots as well as the swarm density. The connectivity metric becomes larger when the robots move closer to each other.

To evaluate the efficiency of our algorithm, the connectivity performance of the swarm for morphogen diffusion-based target following with or without coordinated motion planning is compared. For maintaining connectivity, a robot makes a turn if the number of neighbors it has decreases below a certain threshold. As mentioned before, a robot updates its morphogen concentration using information only from the robots with a lower morphogen concentration. Consequently, the robot's morphogen concentration stops changing if no robot has a lower level of concentration. In the algorithm, if there is no change in the morphogen concentration after a motion step, a robot will move randomly. Hence, if a robot is disconnected from the swarm, it starts to do a random walk till it finds a robot with a lower target morphogen concentration.

The snapshots of the experimental comparative results for the following two targets are shown in Fig. 7.37. It can be seen from Fig. 7.37a that with the coordination movement, the robots are able to maintain connectivity even if no motion strategy

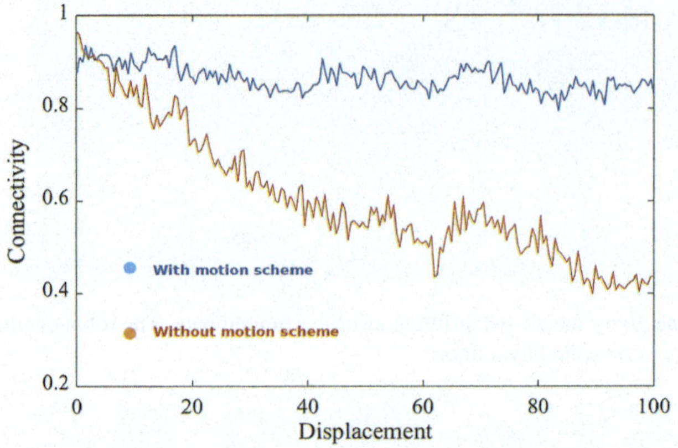

Fig. 7.38 Comparison of the connectivity of two swarms with or without motion scheme versus the displacement of the target

is specified for missing robots. However, there are still collisions between robots as expected, because of the short communication range. Nevertheless, the motion scheduling strategy successfully reduces collisions and jams. Moreover, the edge robots tend to move ahead of their front neighbors by moving in the direction perpendicular to the target direction because of the outward coherence force, resulting in a better flocking and more space for the inner robot of the swarm to move by preventing the swarm from becoming elongated.

By contrast, collision jams occur very quickly after the initiation of the target following without the localization and motion scheduling. The robots in the collision jams lag behind the others, leading to disconnection. Once a group of robots are disconnected from the rest, the robots with the highest target morphogen concentration will move randomly and the others will follow it. As we can see from Fig. 7.37b, there are many robots disconnected from the swarm during target following.

Finally, the change of the connectivity metric of the swarm during the target following process is plotted in Fig. 7.38. The experimental results confirm that the proposed algorithm for self-organizing coordinated motion performs much better than the algorithm without using a motion scheme. From these results, we can see that the algorithm without motion scheduling loses its connectivity continuously during the flocking. By contrast, the swarm is able to maintain the connectivity within a certain range when the motion coordination is implemented.

7.6.5 Regulated Morphogen Gradients for Target Surrounding and Adaptive Shape Formation

7.6.5.1 Motivations

In this section, the algorithm for surrounding static targets using minimalist robots with limited omnidirectional sensing capability is described [383]. Assume that the targets are directly detectable only by the robots that are within the communication range of the target and the robot that receives this signal will pass it to its neighbors. This way, it is possible to simulate morphogen diffusion that allows the robots to have a rough estimation of their distance to the target. The desired behavior is to enclose a target or targets evenly from all sides, which can be realized by forming a circular shape swarm. To this end, the robots need to be able to identify the positions on the edge of a swarm where more robots should aggregate using a second morphogen regulated by the target morphogen.

7.6.5.2 Target Morphogen Diffusion and Regulation

Here we assume that morphogen gradients emitted by targets can be detected and interpreted by the robots in the neighborhood of the targets. These robots can also pass morphogen concentration information onto their neighboring robots and this process continues until the diffusion process stabilizes. The following equation is to simulate morphogen diffusion:

$$\frac{C^i}{dt} = \alpha \sum_{j}^{N^i} \left(\frac{R}{d^{ij}} \right) \cdot C_j - \gamma \cdot C^i \qquad \begin{array}{c} \alpha, \gamma > 0, \\ 0 < C^i < C_{\mathrm{MAX}}, \end{array} \tag{7.74}$$

where C^i represents the morphogen concentration of robot i, C_{MAX} is the maximum concentration, d^{ij} indicates the distance between robot i and robot j, R is the radius of each robot for the purpose of normalization and non-dimensionalization, N^i is the number of neighbors of robot i, and α and γ denote the diffusion rate and decay rate, respectively. The morphogen concentration in an aggregation with a morphogen source in the middle is shown in Fig. 7.39. The level of the morphogen concentration is the highest in the source location (C_{MAX}). The morphogen concentration of the robots in terms of their distances to the source at different time instances starting from the beginning of diffusion is plotted in Fig. 7.39b. By choosing proper values for α and γ, a gradient of morphogen concentration can form from the source toward the boundary of the swarm aggregation, which reaches a steady state within a short period of time.

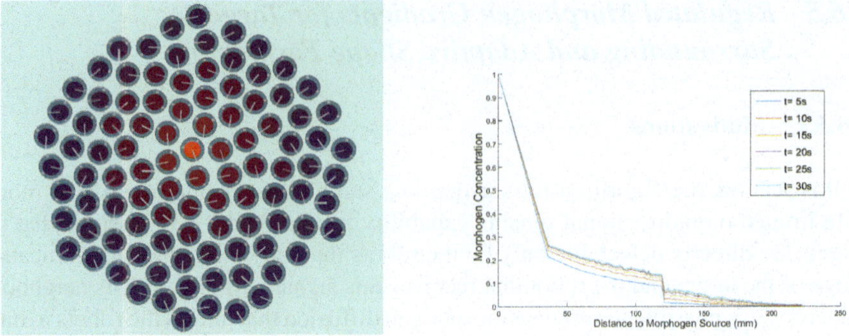

(a) (b)

Fig. 7.39 Morphogen diffusion from a source in the middle of an aggregation. **a** The intensity of red color indicates the level of morphogen concentration. **b** Morphogen concentration in terms of distance to the source

7.6.5.3 Target Tracking

In [332], a target tracking algorithm using morphogen diffusion for minimalist robots without directional sensing and self-localization is suggested. Based on the calculated morphogen concentration in the previous positions, a robot is able to decide in which direction and how much it should rotate in order to be aligned with the direction of the increasing or decreasing morphogen concentration. The direction of morphogen gradients can be estimated using a number of previous estimated positions in the robot's coordinate system. A robot can estimate its position in its local coordinate system as follows:

$$\mathbf{P}_{t+1} = \mathbf{P}_t + \hat{\boldsymbol{\theta}}_t V dt, \tag{7.75}$$

where $\mathbf{P}_t$ is the current position, $\mathbf{P}_{(t+1)}$ is the next position, V is the velocity of the robot, $\hat{\boldsymbol{\theta}}_t$ is a unit vector determining the direction of the movement, and dt is the change of time. In this way, a robot can approximately self-localize its position. The direction of the morphogen gradient can be estimated using the following equations:

$$\Delta \mathbf{C}_{t,i} = \frac{\mathbf{P}_{t,i} - \mathbf{P}_{t-1,i}}{||\mathbf{P}_{t,i} - \mathbf{P}_{t-1,i}||}(C_{\mathbf{P}_{t,i}} - C_{\mathbf{P}_{t-1,i}}), \tag{7.76}$$

$$\mathbf{C}_{T,i} = \sum_{T=t-N_S}^{t} \Delta \mathbf{C}_{T,i}, \tag{7.77}$$

$$\hat{\mathbf{C}}_{T,i} = \frac{\mathbf{C}_{T,i}}{||\mathbf{C}_{T,i}||}, \tag{7.78}$$

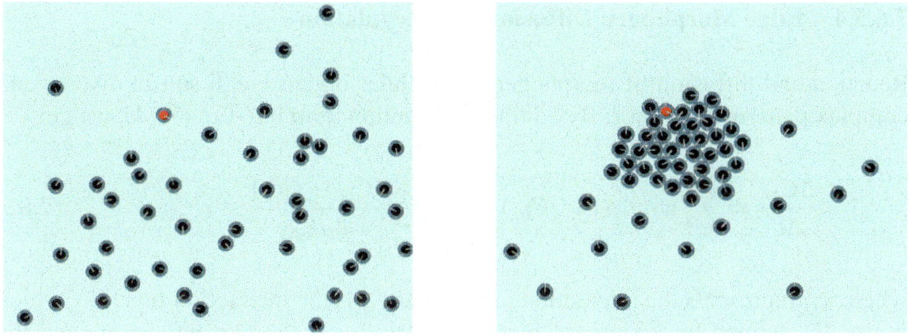

(a) (b)

Fig. 7.40 Tracking a target does not always end up with fully surrounding the target

where $\mathbf{P}_{t,i}$ represents the position of robot i at time-step t, $C_{\mathbf{P}_{t,i}}$ is the morphogen concentration measured at position $\mathbf{P}_{t,i}$, N_s is the number of the previous time-steps taken into account to evaluate the morphogen gradient, $||.||$ returns the magnitude of a vector, and $\hat{\mathbf{C}}_{T,i}$ is a unit vector indicating the orientation of the morphogen concentration.

At each iteration, a robot updates its current direction using Eq. (7.78):

$$\Delta\boldsymbol{\theta}_t = \Delta\hat{\mathbf{C}}_t - \hat{\boldsymbol{\theta}}_{t-1}, \tag{7.79}$$

$$sat(\Delta\theta_t) = \begin{cases} ||\Delta\boldsymbol{\theta}_t|| = \Delta\boldsymbol{\theta}_{max}, & if||\Delta\boldsymbol{\theta}_t|| > \Delta\theta_{max} \\ \Delta\theta_t, & otherwise \end{cases}, \tag{7.80}$$

$$\hat{\boldsymbol{\theta}}_t = \hat{\boldsymbol{\theta}}_{t-1} + sat(\Delta\theta_t). \tag{7.81}$$

In experiments, we need to take limitations of the rotational speed into account, i.e., the change in direction will be bounded if the change in direction in one step of the simulation is greater than $\Delta\theta_{\text{max}}$, referring to Eq. (7.80). The same mechanism is used for approaching a diffusive target. Target tracking only cannot guarantee a complete and robust enclosure of the target on all sides. As shown in Fig. 7.40, tracking and reaching a target may not lead to a reliable surrounding of the target when the distribution of the robots around the target is extremely uneven. However, an even surrounding of the target is non-trivial for robots with only limited local awareness. In the following, it will be shown that it is possible to guide robots to the parts of a swarm where insufficient robots are aggregated by taking advantage of morphogen diffusion.

7.6.5.4 Edge Morphogen Diffusion and Regulation

Reaction and diffusion of morphogens in cellular organs can result in diverse and complex patterns [77, 280]. By adding the reaction term into Eq. (7.74), we get

$$\frac{\Delta C_{bi}}{\Delta t} = \sum_j w_{ij} f_{ij}(C_{bi}, C_{bj}) + D_i \sum_{b'=1}^{N_b} \frac{(C_{bi} - C_{b'i})}{n_i d_{bb'}} - r_i C_{bi}, \tag{7.82}$$

where w_{ij} represents the interaction matrix and f_{ij} is the interaction function, which is usually defined by a sigmoid function. Turing was the first who explained the reaction-diffusion of two morphogens [431], where one morphogen serves as an activator, while the other inhibits the expression of the activator. Turing's reaction-diffusion model has found many successful applications in the field of distributed control of self-organizing systems [310]. Inspired by Turing's reaction-diffusion model, a morphogen diffusion for the target surrounding is defined in Eq. (7.74), where the morphogen from the diffusive target in the middle of the aggregation inhibits the production of the second morphogen in the edge of aggregation. To achieve a circular-shape aggregation around the target, a morphogen, named edge morphogen, is introduced, which is produced and diffused only by the robots at the boundary of the aggregation, called edge robots. The diffusion-reaction equation of the edge morphogen can then be defined, which is regulated by the target morphogen:

$$\begin{aligned} \frac{dC_E^i}{dt} &= \varphi \cdot sig\left(C_{MAX} - C_T^i, k, m\right) - \omega \cdot sig\left(C_T^i, k, m\right) + \\ &\quad \alpha \sum_j^{N^i} \left(\frac{R}{d^{ij}}\right) \cdot \left(\frac{C_E^j - C_E^i}{C_{MAX}}\right) \cdot C_E^j - \gamma_E \cdot C_E^i \end{aligned} \tag{7.83}$$

$$sig(x, k, m) = \frac{2}{\pi} atan\left(k \cdot (x + m)\right), \tag{7.84}$$

where C_T and C_E represent the concentration of the target morphogen and edge morphogen, respectively. From the above equation, we can observe that the production rate of the edge morphogen increases by decreasing the target morphogen. Figure 7.41 illustrates the result of diffusion and regulation of the edge morphogen for an irregular-shape aggregation.

If a robot does not find another robot in its neighborhood with a higher edge morphogen, it will identify itself as an edge target point and set a certain bit in its message to inform the free robots that have not yet joined the swarm. The free robots perform target tracking until they join the aggregation around the targeted points. If a robot is close to the aggregation but there is no edge target point in its neighborhood, it starts to perform edge-following before reaching an edge target point.

To reduce collision, a robot stops moving if it has a robot in its neighborhood in waiting status and closer to the swarm. It means that the first layer of waiting robots

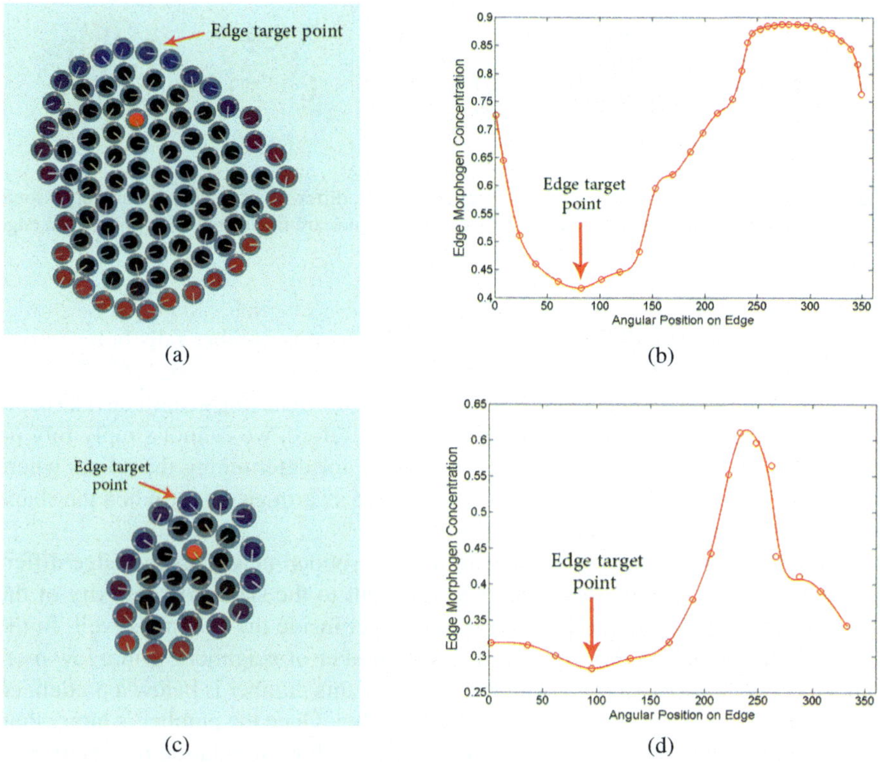

Fig. 7.41 Edge morphogen diffusion from a target at the middle of an aggregation indicated by red color. **a** Edge morphogen concentration for an aggregation of 100 robots. The edge robots' colors indicate the concentration of the edge morphogen, blue for low and red for higher concentration. **b** The curve represents the edge morphogen concentration of each robot in terms of its angular position which is defined as the angle that it makes with the target. **c** Edge morphogen concentration of edge robots for an aggregation of 30 robots, and **d** is its associated edge morphogen concentration

generates another layer of waiting robots. But this second layer does not change their status into waiting to avoid having the whole swarm stop moving. In this way, there are two layers of the robots that stop moving and allow robots in edge-following status joint the aggregation.

7.6.5.5 Edge Differentiation Based on Morphogen Concentration

In the algorithm, the robots at the boundary of the swarm should differentiate into edge robots to allow the production and diffusion of the edge morphogen for indicating where on the boundary a free robot should join. When an edge robot is covered by another layer of robots, it must differentiate back into a normal aggregated robot. It has been proposed that the number of neighbors or neighbors' neighbors is utilized

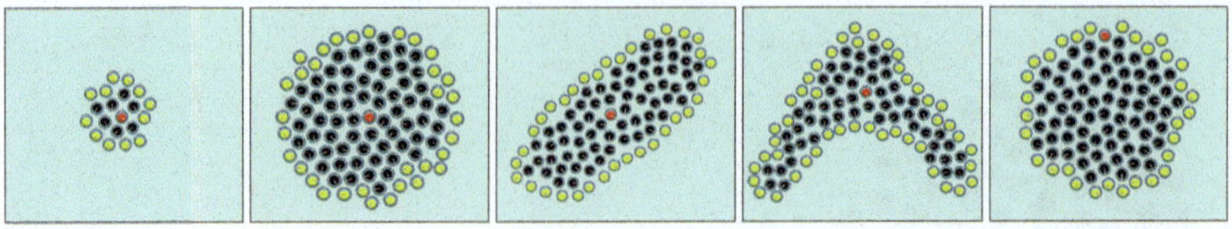

Fig. 7.42 Edge differentiation for robot aggregations with different sizes, shapes, and target locations. The red circle indicates the target, and the yellow ones are the agents differentiated to edge

as a criterion for edge differentiation [332, 347]. If this number falls below a certain threshold, a robot will differentiate into an edge robot. However, this threshold depends on the swarm size. Moreover, for a robot swarm having a non-convex shape, the above method may fail to work properly. Therefore, we cannot simply rely on the number of neighbors or neighbors' neighbors for determining the points where free robots should join as the size of the enclosure swarm grows, or when the shape is non-convex at some points of growth.

Using the target morphogen concentration of robots' neighbors, an edge differentiation criterion is suggested, which is invariant to the size and convexity of the shape of the swarm, and the location of the target inside the swarm as well. In the proposed algorithm [383], a robot counts the number of neighbors with a low morphogen concentration within a certain distance. If this number is below a predefined threshold, the robot differentiates into an edge robot. Once the number is larger than the threshold again, the robot is no longer an edge robot. Simulation results of edge differentiation for swarms of different sizes, shapes, and target locations are shown in Fig. 7.42. It can be seen from these results that the simulated kilobots identify themselves as edge robots if there are four robots within seven centimeters with a lower target morphogen concentration.

7.6.5.6 Results and Discussions

In the simulations, 100 simulated kilobots are scattered unevenly over a confined area with the dimensions of 120 by 120 cm. Figure 7.43 depicts the snapshots of the process of surrounding a single target. The results show that the algorithm is able to regulate the circular shape of the aggregation around the target.

In addition, the results of a multiple target enclosure are shown in Fig. 7.44, where the targets are too far away to be surrounded by a single swarm. The enclosure of the targets may end up with different sizes either due to the randomness in the initialization of the free robots, or due to the different locations of the targets in the field. However, when the targets are close enough to be surrounded by a single swarm, the final shape of the aggregation ends up with a single circular shape, as shown in Fig. 7.45. The diffusion of the edge morphogen through the boundary of all robot swarms regulates their growth. Finally, an experiment using 50 Kilobots for the

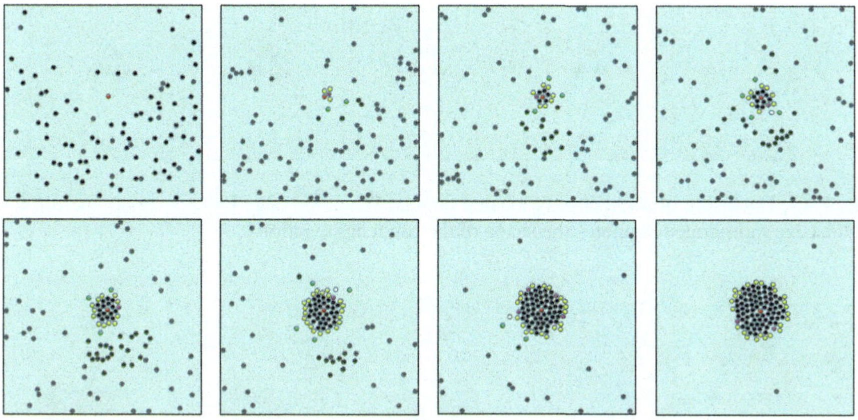

Fig. 7.43 Target enclosure over time by 100 robots. Time increases from up-left to bottom-right. The robots show yellow color once they differentiate into an edge robot, pink color for being edge target, green while doing edge-following, and white for direct target tracking

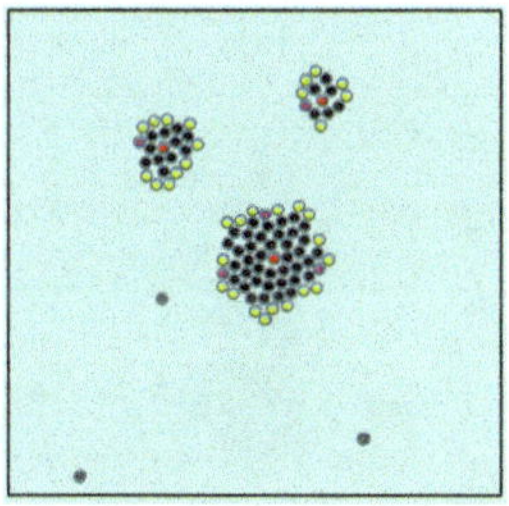

Fig. 7.44 Multi-target enclosure with three separated targets. If the edges of robot swarms do not overlap, the swarm may end up with different sizes

enclosure of one target. The experimental results are shown in Fig. 7.46. The robots are initialized by scattering unevenly over the field. The field itself is confined by barriers for kilobots. However, kilobots easily get stuck on the boundaries, especially in the corners due to their limited mobility. In this case, we need to rotate the stuck robots on the boundaries to let them move back into the field. Hence, the barriers are not covered in the photos and footages. The snapshots of the enclosure procedure are shown in Fig. 7.46a, from which we can conclude that the proposed algorithm works properly since the robot swarm successfully surrounds the target with a fairly evenly distributed circular shape. For comparison, the shape adaptation procedure is removed from the enclosure algorithm and conducted the experiment again. The results, as depicted in Fig. 7.46b, show that the final shape of aggregation is less regular and the enclosure around the target is poor.

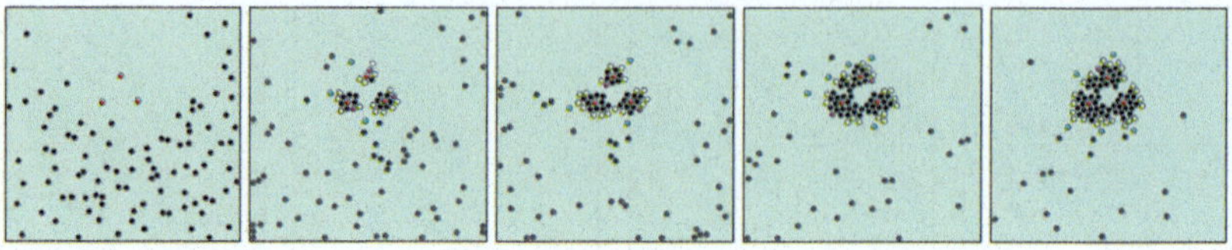

Fig. 7.45 Enclosure of three closely located targets. When the edge layers overlap, the diffusion of the edge morphogen regulates the shape of the robot aggregation

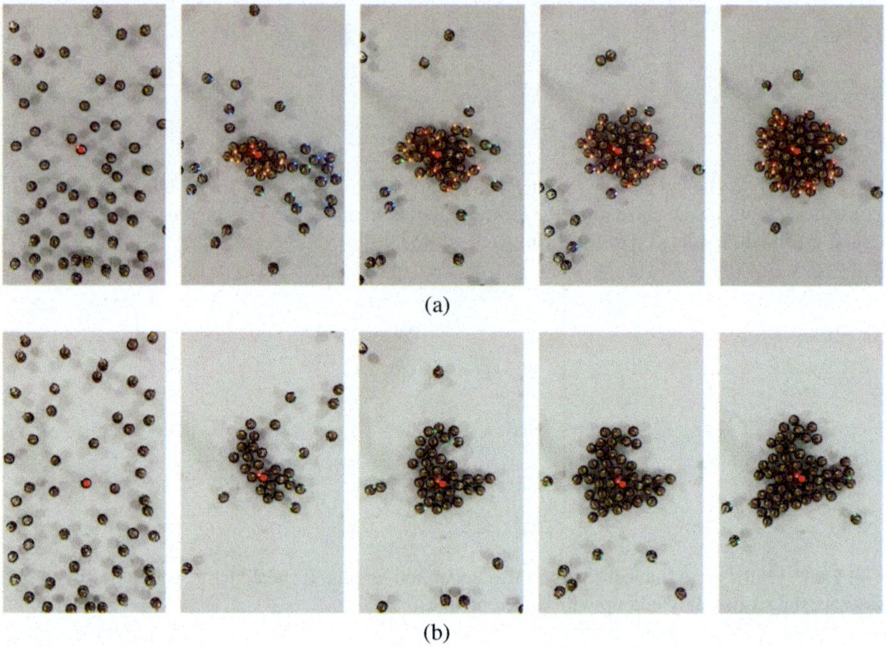

Fig. 7.46 Growth adaptation of the robot aggregation. **a** The edge morphogen concentration of the edge robots increases as the aggregation grows. **b** Distance from the edge robots to the target

7.7 Summary

This chapter presents the application of the genetic and cellular mechanisms to self-organizing swarm robot, which is known as morphogenetic robotics. The main idea is to treat the biological morphogenesis as a self-organizing process that is robust to environmental changes. Evolutionary algorithms are adopted to evolve single gene regulatory network or a hierarchical gene regulatory network for pattern formation, ranging from boundary coverage to region coverage. In addition, it is shown that the hierarchical gene regulatory network can also be evolved using basic regulatory motifs. Finally, algorithms for self-organizing minimalist robots that do not have self-localization and directional sensing capabilities are presented based on morphogen diffusion. It is shown that even with very basic diffusion mechanisms, it is possible to possible to guide minimalist robots to follow targets while maintaining a swarm or to surround targets.

Chapter 8
Towards Evolutionary Developmental Systems

Deep learning [240] has achieved tremendous successes over the past decade in many fields, ranging from traditional machine learning tasks such as computer visions, to go playing [391], to protein structure prediction [103], and to the currently extremely popular chatbot ChatGPT [434], one powerful large language model. Although many of them are in principle inspired by information processing in the human brain, deep learning models still lack several fundamental properties of human intelligence, such as learning many tasks based on small data, planning, anticipation, and learning to learn, i.e., autonomous learning.

In this monograph, we have presented preliminary results on modeling the evolution and development of primitive nervous systems and morphology, highlighting the findings in computational systems showing the genetic and environmental impacts on neural and morphological evolution and development. These results also reveal that adaptation at both genotypic and phenotypic levels is indispensable for resolving the robustness and evolvability trade-off, and in maintaining the adaptation capability, particularly in dynamic environments. We also show that energy minimization is one main principle behind the organization of nervous systems and there is a close coupling between brain, body, and environment in evolution and development. In addition, we also show that genetic and cellular mechanisms underlying biological development open up rich opportunities for solving engineering problems such as self-organization of swarm robots.

Considering the fact that natural intelligence has been shaped by three main adaptation mechanisms in nature, namely evolution, development, and learning, we believe the evolutionary developmental approach to artificial intelligence is an important path to go for sustained future development of artificial intelligence, in particular for artificial general intelligence. Although various methodologies in artificial life [4] or artificial intelligence can be used for studying evolutionary developmental systems, evolutionary developmental robotics may provide an effective platform for studying the modeling of the development of cognitive behaviors in living systems, such as language, emotion, curiosity, anticipation, and social skills, together with the morphological development in an embedded environment. Clearly, developmental robotics focusing on cognitive development only, which is also the typical approach

Y. Jin, *Computational Evolution of Neural and Morphological Development*, Natural Computing Series, https://doi.org/10.1007/978-981-99-1854-6_8

in most artificial general intelligence research, can be challenged both from ontogenetic and phylogenetic point of view. First, the development of biological organisms consists of both physical and mental development. Thus, developmental robotics that concentrates on mental development is incomplete. Ontogenetically, mental development is based on and closely coupled with the physical development of an organism, including the development of both the body plan and the nervous systems. Such close coupling has been confirmed by findings both in neuroscience [398] and in computational systems (as reported in Chap. 6), suggesting that activity-dependent and activity-independent development of neural networks cannot be separated.

Evolutionary robotics [306] applies evolutionary algorithms to the automatic design of neural controllers for autonomous robots. Unfortunately, the role of development has largely been neglected. Thus, integrating research on developmental robotics, including epigenetic and morphogenetic, and evolutionary robotics is indispensable. Developmental plasticity cannot only bias evolution, but also enhance evolvability by maintaining genetic diversity in changing environments and resolving robustness-variability trade-off. For example, it has been shown that an evolutionary perspective on autonomous learning might bring about new insight into understanding the driving force behind mental development [390].

Biological evidence also suggests that autonomous mental development is heavily driven by intrinsic motivational systems [85]. Such findings have triggered one key research area in epigenetic robotics that studies the role of intrinsic motivation systems in mental development [317, 353]. Typically, an intrinsic motivation system is predefined in robotic systems; however, autonomous mental development in living system has not been hard-wired from scratch. Instead, it has been gradually shaped by a brain-body co-evolution embedded in a changing environment.

The morphogenetic self-organization and multi-cellular developmental models introduced in this book address the first challenge in developmental robotics to a certain extent by integrating mental and physical development. Nevertheless, the evolutionary origin that accounts for both physical and mental development is still missing. In addition, in morphogenetic robotics, the gene regulatory network that governs the neural and morphological development should not be designed manually. Thus, the role of evolution becomes increasingly important in studying physical and mental development in robotics systems [311].

Interestingly, the past decade has also witnessed rapid technical and theoretical advances in evolutionary developmental biology [295, 300, 394], often known as evo-devo, in understanding molecular and cellular mechanisms that control the biological morphogenesis. Evolutionary developmental biology has helped and will continue to help us gain deep insights into human cognitive development [140], including emotion [159], and pathology [87]. New findings in developmental biology, neuroscience, systems biology, and brain science will also help us to develop human-like intelligent systems. As the role of genes is too important to be neglected in cognitive development [335], it is high time that the evolutionary developmental approach be adopted for shaping the future of artificial intelligence.

Going from studying cognitive development in epigenetic robotics to evolutionary developmental robotics taking into account both neural and morphological development is non-trivial. Many important questions remain to be answered. For example, how to realize physical development in robotic systems? Must we use hardware (real robotic systems) for research in evo-devo-robo? Another question is, given two different time scales in evolution and development, how much resources should be devoted to evolution and how much to development? How to provide a physical environment that is complex enough to enable the system to evolve and develop behaviors of cognitive significance? Many other questions remain elusive even in computational systems, e.g., how evolution and development have contributed to modularity and connectivity of advanced nervous systems such as the brain? How did evolutionary developmental systems canalize and thus reuse environmental constraints without explicitly encoding them in the DNA?

To conclude, the proposal to move from pure cognitive models including deep learning models toward evolutionary developmental systems may open up abundant new opportunities, although it might have raised more questions than answers in artificial general intelligence research.

References

1. https://github.com/ajhalme/kbsim, 2012
2. A. Aalto, L. Viitasaari, P. Ilmonen, L. Mombaerts, J. Goncalves, Gene regulatory network inference from sparsely sampled noisy data. Nat. Commun. **11**, 3493 (2020)
3. W.C. Abraham, M.F. Bear, Metaplasticity: The plasticity of synaptic plasticity. Trends Neurosci. **19**(4), 126–130 (1996)
4. C. Adami, A brief history of artificial intelligence research. Artif. Life **27**(2), 131–137 (2021)
5. J. Aguilar-Rodríguez, J.L. Payne, *Robustness and Evolvability in Transcriptional Regulation.* (2021)
6. R. Albert, A. Barabási, Statistical mechanics of complex networks. Rev Mod. Phys. **74**(1), 47–97 (2002)
7. N.E.E. Allenby, E. Laing, G. Bucca, A.M. Kierzek, C.P. Smith, Diverse control of metabolism and other cellular processes in streptomyces coelicolor by the phop transcription factor: Genome-wide identification of in vivo targets. Nucleic Acids Res. (2012)
8. J. Allman, *Evolving Brains.* (Scientific American Library, 1999)
9. U. Alon, *An Introduction to Systems Biology: Design Principles of Biological Circuits.* (Chapman & Hall, 2006)
10. U. Alon, Network motifs: Theory and experimental approaches. Nat. Rev. Genet. **8**, 450–461 (2007)
11. A.S. Alvarado, P.A. Tsonis, Bridging the regeneration gap: Genetic insights from diverse animal models. Nat. Rev. Genet. **7**, 873–884 (2006)
12. T. Andersen, R. Newman, T. Otter, Shape homeostasis in virtual embryos. Artif. Life **15**(2), 161–183 (2009)
13. S. Ando, H. Iba, Inference of gene regulatory model by genetic algorithms, in *Evolutionary Computation, 2001. Proceedings of the 2001 Congress on*, vol. 1. (2001), pp. 712 –719
14. O. Antonios, P. Maja, P. Ioannis, Sparse b-spline polynomial descriptors for human activity recognition. Image Vis. Comput. **2**, 1814–1825 (2009)
15. L. Hilary, J.B. Ashe, The interpretation of morphogen gradients. Development **133**(3), 385–394 (2006)
16. S. Babinec, J. Pospichal, Improving the prediction accuracy of echo state neural networks by anti-oja's learning, in *Proceedings of the 17th International Conference on Artificial Neural Networks*, ICANN'07. (Springer-Verlag, 2007), pp. 19–28
17. M.M. Babu, N.M. Luscombe, L. Aravind, M. Gerstein, S.A. Teichmann, Structure and evolution of transcriptional regulatory networks. Curr. Opin. Struct. Biolo. **14**(3), 283–291 (2004)

Y. Jin, *Computational Evolution of Neural and Morphological Development*, Natural Computing Series, https://doi.org/10.1007/978-981-99-1854-6

18. T. Bäck, *Evolutionary Algorithms in Theory and Practice: Evolution Strategies, Evolutionary Programming, Genetic Algorithms*. (Oxford University Press, 1996)
19. T. Bäck, H.-P. Schwefel, An overview of evolutionary algorithms for parameter optimization. Evol. Comput. **1**(1), 1–23 (1993)
20. R. Baddeley, L.F. Abbott, M.C.A. Booth, F. Sengpiel, T. Freeman, E.A. Wakeman, E.T. Rolls, Responses of neurons in primary and inferior temporal visual cortices to natural scenes. Proc. R. Soc. B-Biol. Sci. **264**(1389), 1775–1783 (1997)
21. H.B. Barlow, P. Foldiak, Adaptation and decorrelation in the cortex, in *The Computing Neuron*. ed. by R. Durbin, C. Miall, G. Mitchison (Addison-Wesley, Wokingham, England, 1989), pp.54–72
22. L. Barnett, *Evolutionary Search on Fitness Landscapes with Neutral Networks*. PhD thesis, School of Cognitive and Computer Science (University of Sussex, Falmer, Brighton, 2003)
23. D. Basanta, M. Miodownik, B. Baum, The evolution of robust development and homeostasis in artificial organisms. PLoS Comput. Biol. **4**(3), e1000030 (2008)
24. M. Beauregard, P.J. Kennedy, Robust simulation of lamprey tracking, in *Parallel Problem Solving from Nature - PPSN IX*, (Springer-Verlag, Berlin, 2006), pp. 641–650
25. R.D. Beer, Beyond control: The dynamics of brain-body-environment interaction in motor systems, in *Progress in Motor Control V: A Multidisciplinary Perspective* (2006)
26. R.D. Beer, J.C. Gallagher, Evolving dynamical neural networks for adaptive behaviour. Adapt. Behav. **1**(1), 91–122 (1992)
27. G. Bellec, F. Scherr, A. Subramoney, E. Hajek, D. Salaj, R. Legenstein, W. Maass, A solution to the learning dilemma for recurrent networks of spiking neurons. Nat. Commun. **11**(1), 3625 (2020)
28. L.V. Beloussov, Integrating self-organization theory into an advanced course on morphogenesis at Moscow State University. Int. J. Dev. Biol. **47**, 177–181 (2003)
29. C. Belta, V. Kumar, Abstraction and control for groups of robots. IEEE Trans. Robot. **20**(5), 865–875 (2004)
30. H. Ben-Amor, S. Cadau, A. Elena, D. Dhouailly, J. Demongeot, Regulatory networks analysis: Robustness in morphogenesis, in *2009 Int. Conf. on Advanced Information Networking and Application Workshops* (Bradford, UK, 2009), pp. 924–928
31. P. Bentley, S. Kumar, Three ways to grow designs: A comparison of embryogenies for an evolutionary design problem, in *Proceedings of the Genetic and Evolutionary Computation Conference*, eds. by W. Banzhaf, J. Daida, A.E. Eiben, M.H. Garzon, V. Honavar, M. Jakiela, R.E. Smith, vol. 1 (Morgan Kaufmann, Orlando, Florida, USA, 1999), pp. 35–43
32. P.J. Bentley, Representations are more important than algorithms: Why evolution needs embryology, in *Proceedings of the Fourth International Conference on Adaptive Computing in Design and Manufacture* (Plymouth, UK, 2000)
33. N. Beume, B. Naujoks, M. Emmerich, SMS-EMOA: Multiobjective selection based on dominated hypervolume. Eur. J. Oper. Res. **181**(3), 1653–1669 (2007)
34. G.Q. Bi, M.M. Poo, Synaptic modifications in cultured hippocampal neurons: Dependence on spike timing, synaptic strength, and postsynaptic cell type. J. Neurosci. **18**(24), 10464–10472 (1998)
35. E.L. Bienenstock, L.N. Cooper, P.W. Munro, Theory for the development of neuron selectivity: Orientation specificity and binocular interaction in visual cortex. J. Neurosci. **2**(1), 32–48 (1982)
36. C.D. Bishop, D.F. Erezyilmaz, T. Flatt, C.D. Georgiou, M.G. Hadfield, A. Heyland, J. Hodin, M.W. Jacobs, S.A. Maslakova, A. Pires, A.M. Reitzel, S. Santagata, K. Tanaka, J.H. Youson, What is metamophosis? Integr. Comp. Biol. **46**, 655–661 (2006)
37. C.M. Bishop, *Neural Networks for Pattern Recognition* (Oxford University Press, 1995)
38. B.J. Blaiszik, S.L.B. Kramer, S.C. Olugebefola, J.S. Moore, N.R. Sottos, S.R. White, Self-healing polymers and composites. Ann. Rev. Mater. Res. **40**, 179–211 (2010)
39. J. Blynel, D. Floreano, Levels of dynamics and adaptive behavior in evolutionary neural controllers, in *Proceedings of the Seventh International Conference on Simulation of Adaptive Behavior on From Animals to Animats* (2002)

40. J.C. Bongard, C. Paul, Investigating morphological symmetry and locomotive efficiency using virtual embodied evolution, in *From Animals to Animats 6: Proceedings of the Sixth International Conference on Simulation of Adaptive Behavior* (MIT Press, 2000)
41. C.P. Bowers, Formation of modules in a computational model of embryogeny, in *Proceedings of the 2005 Congress on Evolutionary Computation* (2005)
42. G. Box, Evolutionary operation: A method for increasing industrial productivity. Appl. Stat. **6**, 81–101 (1957)
43. C.A. Brackley, N. Gilbert, D. Michieletto, A. Papantonis, M.C.F. Pereira, P.R. Cook, D. Marenduzzo, Complex small-world regulatory networks emerge from the 3d organisation of the human genome. Nat. Coomun. **12**, 5756 (2021)
44. M. Brenn, Dreidimensionale Topologieoptimierung mit Hilfe eines Zellwachstummodells. Master's Thesis (University of Darmstadt, 2008)
45. D. Brockhoff, T. Friedrich, N. Hebbinghaus, C. Klein, F. Neumann, E. Zitzler, Do additional objectives make a problem harder? in *Proceedings of the 9th Annual Conference on Genetic and Evolutionary Computation*, GECCO '07 (ACM, New York, NY, USA, 2007), pp. 765–772
46. J.T. Brown, A.D. Randall, Activity-dependent depression of the spike after-depolarization generates long-lasting intrinsic plasticity in hippocampal CA3 pyramidal neurons. J. Physiol. (Lond.) **587**(6), 1265–1281 (2009)
47. V.D. Buonomano, W. Maass, State-dependent computations: Spatiotemporal processing in cortical networks. Nat. Rev. Neurosci. **10**(2), 113–25 (2009)
48. D. Bush, Y. Jin, Calcium control of triphasic hippocampal stdp. J. Comput. Neurosci. **33**, 495–514 (2012)
49. D. Bush, Y. Jin, Calcium control of triphasic hippocampal stdp. J. Comput. Neurosci. **33**(3), 495–514 (2012)
50. I. Laptev, C. Schuldt, B. Caputo, Recognizing human actions: A local SVM approach. Proc. Int. Conf. Pattern Recognit. 32–36 (2004)
51. A. Cangelosi, D. Parisi, S. Nolfi, Cell division and migration in a 'genotype' for neural networks. Networks Comput. Neural Syst. **5**, 479–515 (1994)
52. R.A. Caruana, J.D. Schaffer, Representation and hidden bias: Gray versus binary coding in genetic algorithms, in *Proceedings of 5th Conference on Machine Learning*, ed. by J. Laird (Morgan Kaufmann, Ann Arbor, MI, 1988), pp. 153–161
53. E.C. Charnov, S.K.M. Ernest, The offspring-size/clutch-size trade-off in mammals. Am. Naturalist **167**(4), 578–582 (2006)
54. C.C. Cheah, S.P. Hou, J.J.E. Slotine, Region-based shape control for a swarm of robots. Automatica **45**, 2406–2411 (2009)
55. B.-S. Chen, C.-Y. Hsu, J.-J. Liou, Robust design of biological circuits: Evolutionary systems biology approach. J. Biomed. Biotechnol. (2011)
56. L. Chen, Computational systems biology on networksand dynamics. Optim. Syst. Biol. 5–12 (2007)
57. R. Cheng, Y. Jin, M. Olhofer, B. Sendhoff, A reference vector guided evolutionary algorithm for many-objective optimization. IEEE Trans. Evol. Comput. **20**(5), 773–791 (2016)
58. J. Chiquet, Y. Grandvalet, C. Ambroise, Inferring multiple graphical structures. Stat. Comput. **21**(4), 537–553 (2011)
59. A. Chowdhury, M. Chetty, X. Vinh, On the reconstruction of genetic network from partial microarray data, in *Neural Information Processing*. ed. by T. Huang, Z. Zeng, C. Li, C. Leung, L. Notes, in Computer Science, vol. 7663, (Springer, Berlin, 2012), pp.689–696
60. J. Chrol-Cannon, Y. Jin, Computational modeling of neural plasticity for self-organization of neural networks. Biosystems **125**, 43–54 (2014)
61. D. Chu, Evolving genetic regulatory networks for systems biology, in *Congress on Evolutionary Computation* (IEEE Press, 2007), pp. 875–882
62. S.-J. Chung, J.-J. Slotine, On synchronization of coupled hopf-kuramoto oscillators with phase delays, in *Submitted to IEEE Transactions on Automatic Control* (2010). arXiv:1004.5366v1
63. S. Ciliberti, O.C. Martin, A. Wagner, Robustness can evolve gradually in complex regulatory gene networks with varying topology. PLoS Comput. Biol. **3**(2), e15 (2007)

64. A. Citri, R.C. Malenka, Synaptic plasticity: Multiple forms, functions, and mechanisms. Neuropsychopharmacology **33**(1), 18–41 (2007)
65. D.E. Clyde, M.S.G. Corado, X. Wu, A. Pare, D. Papatsenko, S. Small, A self-organizing system of repressor gradients establishes segmental complexity in drosophila. Nature 849–853 (2003)
66. P. Coates, Using genetic programming and L-systems to explore 3D design worlds, in *CAAD-Future'97*, ed. by R. Junge (Kluwer Academic Press, Munich, 1997)
67. C.A. Coello Coello, An updated survey of evolutionary multi-objective optimization techniques: State of art and future trends, in *Proceedings of 1999 Congress on Evolutionary Computation* (IEEE Press, Washington D.C., 1999), pp. 3–13
68. J. Cohen, I. Stewart, *The Collapse of Chaos*. (Penguin Books, 1994)
69. M.B. Cooper, M. Loose, J.F.Y. Brookfield, Evolutionary modelling of feed forward loops in gene regulatory networks. Biosystems **91**(1), 231–244 (2008)
70. C. Joseph, C.B. Bhushan, Bioinspired self-healing materials: Lessons from nature. Beilstein J. Nanotechnol. **9**, 907–935 (2018)
71. F. Crick, Central dogma of molecular biology. Nature **227**, 561–563 (1970)
72. A. Crombach, P. Hogeweg, Evolution of evolvability in gene regulatory networks. PLoS Comput. Biol. **4**(7), e1000112 (2008)
73. A. Crumire, M. Sablik, Positive circuits and d-dimensional spatial differentiation: Application to the formation of sense organs in drosophila. Biosystems **94**(1–2), 102–108 (2008)
74. D. Cvetkovic, I. Parmee, Use of preferences for GA-based multi-objective optimization, in *Proceedings of 1999 Genetic and Evolutionary Computation Conference* (Morgan Kaufmann, San Francisco, California, 1999), pp. 1504–1510
75. R. Darshan, A. Leblois, D. Hansel, Interference and shaping in sensorimotor adaptations with rewards. PLoS Comput. Biol. **10**(1), e1003377 (2014)
76. D. Dasgupta, Incorporate redundancy and gene activation mechanisms in genetic search for adapting to non-stationary environments, in *Practical Handbook of Genetic Algorithms*, ed. by L. Chambers (CRC Press, 1995)
77. J. Davies, *Mechanisms of Morphogenesis* (Elsevier Academic Press, 2005)
78. P. Dayan, L.F. Abbott, *Theoretical Neuroscience: Computational and Mathematical Modeling of Neural Systems* (The MIT Press, Cambridge, MA, 2001)
79. H. de Jong, Modeling and simulation of genetic regulatory systems: A literature review. J. Comput. Biol. **9**(1), 67–103 (2002)
80. K. De Jong, *Analysis of the Behavior of a Class of Genetic Adaptive Systems*. PhD thesis (University of Michigan, Ann Arbor, 1975)
81. K. Deb, S. Agrawal, A. Pratab, T. Meyarivan, A fast elitist non-dominated sorting genetic algorithm for multi-objective optimization: NSGA-II, in *Proceedings of the Parallel Problem Solving from Nature VI Conference*, eds. by M. Schoenauer, K. Deb, G. Rudolph, X. Yao, E. Lutton, J.J. Merelo, H.-P. Schwefel. Lecture Notes in Computer Science No. 1917 (Springer, Paris, France, 2000), pp. 849–858
82. K. Deb, S. Agrawal, A. Pratap, T. Meyarivan, A fast elitist non-dominated sorting genetic algorithm for multi-objective optimization: NSGA-II, in *Parallel Problem Solving from Nature*, vol. VI (2000), pp. 849–858
83. K. Deb, A. Pratap, S. Agarwal, T. Meyarivan, A fast and elitist multiobjective genetic algorithm: Nsga-ii. IEEE Trans. Evol. Comput. **6**(2), 182–197 (2002)
84. K. Deb, *Multi-Objective Optimisation using Evolutionary Algortihms*, 1st edn. (Wiley, Kanpur, India, 2001)
85. E.L. Deci, R.M. Ryan, *Intrinsic Motivation and Self-Determination in Human Behavior* (Springer, Berlin, 1985)
86. L Diambra, L. da Fontoura Costa, Pattern formation in a gene network model with boundary shape dependence. Phys. Rev. E **73**(031917) (2006)
87. R. Diogo, Introduction to evolutionary developmental pathology, or evo-devo-path: On neo-darwinism, natural mutants, hopeful monsters, syndromes, genomics, variations, humans, apes, chameleons, and dinosaurs. Curr. Mol. Biol. Rep. **6**(4), 11–15 (2020)

88. P. Dollár, V. Rabaud, G. Cottrell, S. Belongie, Behavior recognition via sparse spatio-temporal features. Proc. VS-PETS 65–72 (2005)
89. R. Doursat, The growing canvas of biological development: Multiscale pattern generation on an expanding lattice of gene regulatory nets, in *Proceedings of the Sixth International Conference on Complex Systems* (2008), pp. 205–210
90. K.L. Downing, Supplementing evolutionary developmental systems with abstract models of neurogenesis, in *Genetic and Evolutionary Computation Conference* (2007), pp. 990–996
91. K.L. Downing, Supplementing evolutionary developmental systems with abstract models of neurogenesis, in *Proceedings of the 9th Annual Conference on Genetic and Evolutionary Computation* (ACM, 2007), pp. 990–996
92. R.G.L. D'Souza, K.C. Sekaran, A. Kandasamy, Reconstruction of gene networks using phenomic algorithms. Int. J. Artif. Intell. Appl. **1**(2) (2010)
93. D. Zhouwei, H. Fang, X. Jian, Snake-worm: A bi-modal locomotion robot. J. Bionic Eng. **19**, 1272–1289 (2022)
94. D.H. Sharp, E. Mjolsness, J. Reinitz, A connectionist model of development. J. Theoret. Biol. **18**, 429–453 (1991)
95. D.J. Earl, M.W. Deem, Evolvability is a selectable trait. PNAS **101**, 11531–11536 (2004)
96. P. Eggenberger, Evolving morphologies of simulated 3d organisms based on differential gene expression, in *Proceedings of the 4th European Conference on Artificial Life* (1997)
97. P. Eggenberger, Evolving morphologies of simulated 3d organisms based on differential gene expression, in *Proceedings of the Fourth European Conference on Artificial Life* (MIT Press, 1997), pp. 205–213
98. P. Eggenberger Hotz, G. Gómez, R. Pfeifer, Evolving the morphology of a neural network for controlling a foveating retina–and its test on a real robot, in *Proceedings of the 8th International Conference on Artificial Life (ALife 03)* (2003), pp. 243–251
99. O. Ekeberg, An integrated neuronal and mechanical model of fish swim, in *Proc. Computation and Neural Systems (CNS)* (Klumer, 1993)
100. Ö. Ekeberg, A combined neuronal and mechanical model of fish swimming. Biol. Cybern. **69**, 363–374 (1993)
101. T. Elsken, J.H. Metzen, F. Hutter, Neural architecture search: A survey. J. Mach. Learn. Res. **20**(1), 1997–2017 (2019)
102. D.H. Erwin, E.H. Davidson, The evolution of hierarchical gene regulatory networks. Nat. Rev. Genet. **10**, 141–148 (2009)
103. J. Jumper et al., Highly accurate protein structure prediction with AlphaFold. Nature **596**, 583–589 (2021)
104. D. Federici, Increasing evolvability for developmental programs, in *GECCO Workshop on Regeneration and Learning in Developmental Systems*, ed. by J. Miller (2004)
105. D. Federici, T. Ziemke, Why are developing organisms also fault-tolerant, in *Proceedings of the 9th International Conference on Simulation of Adaptive Behavior* (2006), pp. 449–459
106. D. Federici, Evolving developing spiking neural networks, in *Congress on Evolutionary Computation* (2005), pp. 543–550
107. D. Federici, K. Downing, Evolution and development of a multicellular organism: Scalability, resilience, and neutral complexification. Artif. Life **12**(3), 381–409 (2006)
108. D.E. Feldman, Timing-based LTP and LTD at vertical inputs to layer ii/iii pyramidal cells in rat barrel cortex. Neuron **27**(1), 45–56 (2000)
109. F. Fioravanti, M. Helmer-Citterich, E. Nardelli, Modeling gene regulatory network motifs using statecharts. BMC Bioinform. **13**(4), 1–12 (2012)
110. S. Flavell, M.E. Greenberg, Signaling mechanisms linking neuronal activity to gene expression and plasticity of the nervous system. Annu. Rev. Neurosci. **31**, 563–590 (2008)
111. T.C. Fogarty, Varying the probability of mutation in the genetic algorithm, in *Proceedings of International Conference on Genetic Algorithms*, ed. by J.D. Schaffer (Morgan Kanfmann, Fairfax, VA, 1989), pp. 104–109
112. D. Fogel, *Evolving Artificial Intelligence*. PhD Thesis (University of California, San Diego, 1992)

113. L. Fogel, J. Owens, M. Walsh, *Artificial Intelligence through Simulated Evolution* (John Wiley, New York, 1966)
114. C.M. Fonseca, P.J. Flemming, On the performance assessment and comparison of stochastic multi-objective optimizers, in *Parallel Problem Solving from Nature*, vol. IV, ed. by H.-M. Voigt, W. Ebeling, I. Rechenberg, H.-P. Schwefel (Springer, Berlin, 1996), pp.584–593
115. P. François, V. Hakim, Design of genetic networks with specified functions by evolution in silico. Proc. Nat. Acad. Sci. U.S.A. **101**(2), 580–585 (2004)
116. P. Francois, V. Hakim, Design of genetic networks with specified functions by evolution in silico. Proc. Nat. Acad. Sci. U.S.A. **101**(2), 580–585 (2007)
117. R.M. French, Catastrophic forgetting in connectionist networks. Trends Cogn. Sci. **3**(4), 128–35 (1999)
118. G. Friedmann, Digital simulation of an evolutionary process. Gen. Syst. Yearb. **4**, 171–184 (1959)
119. S. Fujita, E. Kuranaga, Y. Nakajima, Regeneration potential of jellyfish: Cellular mechanisms and molecular insights. Genes **12**(5), 758 (2006)
120. G. Gallego, T. Delbrück, G. Orchard, C. Bartolozzi, B. Taba, A. Censi, S. Leutenegger, A.J. Davison, J. Conradt, K. Daniilidis, D. Scaramuzza, Event-based vision: A survey. IEEE Trans. Pattern Anal. Mach. Intell. 1 (2020)
121. G. Gardi, S. Ceron, W. Wang, K. Petersen5, M. Sitti, Microrobot collectives with reconfigurable morphologies, behaviors, and functions. Nat. Commun. **13**, 2239 (2022)
122. T.S. Gardner, D. di Bernardo, D. Lorenz, J.J. Collins, Inferring genetic networks and identifying compound mode of action via expression profiling. Science **301**(5629), 102–105 (2003)
123. N. Geard, J. Wiles, Investigating ontogenetic space with developmental cell lineages, in *Proceedings of the Tenth International Conference on the Simulation and Synthesis if Living Systems* (2006), pp. 56–62
124. N. Geard, J. Wiles, A gene network model for developing cell lineages. Artif. Life **11**(3), 249–267 (2005)
125. F. Geier, J. Timmer, C. Fleck, Reconstructing gene-regulatory networks from time series, knock-out data, and prior knowledge. BMC Syst. Biol. 1 (2007)
126. A. Geoffrion, J. Dyer, A. Feinberg, An interactive approach for multi-criterion optimization. Manage. Sci. **19**, 357–368 (1972)
127. M.-O. Gewaltig, M. Diesmann, NEST (Neural Simulation Tool). Scholarpedia **2**(4), 1430 (2007). Source code available at: http://www.nest-initiative.org/
128. M. Ghildiyal, P.D. Zamore, Small silencing rnas: An expanding universe. Nature **10**, 94–108 (2009)
129. A. Ghysen, The origin and evolution of the nervous system. Int. J. Develop. Biol. **47**, 555–562 (2003)
130. S.F. Gilbert, *Developmental Biology*, 7th edn. (Sinauer Associates, 2003)
131. S.F. Gilbert, *Developmental Biology* (Sinauer Associates, 2003)
132. G.B. Gillis, Angulliform swimming since sir james gray. Undulatory locomotion in elongate aquatic vertebrates. Am. Zool. **36**, 656–665 (1996)
133. C.M. Glen, M.L. Kemp, E.O. Voit, Agent-based modeling of morphogenetic systems: Advantages and challenges. PLOS Comput. Biol. **15**(3), e1006577 (2019)
134. D. Goldberg, *Genetic Algorithms in Search, Optimization and Machine Learning* (Addison-Wesley, Reading, MA, 1989)
135. D. Goldberg, K. Deb, B. Korb, Messy genetic algorithms: Motivation, analysis, and first results. Complex Syst. **3**, 493–530 (1989)
136. D.E. Goldberg, K. Deb, *A comparative analysis of selection schemes used in genetic algorithms, in Foundations of Genetic Algorithms* (Morgan Kaufmann, San Mateo, 1990)
137. A. Goldbeter, *Biochemical Oscillations and Cellular Rhythms*, vol. 1. (Cambridge University Press, April 1997)
138. D. Gonze, Coupling oscillations and switches in genetic networks. Biosystems **99**(1), 60–69 (2010)

139. M. Graupner, N. Brunel, Calcium-based plasticity model explains sensitivity of synaptic changes to spike pattern, rate, and dendritic location. Proc. Nat. Acad. Sci. **109**(10), 3991–3996 (2012)
140. P.E. Griffiths, Evo-devo meets the mind: Towards a developmental evolutionary psychology, in *Integrating Development and Evolution*, eds. by R. Sanson, R.N. Brandon (Cambridge University Press, 2007)
141. F. Gruau, Genetic synthesis of boolean neural networks with a cell rewriting developmental process, in *Proceedings of International Workshop on Combination of Genetic Algorithms and Neural Networks*, eds. by D. Whitley, J. Schaffer (IEEE Press, 1992), pp. 55–73
142. F. Gruau, *Neural Network Synthesis using Cellular Encoding and the Genetic Algorithm.* PhD Thesis (Ecole Normale Supérieure de Lyon and Centre d'études nucléaire de Grenoble, France, 1994)
143. R. Gruna, *Analysis of Redundant Genotype-phenotype Mappings–Investigation of the Effect of Neutrality on Evolvability and Robustness.* Master's Thesis (AIFB, Universität Karlsruhe, 2007)
144. H. Guo, Y. Meng, Y. Jin, Self-adaptive multi-robot construction using gene regulatory networks, in *IEEE Symposium on Artificial Life* (2009), pp. 53–60
145. H. Hache, H. Lehrach, R. Herwig, Reverse engineering of gene regulatory networks: A comparative study. EURASIP J. Bioinform. Syst. Biol. **8**(1–8), 12 (2009)
146. T. Hackworth, Genetic algorithms: Some effects of redundancy in chromosome. Available at: http://www.cs.ucl.ac.uk/staff/P.Bentley/birkbeck/tim.txt (1999)
147. R. Hadsell, D. Rao, A.A. Rusu, R. Pascanu, Embracing change: Continual learning in deep neural networks. Trends Cogn. Sci. **24**(12), 1028–1040 (2020)
148. J. Hallinan, *Gene Networks and Evolutionary Computation* (John Wiley & Sons, Inc., 2007), pp. 67–96
149. M.S. Halvagal, F. Zenke, The combination of hebbian and predictive plasticity learns invariant object representations in deep sensory networks. *bioRxiv* (2022)
150. J. Handl, D. Kell, J. Knowles, Multi-objective optimization in computational biology and bioinformatics. ACM/IEEE Trans. Comput. Biol. Bioinform. **4**, 279–292 (2007)
151. J. Handl, D.B. Kell, J. Knowles, Multiobjective optimization in bioinformatics and computational biology. IEEE/ACM Trans. Comput. Biol. Bioinform. **4**(2), 279–292 (2007)
152. J. Handl, S.C. Lovell, J. Knowles, Investigations into the effect of multiobjectivization in protein structure prediction, in *Proceedings of the 10th international conference on Parallel Problem Solving from Nature: PPSN X* (Springer-Verlag, Berlin, Heidelberg, 2008), pp. 702–711
153. N. Hansen, A. Ostermeier, Completely derandomized self-adaptation in evolution strategies. Evol. Comput. (2000). To appear
154. N. Hansen, A. Ostermeier, Completely derandomized self-adaptation in evolution strategies. Evol. Comput. **9**(2), 159–195 (2001)
155. N. Hansen, A. Ostermeier, A. Gawelczyk, On the adaptation of arbitrary normal mutation distributions in evolution strategies: The generating set adaptation, in *Proceedings of International Conference on Genetic Algorithms*, ed. by L.J. Eshelman (Morgan Kaufmann, 1995), pp. 57–64
156. S. Harding, W. Banzhaf, Artificial development, in *Organic Computing*, ed. by R.W. Würz, Chapter 9 (Springer, 2008), pp. 201–219
157. D.O. Hebb, *The Organization of Behavior* (Wiley, New York, 1949)
158. G. Hennequin, W. Gerstner, J.-P. Pfister, Stdp in adaptive neurons gives close-to-optimal information transmission. Front. Comput. Neurosci. **4**(143) (2010)
159. J. Luis Hernández-Ochoa, F. Vergara-Silva, Is it necessary to integrate evo-devo to the analysis and construction of artificial emotional systems? Front. Ecol. Evol. **16**, 728829 (2022)
160. F. Herrera, M. Lozano, A.M. Sánchez, A taxonomy for the crossover operator for real-coded genetic algorithms: An experimental study. Int. J. Intell. Syst. **18**(3), 309–338 (2003)
161. A. Hesketh, C. Hill, J. Mokhtar, G. Novotna, N. Tran, M. Bibb, H.-J. Hong, Genome-wide dynamics of a bacterial response to antibiotics that target the cell envelope. BMC Genom. **12**, 11 (2011)

162. C. Higuera, A.F. Villaverde, J.R. Banga, J. Ross, F. Morán, Multi-criteria optimization of regulation in metabolic networks. PLoS ONE **7**(7), e41122 (2012)
163. T.K. Ho, Random decision forests, in *Proceedings of 3rd International Conference on Document Analysis and Recognition*, vol. 1 (1995), pp. 278–282
164. D.C. Hoaglin, F. Mosteller, J.W. Tukey, *Understanding Robust and Exploratory Data Analysis* (Wiley, 1983)
165. T. Hohm, E. Zitzler, Multiobjectivization for parameter estimation: A case-study on the segment polarity network of drosophila, in *GECCO '09: Genetic and Evolutionary Computation Conference (GECCO 2009)*. ed. by F. Rothlauf et al. (ACM, New York, NY, USA, 2009), pp.209–216
166. J. Holland, *Adaptation in Natural and Artificial Systems* (University of Michigan Press, Ann Arbor, Michigan, 1975)
167. P. Holmes, R.J. Full, D. Koditschek, J. Guckenheimer, The dynamics of legged locomotion: Models, analyses, and challenges. SIAM Rev. **48**(2), 1–13 (2006)
168. A. Holtmaat1, K. Svoboda, Experience-dependent structural synaptic plasticity in the mammalian brain. Nat. Rev. Neurosci. **10**, 647–658 (2009)
169. J. Horn, N. Nafpliotis, D.E. Goldberg, A niched Pareto genetic algorithm for multi-objective optimization, in *Proceedings of First IEEE International Conference on Evolutionary Computation*, vol. 1 (IEEE Press, Piscataway, New Jersey, 1994), pp. 82–87
170. G.S. Hornby, J.B. Pollack, Creating high-level components with a generative representation for body-brain evolution. Artif. Life **8**, 223–246 (2002)
171. S.P. Hou, C.C. Cheah, Multiplicative potential energy function for swarm control, in *IEEE/RSJ International Conference on Intelligent Robots and Systems* (St. Louis, USA, Oct. 2009), pp. 4363–4368
172. Howard, M. Mataric, S. Sukhatme, Mobile sensor network deployment using potential fields: A distributed scalable solution to the area coverage problema, in *6th International Symposium on Distributed Autonomous Robotics Systems* (2002), pp. 299–308
173. M. Hsieh, V. Kumar, L. Chaimowicz, Decentralized controllers for shape generation with robotic swarms. Robotica **26**(5), 691–701 (2008)
174. Y. Hua, Q. Liu, K. Hao, Y. Jin, A survey of evolutionary algorithms for multi-objective optimization problems with irregular pareto fronts. IEEE/CAA J. Autom. Sinica **8**(2), 303–318 (2021)
175. Y. Huang, Y. Jin, K. Hao, Decision-making and multi-objectivization for cost sensitive robust optimization over time. Knowl.-Based Syst. **199**, 105857 (2020)
176. H. Iba, A. Mimura, Inference of a gene regulatory network by means of interactive evolutionary computing. Inf. Sci. **145**(3–4), 225–236 (2002)
177. A. Ijspeert, J. Kodjabachian, Evolution and development of a central pattern generator for the swimming of a lamprey (1998)
178. B. Illing, J. Ventura, W.G. Guillaume Bellec, Local plasticity rules can learn deep representations using self-supervised contrastive predictions. arXiv:2010.08262 (2022)
179. H. Ishibuchi, R. Imada, Y. Setoguchi, Y. Nojima, Reference point specification in Hypervolume calculation for fair comparison and efficient search, in *Proceedings of the Genetic and Evolutionary Computation Conference* (ACM, New York, NY, USA, 2017), pp. 585–592
180. H. Ishibuchi, H. Masuda, Y. Tanigaki, Y. Nojima, Modified distance calculation in generational distance and inverted generational distance, in *Proceedings of the International Conference on Evolutionary Multi-criterion Optimization* (2015), pp. 110–125
181. E.M. Izhikevich, Simple model of spiking neurons. IEEE Trans. Neural Networks **14**(6), 1569–72 (2003)
182. E.M. Izhikevich, Which model to use for cortical spiking neurons? IEEE Trans. Neural Networks **15**(5), 1063–70 (2004)
183. E.M. Izhikevich, Solving the distal reward problem through linkage of stdp and dopamine signaling. BMC Neurosci. **8**(Suppl 2), 1–2 (2007)
184. E.J. Izquierdo, C.T. Fernando, The evolution of evolvability in gene transcription networks, in *Proceedings of Alife XI* (2008), pp. 265–273

185. H. Jaeger, The 'echo state' approach to analysing and training recurrent neural networks. Technical Report 148, German Nat. Res. Center Inf. Technol., Sankt Augustin, Germany (2001)
186. H. Jaeger, Reservoir riddles: suggestions for echo state network research, in *Neural Networks, 2005. IJCNN '05. Proceedings. 2005 IEEE International Joint Conference on*, vol. 3 (2005), pp. 1460–1462
187. H. Jaeger, *Discovering Multiscale Dynamical Features with Hierarchical Echo State Networks*. Technical Report (Jacobs University Bremen, 2007)
188. H. Jahnsen, R. Llinas, Electrophysiological properties of guinea-pig thalamic neurones: An in vitro study. J. Physiol. **349**(1), 205–226 (1984)
189. T.M. Jensen, Helper-objectives: Using multi-objective evolutionary algorithms for single-objective optimisation. J. Math. Modell. Algorithms **3**, 323–347 (2004). https://doi.org/10.1023/B:JMMA.0000049378.57591.c6
190. H. Jhuang, T. Serre, L. Wolf, T. Poggio, *A biologically inspired system for action recognition* (Proc. IEEE Int. Conf. Comput, Vis, 2007)
191. Y. Jin, *Advanced Fuzzy Systems Design and Applications* (Physica Verlag/Springer Verlag, Heidelberg, 2003)
192. Y. Jin, R. Gruna, I. Paenke, B. Sendhoff, Multi-objective optimization of robustness and innovation in redundant genetic representations, in *IEEE Symposium on Multi-Criteria Decision-Making* (2009), pp. 38–45
193. Y. Jin, Y. Meng, B. Sendhoff, Influence of regulation logic on the easiness of evolving sustained oscillation for gene regulatory networks, in *IEEE Symposium on Artificial Life* (IEEE Press, 2009), pp. 61–68
194. Y. Jin, T. Okabe, B. Sendhoff, Adapting weighted aggregation for multi-objective evolution strategies, in *Proceedings of First International Conference on Evolutionary Multi-Criterion Optimization*, Leture Notes in Computer Science (Springer, Zurich, March 2001), pp. 96–110
195. Y. Jin, Surrogate-assisted evolutionary computation: Recent advances and future challenges. Swarm Evol. Comput. **1**(2), 61–70 (2011)
196. Y. Jin, R. Gruna, I. Paenke, B. Sendhoff, Evolutionary multi-objective optimization of robustness and innovation in redundant genetic representations, in *IEEE Symposium on Computational Intelligence in Multi-Criteria Decision-Making* (2009), pp. 38–45
197. Y. Jin, H. Guo, Y. Meng, Robustness analysis and failure recovery of a bio-inspired self-organizing multi-robot system, in *Third IEEE International Conference on Self-Adaptive and Self-organizing Systems* (IEEE Press, 2009), pp. 154–164
198. Y. Jin, Y. Meng, Morphogenetic robotics: A new paradigm for designing self-organizing, self-reconfigurable and self-adaptive robots, in *Morphogenetic Engineering: Toward Programmable Complex Systems* (Springer, 2010). To appear
199. Y. Jin, Y. Meng, Emergence of robust regulatory motifs from in silico evolution of sustained oscillation. BioSystems **103**(1), 38–44 (2011)
200. Y. Jin, Y. Meng, Morphogenetic robotics: An emerging new field in developmental robotics. IEEE Trans. Syst. Man Cybern. Part C: Appl. Rev. **41**(2), 145–160 (2011)
201. Y. Jin, Y. Meng, H. Guo, A morphogenetic self-organization algorithm for swarm robotic systems using relative position information, in *UK Workshop on Computational Intelligence* (2010)
202. Y. Jin, Y. Meng, B. Sendhoff, Influence of regulation logic on the easiness of evolving sustained oscillation for gene regulatory networks, in *Artificial Life, 2009. ALife '09. IEEE Symposium on* (2009), pp. 61 –68
203. Y. Jin, L. Schramm, B. Sendhoff, *A gene regulatory model for the development of primitive nervous systems, in INNS-NNN Symposia on Modeling the Brain and Nervous Systems* (Springer, LNCS, 2009), pp.48–55
204. Y. Jin, B. Sendhoff, A systems approach to evolutionary multi-objective structural optimization and beyond. IEEE Comput. Intell. Mag. **4**(3), 62–76 (2009)
205. Y. Jin, J. Trommler, A fitness-independent evolvability measure for evolutionary developmental systems, in *IEEE Symposium on Computational Intelligence in Bioinformatics and Computational Biology* (2010), pp. 69–76

206. M. Joachimczak, B. Wròbel, Complexity of the search space in a model of artificial evolution of gene regulatory networks controlling 3d multicellular morphogenesis. Adv. Complex Syst. **12**, 347–369 (2009)
207. M. Joachimczak, B. Wròbel, Evolution of the morphology and patterning of artificial embryos: Scaling the tricolour problem to the third dimension, in *Proceedings of the European Conference on Artificial Life (ECAL)* (2009), pp. 33–40
208. B. Jones, Y. Jin, X. Yao, S. Sendhoff, *Evolution of neural organization in the hydramat-a computational model of a radially-symmetric organism* (ACM/IEEE Trans. Comput, Biol, Bioinform, 2008). ((In revision))
209. B. Jones, Y. Jin, B. Sendhoff, X. Yao, Emerged optimal distribution of computational workload in the evolution of an undulatory animat, in *The 11th International Conference on Simulation of Adaptive Behaviors (SAB 2010)* (1998)
210. B. Jones, Y. Jin, B. Sendhoff, X. Yao, Evolving functional symmetry in a three dimensional model of an elongated organism, in *Proceedings of the 11th International Conference on the Simulation and Synthesis of Living Systems (ALife)* (MIT Press, 2008), pp. 305–312
211. B. Jones, A. Soltoggio, B. Sendhoff, X. Yao, Evolution of neural symmetry and its coupled alignment to body plan morphology, in *Proceedings of the 13th Annual Conference on Genetic and Evolutionary Computation* (2011), pp. 235–242
212. B.H.D. Jones, Investigating the emergence of modularity in a model of artificial neural network morphogenesis. Master's Thesis, Computer Science (University of Birmingham, 2005)
213. B.H.D. Jones, *The Evolutionary Emergence of Neural Organisation in Computational Models of Primitive Organisms*. PhD Thesis (University of Birmingham, 2010)
214. P. Joshi, J. Triesch, Optimizing generic neural microcircuits through reward modulated stdp, in *Artificial Neural Networks ICANN 2009*. ed. by C. Alippi, M. Polycarpou, C. Panayiotou, G. Ellinas, L. Notes, in Computer Science, vol. 5768, (Springer, Berlin Heidelberg, 2009), pp.239–248
215. J.W. Kalat, *Biological Psychology* (Wadsworth Publishing, 2008)
216. G. Karlebach, R. Shamir, Modelling and analysis of gene regulatory networks. Nat. Rev. Mol. Cell Biol. **9**, 770–780
217. M. Khammash, H. El-Samad, Systems biology: From physiology to gene regulation. Control Syst. IEEE **24**(4), 62–76 (2004)
218. J. Kirkpatrick, R. Pascanu, N. Rabinowitz, R. Hadsell, Overcoming catastrophic forgetting in neural networks. PNAS **13**(114), 521–3526 (2017)
219. H. Kitano, Designing neural networks using genetic algorithms with graph generation system. Complex Syst. **4**, 461–476 (1990)
220. H. Kitano, Biological robustness. Nat. Rev. Genet. **5**(11), 826–37 (2004)
221. H. Kitano, Biological robustness. Nat. Rev. Genet. **5**, 826–837 (2004)
222. H. Kitano, Designing neural networks using genetic algorithms with graph generation system. Complex Syst. J. **4**, 461–476 (1990)
223. H. Kitano, A simple model of neurogenesis and cell differentiation based on evolutionary large-scale chaos. Artif. Life **2**(1), 79–99 (1995)
224. H. Kitano, Computational systems biology. Nature **420**, 206–210 (2002)
225. J.F. Knabe, C.L. Nehaniv, M.J. Schilstra, T. Quick, Evolving biological clocks using genetic regulatory networks, in *Artificial Life X* (MIT Press, 2006), pp. 15–21
226. J.F. Knabe, M.J. Schilstra, C. Nehaniv, Evolution and morphogenesis of differentiated multicellular organisms: Autonomously generated diffusion gradients for positional information, in *Proceedings of the 11th International Conference on the Simulation and Synthesis of Living Systems (ALife)* (2008)
227. J.D. Knowles, R.A. Watson, D. Corne, Reducing local optima in single-objective problems by multi-objectivization, in *Proceedings of the First International Conference on Evolutionary Multi-Criterion Optimization*, EMO '01 (Springer-Verlag, London, UK, UK, 2001), pp. 269–283
228. O. Koizumi, Developmental neurobiology of hydra, a model animal of cnidarians. Can. J. Zool. **80**(10), 1678–1689 (2002)

229. T. Kowaliw, N. Bredeche, S. Chevallier, R. Doursat, Artificial neurogenesis: An introduction and selective review, in *CStudies in Computational Intelligence*, ed. by T. Kowaliw et al (Springer, 2014), pp. 1–60
230. J. Koza, *Genetic Programming: On the Programming of Computers by Means of Natural Selection* (The MIT Press, Cambridge, MA, 1992)
231. A. Krishnan, M. Tomita, A. Giuliani, Evolution of gene regulatory networks: Robustness as an emergent property of evolution. Phys. A **387**, 2170–2186 (2008)
232. M. Kudo, J. Toyama, M. Shimbo, Multidimensional curve classification using passing-through regions. Pattern Recog. Lett. **20**(11–3), 1103–1111 (1999)
233. Y. Kuniyoshi, M. Lungarella, *Developmental cognitive robotics: An emerging new field, in Workshop on Cognitive Developmental Robotics* (Japan, Kyoto, 2006), pp.3–5
234. P. Dwight Kuo, W. Banzhaf, A. Leier, Network topology and the evolution of dynamics in an artificial genetic regulatory network model created by whole genome duplication and divergence. BioSystems **85**(3), 177–200 (2006)
235. Y.-K. Kwon, K.-H. Cho, Analysis of feedback loops and robustness in network evolution based on boolean models. BMC Bioinform. 430 (2007)
236. Y.-K. Kwon, K.-H. Cho, Coherent coupling of feedback loops: A design principle of cell signaling networks. Bioinformatics **24**(17), 1926–1932 (2008)
237. Y.-K. Kwon, K.-H. Cho, Quantitative analysis of robustness and fragility in biological networks based on feedback dynamics. Bioinformatics **7**(24), 987–994 (2008)
238. M.G. Lamarche, B.L. Wanner, S. Crépin, J. Harel, The phosphate regulon and bacterial virulence: A regulatory network connecting phosphate homeostasis and pathogenesis. FEMS Microbiol. Rev. **32**(3), 461–473 (2008)
239. W. Langdon, *genetic Programming and Data Structures: Genetic Programming + Data Structures = Automatic Programming* (Kluwer, Boston, MA, 1998)
240. Y. LeCun, Y. Bengio, G. Hinton, Deep learning. Nature **521**, 436–444 (2015)
241. D.-T. Lee, A.K. Lin, Generalized delaunay triangulation for planar graphs. Discrete Comput. Geomet. **1**(3), 201–217 (1986)
242. J.A. Lee, J. Sitte, Morphogenetic evolvable hardware controllers for robot walking, in *2nd International Symposium on Autonomous Minirobots for Research and Edutainment* (2003)
243. T.J. Lee, D. Tu, C.M. Tan, L. You, Modeling cellular networks, in *Systems Bioinformatics: An Engineering Case-Based Approach*, Chapter 6. (Artech House, 2007)
244. W.-P. Lee, Y.-T. Hsiao, Inferring gene regulatory networks by incremental evolution and network decomposition, in *Optimization and Systems Biology* (2008), pp. 311–324
245. R.A. Legenstein, D. Pecevski, W. Maass, A learning theory for reward-modulated spike-timing-dependent plasticity with application to biofeedback. PLoS Comput. Biol. **4**(10) (2008)
246. P.K. Lehre, P.C. Haddow, Phenotypic complexity and local variations in neutral degree. BioSystems **87**, 233–242 (2007)
247. D.D. Leitch, *A New Genetic Algorithm for Evolution of Fuzzy Systems*. PhD Thesis (Department of Engineering Science, University Oxford, Oxford, UK, 1995)
248. T. Lenser, T. Hinze, B. Ibrahim, P. Dittrich, Towards evolutionary network reconstruction tools for systems biology, in *Evolutionary Computation, Machine Learning and Data Mining in Bioinformatics*, eds. by E. Marchiori, J.H. Moore, J.C. Rajapakse, vol. 4447. Lecture Notes in Computer Science (Springer, Berlin, 2007), pp. 132–142
249. R.E. Lenski, J.E. Barrick, E. Ofria, Balancing robustness and evolvability. PLOS Biol. **4**(2), e428 (2006)
250. G. Lepousez, A. Nissant, A.K. Bryant, G. Gheusi, C.A. Greer, P.-M. Lledo, Olfactory learning promotes input-specific synaptic plasticity in adult-born neurons. Proc. Nat. Acad. Sci. **111**(38), 13984–9 (2014)
251. M. Levine, E.H. Davidson, Gene regulatory networks for development. Proc. Nat. Acad. Sci. **102**(44), 4936–4942 (2005)
252. B. Li, J. Li, K. Tang, X. Yao, Many-objective evolutionary algorithms: A survey. ACM Comput. Surv. **48**(1), Article No. 13 (2015)

253. C.G. Li, A model of neuronal intrinsic plasticity. IEEE Trans. Auton. Ment. Dev. **3**, 277–284 (2011)
254. H. Li, Y. Cai, J. Hong, X. Peng, H. Cheng, X. Zhu, H. Bingliang, Z. Hao, Z. Fan, VG-Swarm: A vision-based gene regulation network for uavs swarm behavior emergence. IEEE Robot. Autom. Lett. **8**(3), 1175–1182 (2023)
255. J. Li, T. Kim, D. Szymanski, Multi-scale regulation of cell branching: Modeling morphogenesis. Develop. Biol. **451**(1), 40–52 (2019)
256. J. Li, X.-S. Zhang, An optimization model for gene regulatory network reconstruction with known biological information, in *Optimization and Systems Biology*(2007), pp. 35–44
257. Y. Li, E.G. Meloni, W.A. Carlezon, M.R. Milad, R.K. Pitman, K. Nader, V.Y. Bolshakov, Learning and reconsolidation implicate different synaptic mechanisms. Proc. Nat. Acad. Sci. **110**(12), 4798–803 (2013)
258. K.-H. Liang, X. Yao, Y. Liu, C.S. Newton, D. Hoffman, An experimental investigation of selfadaptation in evolutionary programming. LNCS **1447**, 291–300 (1998)
259. H. Liu, K. Simonyan, O. Vinyals, C. Fernando, K. Kavukcuoglu, Hierarchical representations for efficient architecture search, in *International Conference on Learning Representations* (2018)
260. J. Liu, R. Cheng, Y. Jin, Bi-fidelity evolutionary multiobjective search for adversarially robust deep neural architectures. *arXiv preprint* arXiv:2207.05321
261. Y. Liu, Y. Sun, B. Xue, M. Zhang, G.G. Yen, K.C. Tan, A survey on evolutionary neural architecture search. IEEE Trans. Neural Networks Learn. Syst. 1–21 (2021)
262. R.R. Llinás, The intrinsic electrophysiological properties of mammalian neurons: Insights into central nervous system function. Science **242**(4886), 1654–1664 (1988)
263. D.F. Lochtefeld, F.W. Ciarallo, Multiobjectivization via helper-objectives with the tunable objectives problem. Evol. Comput. IEEE Trans. **16**(3), 373–390 (2012)
264. W.-Y. Loh, Classification and regression trees. WIREs Data Min. Knowl. Discov. **1**(1), 14–23 (2011)
265. M. Lukosevicius, H. Jaeger, Reservoir computing approaches to recurrent neural network training. Comput. Sci. Rev. **3**(3), 127–149 (2009)
266. M. Lungarella, G. Metta, G.P.R. Sandini, Developmental robotics: A survey. Conn. Sci. **15**, 151–190 (2003)
267. F. Zambonelli, M. Mamei, M. Vasirani, Experiments in morphogenesis in swarms of simple mobile robots. Appl. Artif. Intell. **18**(9–10), 903–919 (2004)
268. X. Ma, X. Li, Q. Zhang, K. Tang, Z. Liang, W. Xie, Z. Zhu, A survey on cooperative co-evolutionary algorithms. IEEE Trans. Evol. Comput. 1 (2018)
269. W. Maass, T. Natschlager, H. Markram, Real-time computing without stable states: A new framework for neural computation based on perturbations. Neural Comput. **14**(11), 2531–60 (2002)
270. H.A. Maeda, A.R. Fernie, Evolutionary history of plant metabolism. Ann. Rev. Plant Biol. **72**, 31.1–31.32 (2021)
271. H.A. Maeda, A.R. Fernie, The importance of gene duplication and domain repeat expansion for the function and evolution of fertilization proteins. Front. Cell Develop. Biol. **10**, Article 827454 (2022)
272. A.E. Magurran, *Ecological Diversity and its Measurement* (Princeton University Press, 1988)
273. C. Malsburg, Self-organization of orientation sensitive cells in the striate cortex. Kybernetik **14**(2), 85–100 (1973)
274. M. Mamei, M. Vasirani, F. Zambonelli, Experiments of morphogenesis in swarms of simple mobile robots. Appl. Artif. Intell. **18**, 903–919 (2004)
275. A.A. Margolin, I. Nemenman, K. Basso, C. Wiggins, G. Stolovitzky, R.D. Favera, A. Califano, ARACNE: An algorithm for the reconstruction of gene regulatory networks in a mammalian cellular context. BMC Bioinform. **7**(Suppl 1), S7 (2006)
276. H. Markram, J. Luebke, M. Frotscher, B. Sakmann, Regulation of synaptic efficacy by coincidence of postsynaptic aps and epsps. Science **275**(5297), 213–215 (1997)
277. A. Martinoli, *Collective complexity out of individual simplicity* (Artif, Life, 2001)

278. K.E. Mathias, L.D. Whitley, Initial performance comparisons for the delta coding algorithms. Proc. IEEE Conf. Evol. Comput. **1**, 433–438 (1994)
279. M. McCloskey, N. Cohen, Catastrophic interference in connectionist networks: The sequential learning problem. Psychol. Learn. Motiv. **24**, 109–164 (1989)
280. H. Meinhardt, A. Gierer, Pattern formation by local self-activation and lateral inhibition. BioEssays **22**, 753–760 (2000)
281. H. Meinhardt, The radial-symmetric hydra and the evolution of the bilateral body plan: An old body became a young brain. BioEssays **24**, 195–191 (2002)
282. M. Recamonde Mendoza, A.L.C. Bazzan, Evolving random boolean networks with genetic algorithms for regulatory networks reconstruction, in *Proceedings of the 13th Annual Conference on Genetic and Evolutionary Computation*, GECCO '11 (ACM, New York, NY, USA, 2011), pp. 291–298
283. Y. Meng, H. Guo, Y. Jin, A morphogenetic approach to flexible and robust shape formation for swarm robotic systems, in *Robotics and Autonomous Systems* (2010). Submitted
284. Y. Meng, Y. Zhang, Y. Jin, Autonomous self-reconfiguration of modular robots using a hierarchical mechanochemical model. IEEE Comput. Intell. Mag. **6**(1), 43–54 (2011)
285. Y. Meng, Y. Zhang, Y. Jin, A morphogenetic approach to self-reconfigurable modular robots using a hybrid hierarchical gene regulatory network, in *Proceedings of the 12th International Conference on the Synthesis and Simulation of Living Systems (ALife XII)* (Odense, Denmark, 2010), pp. 765–772
286. Y. Meng, Y. Zhang, Y. Jin, Autonomous self-reconfiguration of modular robots by evolving a hierarchical mechanochemical model. IEEE Comput. Intell. Mag. **6**(1), 43–54 (2011)
287. D. Mercatelli, L. Scalambra, L. Triboli, F. Ray, F.M. Giorgi, Gene regulatory network inference resources: A practical overview. BBA Gene Regul. Mech. **1683**, 194430 (2020)
288. M.M. Merzenich, J.H. Kaas, J. Wall, R.J. Nelson, M. Sur, D. Felleman, Topographic reorganization of somatosensory cortical areas 3b and 1 in adult monkeys following restricted deafferentation. Neuroscience **8**(1), 33–55 (1983)
289. M.M. Merzenich, R.J. Nelson, M.P. Stryker, M.S. Cynader, A. Schoppmann, J.M. Zook, Somatosensory cortical map changes following digit amputation in adult monkeys. J. Comp. Neurol. **224**(4), 591–605 (1984)
290. K. Miettinen, F. Ruiz, A.P. Wierzbicki, *Introduction to Multiobjective Optimization: Interactive Approaches* (Springer, Berlin Heidelberg, Berlin, Heidelberg, 2008), pp.27–57
291. J. Miller, *Evolving Developmental Programs for Adaptation, Morphogenesis and Self-repair* (In Proceedings of the European Congress of Artificial Life, ECAL, 2003), pp. 256–265
292. J.F. Miller, Evolving developmental programs for adaptation, morphogenesis, and self-repair, in *Advances in Artificial Life: 7th European Conference on Artificial Life (ECAL 03)* (2003), pp. 256–265
293. J.F. Miller, Evolving a self-repairing, self-regulating, french flag organism, in *Proceedings of the GECCO* (2004), pp. 129–139
294. R. Milo, S. Shen-Orr, S. Itzkovitz, N. Kashtan, D. Chklovskii, U. Alon, Network motifs: Simple building blocks of complex networks. Science **298**(5594), 824–827 (2002)
295. A. Minelli, Grand challenges in evolutionary developmental biology. Front. Ecol. Evol. **2**, 85 (2015)
296. M. Mitchell, *An Introduction to Genetic Algorithms* (The MIT Press, 1999)
297. C. Moeslinger, T. Schmickl, K. Crailsheim, A minimalist flocking algorithm for swarm robots, in *Advances in Artificial Life. Darwin Meets von Neumann* (Springer, 2009), pp. 375–382
298. J.B. Mouret, *Novelty-based multiobjectivization, in New Horizons in Evolutionary Robotics* (Springer, Berlin/Heidelberg, 2011), pp.139–154
299. A. Mukhopadhyay, H.A. Tissenbaum, Reproduction and longevity: Secrets revealed by c. elegans. Trends Cell Biol. **17**(2), 65–71 (2007)
300. G.B. Müller, Evo-devo: Extending the evolutionary synthesis. Nat. Rev. Genet. 943–949 (2007)
301. T. Murata, H. Ishibuchi, M. Gen, Specification of genetic search directions in cellular multi-objective genetic algorithms, in *Proceedings of the First International Conference on Evolutionary Multi-criterion Optimization* (2001), pp. 82–95

302. A.G.O. Mutambara, *Decentralized Estimation and Control for Multisensor Systems* (CRC Press LLC, Boca Raton, Florida, 1998)
303. R. Nagpal, H. Shrobe, J. Bachrach, *Organizing a Global Coordinate System from Local Information on an Ad Hoc Sensor Network. IPSN, LNCS 2634* (Springer-Verlag Berlin Heidelberg, 2003)
304. S.B. Needleman, C.D. Wunsch, A general method applicable to the search for similarities in the amino acid sequence of two proteins. J. Mol. Biol. **48**(3) (1970)
305. G. Nicolis, *Introduction to Nonlinear Science* (Cambridge University Press, 1995)
306. S. Nolfi, D. Parisi, Evolving artificial neural networks that develop in time, in *European Conference on Artificial Life* (1995), pp. 353–367
307. N. Noman, H. Iba, Inferring gene regulatory networks using differential evolution with local search heuristics. IEEE/ACM Trans. Comput. Biol. Bioinform. **4**(4), 634–647 (2007)
308. M.A. Nowak, M.C. Boerlijst, J. Cooke, J.M. Smith, Evolution of genetic redundancy. Nature **388**, 167–171 (1997)
309. H. Oh, Shirazi A. R., Y. Jin, Morphogen diffusion algorithms for tracking and herding using a swarm of kilobots. Soft Comput. **22**(6), 1833–1844 (2018)
310. H. Oh, A.R. Shirazi, C. Sun, Y. Jin, Bio-inspired self-organising multi-robot pattern formation: A review. Robot. Autonom. Syst. **91**(3), 83–100 (2017)
311. H. Oh, Y. Jin, Adaptive swarm robot region coverage using gene regulatory networks, in *Advances in Autonomous Robotics Systems*, vol. 8187. Lecture Notes in Computer Science (Springer, 2014), pp. 197–208
312. E. Oja, Simplified neuron model as a principal component analyzer. J. Math. Biol. **15**(3), 267–273 (1982)
313. T. Okabe, Y. Jin, B. Sendhoff, A critical survey of performance indices for multi-objective optimization, in *Proceedings of the IEEE Congress on Evolutionary Computation (CEC)* (IEEE, 2018), pp. 878–885
314. D.D. O'Leary, S. Sahara, Genetic regulation of arealization of the neocortex. Curr. Opin. Neurobiol. **18**, 90–100 (2008)
315. G. Orchard, A. Jayawant, G.K. Cohen, N. Thakor, Converting static image datasets to spiking neuromorphic datasets using saccades. Front. Neurosci. **9**(437) (2015)
316. A. Ostermeier, A. Gawelczyk, N. Hansen, A derandomized approach to self-adaptation of evolution strategies. Evol. Comput. **2**(4), 369–380 (1994)
317. P-Y. Oudeyer, F. Kaplan, How can we define intrinsic motivation? in *Proceedings of the 8th International Conference on Epigenetic Robotics* (LUCS, Brighton, 2008)
318. P.-Y. Oudeyer, F. Kaplan, What is intrinsic motivation? A typology of computational approaches. Front. Neurorobot. **1**(Article 6), 1–13 (2007)
319. P. Paenke, Y. Jin, J. Branke, Balancing population and individual level adaptation in changing environments. Adap. Behav. **17**(2), 153–174 (2009)
320. S.R. Paladugu, V. Chickarmane, A. Dekard, J.P. Frumkin, M. MaCormarck, H.M. Sauro, In silico evolution of functional modules in biochemical networks. IEEE Proc. Syst. Biol. **153**(4), 223–235 (2006)
321. E. Palsson, H.G. Othmer, A model for individual and collective cell movement in dictyostelium discoideum. PNAS **97**(19), 10448–10453 (2000)
322. C.A. Penfold, D.L. Wild, How to infer gene networks from expression profiles, revisited. *Interface Focus* (2011)
323. X. Peng, S. Zhang, X. Lei, Multi-target trapping with swarm robots based on pattern formation. Int. J. Adv. Robot. Syst. 1–12 (2016)
324. X. Peng, Y. Jin, H. Wang, Multimodal optimization enhanced cooperative coevolution for large-scale optimization. IEEE Trans. Cybern. **49**(9), 3507–3520 (2019)
325. R. Pfeifer, F. Iida, G. Gomez, Morphological computation for adaptive behavior and cognition. Int. Congr. Ser. **1291**, 22–29 (2006)
326. L. Piegl, W. Tiller, *The NURBS Book*, 2nd edn. (Springer, 1997)
327. S. Psujek, J. Ames, R.D. Beer, Connection and coordination: The interplay between architecture and dynamics in evolved model pattern generators. Neural Comput. **18**, 729–747 (2006)

328. S. Psujek, R.D. Beer, Developmental bias in evolution: Evolutionary accessibility of phenotypes in a model of evo-devo systems. Evol. Develop. **10**(3), 375–390 (2008)
329. P. Adrian, Q.S. Bullock, Modelling the evolution of genetic regulatory networks. J. Theoret. Biol. **238**(4), 737–753 (2006)
330. T. Quick, C.L. Nehaniv, K. Dautenhahn, G. Roberts, Evolving embodied genetic regulatory network-driven control systems, in *European Conference on Artificial Life (ECAL'03)*. Lecture Notes in Artificial Intelligence 2801 (Springer, 2003), pp. 266–277
331. Q.V. Le, W.Y. Zou, S.Y. Yeung, A.Y. Ng, Learning hierarchical spatio-temporal features for action recognition with independent subspace analysis, in *Proc. IEEE Conf. Comput. Vis. Pattern Recognit.* (2011)
332. A.R. Shirazi, H. Oh, Y. Jin, Morphogenetic self-organization of collective movement without directional sensing, in *Advances in Autonomous Robotics Systems* (Springer, 2014), pp. 139–150
333. A.F. Ramons, G.C. Innocentini, F.M. Forger, J.E. Hornos, Symmetry in biology: From genetic code to stochastic gene regulation. IET Syst. Biol. **4**(5), 311–329 (2010)
334. C.L. Ramsey, K.A.De Jong, J.J. Grefenstette, A.S. Wu, D.S. Burke, Genome lengthe as an evolutionary self-adaptation, in *Proceedings of 5th Conference on Parallel Problem Solving from Nature*, eds. by A.E Eiben ans, Th. Bäck, M. Schoenauer, H.-P. Schwefel, vol. V (Springer, Amsterdam, Netherlands, 1998), pp. 345–353
335. F. Rasmus, Genes, brain, and cognition: A roadmap for the cognitive scientist. Cognition **101**, 247–269 (2006)
336. R. Ratcliff, Connectionist models of recognition memory: Constraints imposed by learning and forgetting functions. Psychol. Rev. **97**, 285–308 (1990)
337. J. Reisinger, K.O. Stanley, R. Miikkulainen, Towards an empirical measure of evolvability, in *Proceedings of the 2005 Workshops on Genetic and Evolutionary Computation* (ACM New York, NY, USA, 2005), pp. 257–264
338. J.-L. Reyes-Ortiz, L. Oneto, A. Samà, X. Parra, D. Anguita, Transition-aware human activity recognition using smartphones. Neurocomputing **171**, 754–767 (2016)
339. A. Rodríguez-García, A. Sola-Landa, K. Apel, F. Santos-Beneit, J.F. Martín, Phosphate control over nitrogen metabolism in streptomyces coelicolor: Direct and indirect negative control of glnR, glnA, glniI and amtB expression by the response regulator PhoP. Nucleic Acids Res. **37**(10) (2009)
340. F. Rosenblatt, The perceptron: A probabilistic model for information storage and organization in the brain. Psychol. Rev. **65**(6), 386–408 (1958)
341. M. Rubenstein, C. Ahler, R. Nagpal, Kilobot: A low cost scalable robot system for collective behaviors, in *IEEE International Conference on Robotics and Automation (ICRA)* (2012), pp. 3293–3298
342. T. Rudge, N. Geard, Evolving gene regulatory networks for cellular morphogenesis. Recent Adv. Artif. Life 239–252 (2005)
343. G. Rudolph, Convergence analysis of canonical genetic algorithms. IEEE Trans. Neural Networks **5**(1), 96–101 (1994)
344. S.J. Russell, P. Norvig, *Artificial Intelligence: A Modern Approach* (Prentice Hall, 2010)
345. S. Nolfi, D. Floreano, *Evolutionary Robotics* (The MIT Press, 2004)
346. L. Sabattini, C. Secchi, N. Chopra, A. Gasparri, Distributed control of multirobot systems with global connectivity maintenance. Robot. IEEE Trans. **29**(5), 1326–1332 (2013)
347. C.S. Sahin, E. Urrea, M.U. Uyar, Bio-inspired techniques for topology control of mobile nodes, in *Formal and Practical Aspects of Autonomic Computing and Networking: Specification, Development, and Verification* (2011), pp. 217–240
348. C. Salaän, L. Lepiniec, B. Dubreucq, Genetic and molecular control of somatic embryogenesis. Artif. Life **10**(7), 1467 (2021)
349. I. Salazar-Ciudad, J. Garcia-Fernández, R.V. Solé, Gene networks capable of pattern formation: from induction to reaction-diffusion. J. Theoret. Biol. **205**, 587–603 (2000)
350. H.E. Samad, M. Khammash, L. Petzold, D. Gillespie, Stochastic modelling of gene regulatory networks. Int. J. Robust Nonlinear Control

351. R.V. Samonte, E.E. Eichler, Segmental duplications and the evolution of the primate genome. Nat. Rev. Genet. **3**(1), 65–72 (2002)
352. D.H. Sanes, T.A. Reh, W.A. Harris, *Development of the Nervous System* (Academic Press, Amsterdam, 2006)
353. V.G. Santucci, P.-Y. Oudeyer, A. Barto, G. Baldassarre, Editorial: Intrinsically motivated open-ended learning in autonomous robots. Front. Neurorobot. **13** (2020)
354. H. Sayama, Robust morphogenesis of robotic swarms. IEEE Comput. Intell. Mag. **5**(3), 43–49 (2010)
355. J. Schäfer, K. Strimmer, An empirical bayes approach to inferring large-scale gene association networks. Bioinformatics **21**(6), 754–764 (2005)
356. J.D. Schaffer, Multiple objective optimization with vector evaluated genetic algorithms, in *Proceedings of an International Conference on Genetic Algorithms and Their Applications* (1985), pp. 93–100
357. M.J. Schilstra, C.L. Nehaniv, Bio-logic: gene expression and the laws of combinatorial logic. Artif. Life **14**(1), 121–133 (2008)
358. L. Schramm, Y. Jin, B. Sendhoff, Evolution and analysis of genetic networks for stable cellular growth and regeneration. Artif. Life **18**, 1–20 (2012)
359. L. Schramm, *Evolution of Cellular Systems–Optimization of Gene Regulatory Networks for the Development of Morphology and Control*. PhD Dissertation (TU Darmstadt, 2012)
360. L. Schramm, Y. Jin, B. Sendhoff, Emerged coupling of motor control and morphological development in evolution of multi-cellular animats, in *Proceedings of the European Conference on Artificial Life (ECAL)* (2009), pp. 25–32
361. L. Schramm, Y. Jin, B. Sendhoff, Redundancy creates opportunity in developmental representations, in *Proceedings of the IEEE Symposium Series on Computational Intelligence 2011 (IEEE ALIFE 2011)* (2011), pp. 203–210
362. L. Schramm, Y. Jin, B. Sendhoff, Quantitative analysis of redundancy in evolution of developmental systems, in *IEEE Symposium on Computational Intelligence in Bioinformatics and Computational Biology* (IEEE Press, 2012)
363. L. Schramm, V.V. Martins, Y. Jin, B. Sendhoff, Analysis of gene regulatory network motifs in evolutionary development of multicellular organisms, in *Proceedings of the 12th International Conference on the Synthesis and Simulation of Living Systems (ALife XII)* (2010), pp. 133–140
364. L. Schramm, B. Sendhoff, An animat's cell doctrine, in *Proceedings of the European Conference on Artificial Life (ECAL)* (2011), pp. 739–746
365. B. Schrauwen, M. Wardermann, D. Verstraeten, J.J. Steil, D. Stroobandt, Improving reservoirs using intrinsic plasticity. Neurocomputing **71**(7–9), 1159–1171 (2008)
366. C. Schuldt, I. Laptev, B. Caputo, Recognizing human actions: a local svm approach, in *ICPR Proceedings of the 17th International Conference on Pattern Recognition*, vol. 3 (2004), pp. 32–6
367. C. Schuldt, I. Laptev, B. Caputo, Recognizing human actions: a local svm approach, in *Proceedings of the 17th International Conference on Pattern Recognition, 2004. ICPR 2004*, vol. 3 (2004), pp. 32–36
368. H.-P. Schwefel, *Numerical Optimization of Computer Models* (John Wiley, Chichester, 1981)
369. H.-P. Schwefel, *Evolution and Optimum Search* (John Wiley, 1994)
370. M. Sebag, M. Schoenauer, Controlling crossover through inductive learning, in *Parallel Problem Solving from Nature*, eds. by Y. Davidor, H.-P. Schwefel, R. Männer vol. III (Springer, 1994), pp. 209–228
371. B. Sendhoff, *Evolution of Structures: Optimization of Artificial Neural Structures for Information Processing*. PhD thesis (Institut für Neoroinformatik, Ruhr-Univerisität Bochum, Bochum, Germany, 1998)
372. B. Sendhoff, M. Kreutz, Analysis of possible genome-dependance of mutation rates in genetic algorithms, in *Evolutionary Computing– Selected Papers from the 1996 AISB Workshop*, ed. by T.C. Fogarty. Lecture Notes in Computer Science 1143 (1996), pp. 257–269
373. B. Sendhoff, M. Kreutz, W. von Seelen, A condition for the genotype-phenotype mapping: Causality, in *Proceedings of the 7th International Conference on Genetic Algorithms*, ed. by Th. Bäck (Morgan Kaufmann, 1997), pp. 73–80

374. B. Sendhoff, M. Kreuz, Evolutionary optimization of the structure of neural networks using a recursive mapping as encoding, in *Proceedings of International Conference on Artificial Neural Nets and Genetic Algorithms*, eds. by G. Smith, N. Steele, R. Albrecht (Springer, 1998), pp. 370–374
375. M. Sfakiotakis, D.P. Tsakiris, Simuun: A simulation environment for undulatory locomotion. Int. J. Model. Simul. (2006)
376. B. Shahriari, K. Swersky, Z. Wang, R.P. Adams, N. de Freitas, Taking the human out of the loop: A review of bayesian optimization. Proc. IEEE **104**(1), 148–175 (2016)
377. J.C.A. Sharkey, Swarm robotics and minimalism. Conn. Sci. **19**(3), 245–260 (2007)
378. O. Tatyana, Sharpee, Function determines structure in complex neural networks. Proc. Nat. Acad. Sci. **111**(23), 8327–8 (2014)
379. W. Shen, P. Will, A. Galstyan, *Hormone-inspired self-organization and distributed control of robotic swarms* (Autonom, Robot, 2004)
380. S. Shai, Shen-Orr, R. Milo, S. Mangan, U. Alon, Network motifs in the transcriptional regulations network of escherichia coli. Nat. Genet. **31**, 64–68 (2002)
381. J.D. Shepherd, G. Rumbaugh, J. Wu, S. Chowdhury, N. Plath, D. Kuhl, R.L. Huganir, P.F. Worley, Arc mediates homeostatic synaptic scaling of AMPA receptors. Neuron **52**(3), 475–484 (2006)
382. A.R. Shirazi, Y. Jin, A strategy for self-organized coordinated motion of a swarm of minimalist robots. IEEE Trans. Emer. Topics Comput. Intell. **1**(5), 326–338 (2017)
383. A. Ramezan Shirazi, Y. Jin, Regulated morphogen gradients for target surrounding and adaptive shape formation. IEEE Trans. Cogn. Develop. Syst. **13**(4), 818–826 (2021)
384. H.Z. Shouval, S.S. Wang, G.M. Wittenberg, Spike timing dependent plasticity: A consequence of more fundamental learning rules. Front. Comput. Neurosci. **4**(19) (2010)
385. S.Y. Shvartsman, M. Coppey, A.M. Berezhkovskii, Dynamics of maternal morphogen gradients in the drosophila embryo. Curr. Opin. Genet. Dev. **18**(4), 342–347 (2008)
386. K. Simons, E. Karsenti, D. St Johnston, C. Wijer, S. Swarminathan eds. *Self-Organization and Morphogenesis in Biological Systems* (Schloss Ringberg, Tegersee, Germany, 2006)
387. K. Sims, Evolving 3D morphology and behavior by competition. Artif. Life **1** (1994)
388. K. Sims, Evolving 3d morphology and behaviour by competition, in *ALife IV Proceedings* (MIT Press, 1994), pp. 28–39
389. K, Sims, Evolving virtual creatures, in *Computer Graphics, Annual Conference Series* (1994), pp. 15–22
390. S. Singh, R.L. Lewis, A.G. Barto, J. Sorg, Intrinsically motivated reinforcement learning: An evolutionary perspective. IEEE Trans. Autonom. Mental Develop. **2**, 70–82 (2010)
391. S. Singh, A. Okun, A. Jackson, Learning to play go from scratch. Nature **550**, 556–557 (2017)
392. A. Sirbu, H.J. Ruskin, M. Crane, Comparison of evolutionary algorithms in gene regulatory network model inference. BMC Bioinform. **11**, 59 (2010)
393. A. Sîrbu, H.J. Ruskin, M. Crane, Stages of gene regulatory network inference: The evolutionary algorithm role, in *Evolutionary Algorithms*, ed. by E. Kita (InTech, 2011)
394. J.M. Smart, Evolutionary development: A universal perspective, in *Evolution, Development and Complexity*, eds. by G.Y. Georgiev, J.M. Smart, C.L. Flores Martinez, M.E. Price (Springer, 2008), pp. 201–219
395. Q.S. Song, Z.R. Feng, Effects of connectivity structure of complex echo state network on its prediction performance for nonlinear time series. Neurocomputing **73**(10–12), 2177–2185 (2010)
396. S. Song, K.D. Miller, L.F. Abbott, Competitive Hebbian learning through spike-timing-dependent synaptic plasticity. Nat. Neurosci. **3**(9), 919–926 (2000)
397. W. Spears, Recombination parameters, in *Evolutionary Computation*, eds. by Th. Bäck, D.B. Fogel, Z. Michalewicz, vol. 2. (Institut of Physics Publication, Bristol, 2000), pp.152–169
398. N.C. Spitzer, Electrical activity in early neuronal development. Nature **444**, 707–712 (2006)
399. O. Kenneth, Stanley, Risto Miikkulainen, A taxonomy for artificial embryogeny. Artif. Life **9**(2), 93–130 (2003)
400. S.C. Stearns, Trade-offs in life-history evolution. Func. Ecol. **3**, 259–268 (1989)

401. T. Steiner, M. Olhofer, B. Sendhoff, Towards shape and structure opti- mization with evolutionary development. *Proceedings of the ALife X: The Tenth International Conference on the Simulation and Synthesis of Living Systems* (2008), pp. 70–76
402. T. Steiner, J. Trommler, M. Brenn, Y. Jin, B. Sendhoff, Global shape with morphogen gradients and motile polarized cells, in *Congress on Evolutionary Computation* (2009), pp. 2225–2232
403. T. Steiner, Y. Jin, B. Sendhoff, A cellular model for the evolutionary development of lightweight material with an inner structure, in *GECCO'08: Proceedings of the 10th Annual Conference on Genetic and Evolutionary Computation* (2008), pp. 851–858
404. T. Steiner, M. Olhofer, B. Sendhoff, Towards shape and structure optimization with evolutionary development, in *Proceedings of the 10th International Conference on the Simulation and Synthesis of Living Systems (ALife)* (2006), pp. 70–76
405. T. Steiner, L. Schramm, Y. Jin, B. Sendhoff, Emergence of feedback in artificial gene regulatory networks, in *Proceedings of the IEEE Conference on Evolutionary Computation (CEC)* (2007), pp. 867–874
406. T. Steiner, J. Trommler, M. Brenn, Y. Jin, B. Sendhoff, Global shape with morphogen gradients and motile polarized cells, in *Proceedings of the 2009 Congress on Evolutionary Computation* (2009), pp. 2225–2232
407. M. Stemmler, C. Koch, How voltage-dependent conductances can adapt to maximize the information encoded by neuronal firing rate. Nat. Neurosci. **2**, 521–527 (1999)
408. N.R. Stillman, R. Mayor, Generative models of morphogenesis in developmental biology. *Seminars in Cell & Developmental Biology*. https://doi.org/10.1016/j.semcdb.2023.02.001 (2023)
409. Y. Sun, H. Wang, B. Xue, Y. Jin, G.G. Yen, M. Zhang, Surrogate-assisted evolutionary deep learning using an end-to-end random forest-based performance predictor. IEEE Trans. Evol. Comput. **24**(2), 350–364 (2020)
410. R.S. Sutton, A.G. Barto, *Reinforcement Learning: An Introduction* (The MIT Press, 2018)
411. M. Swain, T. Hunniford, J. Mandel, N. Palfreyman, W. Dubitzky, Modeling gene-regulatory networks using evolutionary algorithms and distributed computing, in *Cluster Computing and the Grid, 2005. CCGrid 2005. IEEE International Symposium on*, vol. 1 (2005), pp. 512–519
412. G. Syswerda, Uniform crossover in genetic algorithms, in *Proceedings of International Conference on Genetic Algorithms*, ed. by J.D. Schaffer (Fairfax, VA, 1989), pp. 2–9
413. T. Tanino, M. Tanaka, C. Hijo, An interactive multi-criteria decision making method by using a genetic algorithm, in *Proceedings of 2nd International Conference on Systems Science and Systems Engineering* (1993), pp. 381–386
414. R. Tarjan, Depth-first search and linear graph algorithms. SIAM J. Comput. **1**(2), 146–160 (1972)
415. T. Taylor, C. Massey, Recent developments in the evolution of morphologies and controller for physically simulated creatures. Artif. Life **7**(1), 77–87 (2001)
416. T. Taylor, C. Massey, Recent developments in the evolution of morphologies and controllers for physically simulated creatures. Artif. Life **7**(1) (2001)
417. S.A. Thomas, Y. Jin, Combining genetic oscillators and switches using evolutionary algorithms, in *Computational Intelligence in Bioinformatics and Computational Biology (CIBCB), 2012 IEEE Symposium on* (2012), pp. 28 –34
418. S.A. Thomas, Y. Jin, Evolving connectivity between genetic oscillators and switches using evolutionary algorithms. J. Bioinform. Comput. Biol. **11**(3:1341001) (2013)
419. S.A. Thomas, Y. Jin, Single and multi-objective in silico evolution of tunable genetic oscillators, in *The 7th Int. Conf. on Evolutionary Multi-Criterion Optimization (EMO 2013)* (2013)
420. S.A. Thomas, Y. Jin, Reconstructing gene regulatory networks: Where optimization meets big data. Evol. Intell. **7**(1), 29–47 (2014)
421. S.A. Thomas, Y. Jin, E. Laing, C.P. Smith, Reconstructing regulatory networks in streptomyces using evolutionary algorithms, in *UK Workshop on Computational Intelligence* (2013)
422. S. Thomas, Y. Jin, Single and multi-objective in silico evolution of tunable genetic oscillators, in *Evolutionary Multi-Criterion Optimization*, eds. by R.C. Purshouse, P.J. Fleming, C.M. Fonseca, S. Greco, J. Shaw, vol. 7811. Lecture Notes in Computer Science (Springer Berlin Heidelberg, 2013), pp. 696–709

423. Y. Tian, H. Wang, X. Zhang, Y. Jin, Effectiveness and efficiency of non-dominated sorting for evolutionary multi-and many-objective optimization. Complex Intell. Syst. **3**(4) (2017)
424. Y. Tian, X. Zhang, R. Cheng, Y. Jin, *A Multi-objective Evolutionary Algorithm Based on an Enhanced Inverted Generational Distance Metric* (In Congress on Evolutionary Computation, IEEE, 2016)
425. T. Toyoizumi, J.P. Pfister, K. Aihara, W. Gerstner, Optimality model of unsupervised spike-timing-dependent plasticity: synaptic memory and weight distribution. Neural Comput. **19**(3), 639–671 (2007)
426. J. Triesch, A gradient rule for the plasticity of a neuron's intrinsic excitability, in *15th Int. Conf. Artif. Neural Netw.* (2005), pp. 65–70
427. J. Triesch, Synergies between intrinsic and synaptic plasticity mechanisms. Neural Comput. **19**, 885–909 (2007)
428. T. Yu-Chen Tsai, Y.S. Choi, W. Ma, J.R. Pomerening, C. Tang, J.E. Ferrell, Robust, tunable biological oscillations from interlinked positive and negative feedback loops. Science **321**(5885), 126–129 (2008)
429. T. Yu-Chen Tsai, Y.S. Choi, W. Ma, J.R. Pomerening, C. Tang, J.E. Ferrell, Robust, tunable biological oscillations from interlinked positive and negative feedback loops. Science **321**(5885), 126–129 (2008)
430. G. Tufte, Gene regulation mechanisms introduced in the evaluation criteria for a hardware cellular development system, in *Proceedings of the First NASA/ESA Conference on Adaptive Hardware and Systems* (2006)
431. A. Turing, The chemical basis of morphogenesis. Philos. Trans. R. Soc. London Ser. B. **237**, 37–72 (1952)
432. P.D. Turney, Increasing evolvability considered as a large-scale trend in evolution, in *Proceedings of the 1999 Genetic and Evolutionary Computation Conference Workshop Program (GECCO-99 Workshop on Evolvability)* (1999), pp. 43–46
433. E.D. Vaishnav, C.G. de Boer, J. Molinet, M. Yassour, L. Fan, X. Adiconis, D.A. Thompson, J.Z. Levin, F.A. Cubillos, A. Regev, The evolution, evolvability and engineering of gene regulatory dna. Nature **603**(12), 455–463 (2013)
434. E.A.M. van Dis, J. Bollen, W. Zuidema, R. van Rooij, C.L. Bockting, Chatgpt: Five priorities for research. Nature **614**, 224–226 (2023)
435. M.C.W. van Rossum, G.Q. Bi, G.G. Turrigiano, Stable hebbian learning from spike timing-dependent plasticity. J. Neurosci. **20**(23), 8812–8821 (2000)
436. M.C.W. van Rossum, G.G. Turrigiano, Correlation based learning from spike timing dependent plasticity. Neurocomputing **38–40**, 409–415 (2001)
437. G.P. van Wezel, K.J. McDowall, The regulation of the secondary metabolism of streptomyces: New links and experimental advances. Nat. Prod. Rep. **28**, 1311–1333 (2011)
438. P. Vartholomeos, E. Papadopoulos, Analysis, design and control of a planar micro-robot driven by two centripetal-force actuators, in *IEEE International Conference on Robotics and Automation (ICRA)* (Orlando, Florida, 2006)
439. M. Vignes, J. Vandel, D. Allouche, N. Ramadan-Alban, C. Cierco-Ayrolles, T. Schiex, B. Mangin, S. de Givry, Gene regulatory network reconstruction using bayesian networks, the dantzig selector, the lasso and their meta-analysis. PLoS ONE **6**(12), e29165 (2011)
440. E.J. Visvader, H. Clevers, Tissue-specific designs of stem cell hierarchies. Nat. Cell Biol. **18**(4), 349–355 (2016)
441. E.O. Voit, Model identifictaion: A key challenge is computational systems biology, in *Optimization and Systems Biology* (2008), pp. 1–12
442. A. Wagner, *Robustness and Evolvability in Living Systems* (Princeton University Press, 2007)
443. A. Wagner, Distributed robustness versus redundancy as causes of mutational robustness. BioEssays **27**, 176–188 (2005)
444. G.P. Wagner, G. Booth, H. Bagheri-Chaichian, Genetic measurement theory of epistatic effects. Genetica **102**(103), 569–580 (1998)
445. H. Wang, M. Olhofer, Y. Jin, Mini-review on preference modeling and articulation in multi-objective optimization: Current status and challenges. Complex Intell. Syst. **3**(4), 233–245 (2017)

446. X. Wang, Y. Jin, W. Du, J. Wang, Evolving dual-threshold bienenstock-cooper-munro learning rules in echo state networks. *IEEE Transactions on Neural Networks and Learning Systems*
447. X. Wang, Y. Jin, K. Hao, Echo state networks regulated by local intrinsic plasticity rules for regression. Neurocomputing **351**, 111–122 (2019)
448. X. Wang, Y. Jin, K. Hao, Echo state networks regulated by local intrinsic plasticity rules for regression. Neurocomputing **351**, 111–122 (2019)
449. X. Wang, Y. Jin, K. Hao, Evolving local plasticity rules for synergistic learning in echo state networks. IEEE Trans. Neural Network Learn. Syst. **31**(4), 1363–1374 (2020)
450. Y. Wang, Andrey Minarsky, R. Penner, C. Soule, N. Morozova, Model of morphogenesis. J. Comput. Biol. **27**(9), 1373–1383 (2020)
451. Y. Wang, X.-S. Zhang, L. Chen, Optimization meets systems biology. BMC Syst. Biol. **4**(Suppl 2), 1–4 (2010)
452. W. Weibull, A statistical distribution function of wide applicability. J. Appl. Mech. **13**, 293–297 (1951)
453. J. Weng, J. McClelland, J. Pentland, I. Sporns, O. Stockman, M. Sur, E. Thelen, Autonomous mental development by robots and animals. Science **291**, 599–600 (2003)
454. P.J. Werbos, Generalization of backpropagation with application to a recurrent gas market model. Neural Networks **1**(4), 339–356 (1988)
455. B.A. White, A. Tsourdos, I. Ashokaraj, S. Subchan, R. Zbkowski, Contaminant cloud boundary monitoring using network of UAV sensors. IEEE Sens. J. **8**(10), 1681–1692 (2008)
456. D.J. Whitehead, A. Skusa, P.J. Kennedy, Evaluating an Evolutionary Approach for Reconstructing Gene Regulatory Networks, in *Ninth International Conference on the Simulation and Synthesis of Living Systems (ALIFE9)* (Boston, 2004. MIT Press)
457. A.F.T. Winfield, J. Nembrini, *Emergent Swarm Morphology Control of Wireless Networked Mobile Robots, Morphogenetic Engineering*, vol. 8 (Understanding Complex Systems. Springer-Verlag, Berlin Heidelberg, 2012)
458. G.M. Wittenberg, S.S. Wang, Malleability of spike-timing-dependent plasticity at the ca3-ca1 synapse. J. Neurosci. **26**, 6610–6617 (2006)
459. L. Wolpert, *Principles of Development* (Oxford University Press, 2002)
460. J. Xiong, T. Zhou, Gene regulatory network inference from multifactorial perturbation data using both regression and correlation analyses. PLoS ONE **7**(9), e43819 (2012)
461. F. Xue, Z. Hou, X. Li, Computational capability of liquid state machines with spike-timing-dependent plasticity. Neurocomputing **122**, 324–329 (2013)
462. P. Yang, K. Tang, X. Yao, Turning high-dimensional optimization into computationally expensive optimization. IEEE Trans. Evol. Comput. **22**(1), 143–156 (2018)
463. J. Yin, Y. Meng, Y. Jin, A developmental approach to structural self-organization in reservoir computing. IEEE Trans. Autonom. Mental Develop. **4**(4), 273–289 (2012)
464. K.Y. Yip, R.P. Alexander, K.-K. Yan, M. Gerstein, Improved reconstruction of in silico gene regulatory networks by integrating knockout and perturbation data. PLoS ONE **5**(1), e8121 (2010)
465. G. Yu, Y. Jin, M. Olhofer, Q. Liu, W. Du, Solution set augmentation for knee identification in multiobjective decision analysis. IEEE Trans. Cybern. (2022). https://ieeexplore.ieee.org/document/9613819
466. G. Yu, L. Ma, Y. Jin, W. Du, Q. Liu, H. Zhang, *A survey on knee-oriented multi-objective evolutionary optimization* (IEEE Trans. Evol, Comput, 2022)
467. C.-H. Yuh, H. Bolouri, J. Bower, E. Davidson, A logical model of *cis*-regulatory control in a eukaryotic system, in *Computational Modeling of Genetic and Biochemical Networks*, eds. by J. Bower, H. Bolouri (The MIT Press, 2000), pp. 73–100
468. S. Zhan, J.F. Miller, A.M. Tyrrell, An evolutionary system using development and artificial genetic regulatory networks for electronic circuit design. Biosystems **98**(3), 176–192 (2009)
469. H. Zhang, Y. Jin, R. Cheng, K. Hao, Efficient evolutionary search of attention convolutional networks via sampled training and node inheritance. IEEE Trans. Neural Networks Learn. Syst. **31**(4), 371–385 (2021)
470. J. Zhang, Evolution by gene duplication: an update. Trends Ecol. Evol. **18**(6), 292–298 (2003)

471. Q. Zhang, H. Li, MOEA/D: A multiobjective evolutionary algorithm based on decomposition. IEEE Trans. Evol. Comput. **11**(6), 712–731 (2007)
472. S. Zhang, M. Liu, X. Lei, Y. Huang, F. Zhang, Multi-target trapping with swarm robots based on pattern formation. Robot. Autonom. Syst. **106**, 1–13 (2018)
473. W. Zhang, D.J. Linden, The other side of the engram: Experience-driven changes in neuronal intrinsic excitability. Nat. Rev. Neurosci. **4**, 885–900 (2003)
474. X. Zhang, Y. Tian, R. Cheng, Y. Jin, An efficient approach to non-dominated sorting for evolutionary multi-objective optimization. IEEE Trans. Evol. Comput. **19**(6), 761–776 (2015)
475. X. Zhang, X.-M. Zhao, K. He, L. Le, Y. Cao, J. Liu, J.-K. Hao, Z.-P. Liu, L. Chen, Inferring gene regulatory networks from gene expression data by path consistency algorithm based on conditional mutual information. Bioinformatics **28**(1), 98–104 (2012)
476. M. Zheng, T. Iwasaki, W.O. Friesen, Systems approach to modeling the neuronal cpg for leech swimming, in *Annual International Conference of the IEEE Engineering in Medicine and Biology Society* (2004), pp. 703–706
477. Y. Zhou, Y. Jin, Y. Sun, J. Ding, *Surrogate-assisted cooperative co-evolutionary reservoir architecture search for liquid state machines* (IEEE Trans. Emer. Topics Comput, Intell, 2022)
478. H. Zhu, R. Shyama Prasad Rao, T. Zeng, L. Chen, Reconstructing dynamic gene regulatory networks from sample-based transcriptional data. Nucleic Acids Res. **40**(21), 10657–10667 (2012)
479. H. Zhu, Y. Jin, Multi-objective evolutionary federated learning. IEEE Trans. Neural Networks Learn. Syst. **31**(4), 1310–1322 (2020)
480. E. Zitzler, L. Thiele, Multi-objective evolutionary algorithms: A comparative case study and the strength Pareto approach. IEEE Trans. Evol. Comput. **3**(4), 257–271 (1999)

The manufacturer's authorised representative in the EU is Springer Nature Customer Service Centre GmbH, Europaplatz 3, 69115 Heidelberg, Germany. If you have any concerns regarding our products, please contact ProductSafety@springernature.com

Printed and bound by CPI Group (UK) Ltd, Croydon, CR0 4YY
01/10/2026
02225867-0010